ENERGY FROM THE WEST

D0598070

ENERGY FROM

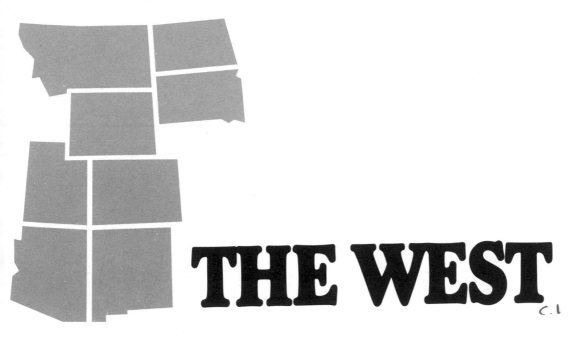

THE WEST

A TECHNOLOGY ASSESSMENT OF WESTERN ENERGY RESOURCE DEVELOPMENT

By Science and
Public Policy Program
University of Oklahoma

Michael D. Devine
Steven C. Ballard
Irvin L. White
Michael A. Chartock
Allyn R. Brosz
Frank J. Calzonetti
Mark S. Eckert

Timothy A. Hall
R. Leon Leonard
Edward J. Malecki
Gary D. Miller
Edward B. Rappaport
Robert W. Rycroft

University of Oklahoma Press : Norman

By the Science and Public Policy Program of the University of Oklahoma

Energy Under the Oceans (Norman, 1973)
North Sea Oil and Gas (Norman, 1973)
Energy Alternatives (Washington, 1975)
Our Energy Future (Norman, 1976)
Energy from the West (Norman, 1981)

Library of Congress Cataloging in Publication Data

Oklahoma. University. Science and Public Policy Program.
  Energy from the West.

  Based largely on the results of a study conducted from 1975 through 1979 for the Office of Energy, Minerals, and Industry of the Environmental Protection Agency.
  Bibliography: p.
  Includes index.
  1. Energy development—The West. 2. Energy policy—The West. 3. Environmental protection—The West. I. Devine, Michael D. II. United States. Environmental Protection Agency. Office of Energy, Minerals, and Industry. III. Title.
TJ163.25.U6035 1981   333.79'0978   80-5936
ISBN 0-8061-1750-8
ISBN 0-8061-1751-6 (pbk.)

Preface

This book is based largely on the results of a "Technology Assessment of Western Energy Resource Development" conducted for the Office of Energy, Minerals, and Industry, part of the Office of Research and Development of the U.S. Environmental Protection Agency (EPA). The study was conducted from 1975 through 1979 by an interdisciplinary research team from the Science and Public Policy Program (S&PP) in the University of Oklahoma. Assistance was provided by two major subcontractors (Radian Corporation, Austin, Texas; and Water Purification Associates, Cambridge, Massachusetts), several other subcontractors, consultants, and an advisory committee.

The overall purpose of the study was to determine what the consequences of western energy development would be and how public and private policymakers could deal with the major problems and issues likely to arise. Our specific objectives were to:

- Identify and describe energy development alternatives;
- Determine and analyze impacts from development;
- Identify and define policy problems and issues; and
- Identify, evaluate, and compare alternative policy responses.

The study focuses on the development of coal, oil shale, uranium, oil, natural gas, and geothermal energy in an eight-state area—Arizona, Colorado, Montana, New Mexico, North Dakota, South Dakota, Utah, and Wyoming. The consequences of development were assessed for the period 1975 to 2000. However, most of the analyses and conclusions are tied to specific levels of development rather than to any particular time period for that development.

All the authors of this book are currently or were formerly associated with the S&PP Program. The primary initiator of the study and project director during its first three years was Irvin L. (Jack) White, assistant director of S&PP and professor of political science. In many respects the completion of this study can be directly attributed to his management efforts and devotion to the idea of interdisciplinary, team research. In particular, we would like to acknowledge his primary contribution to the approach and structure of policy analysis developed in this study. Jack White is now assistant director for energy and mineral resources in the Bureau of Land Management. Michael D. Devine, now director of S&PP and professor of industrial engineering, served as project director during the final stages of this study. Steven C. Ballard, now assistant director of S&PP and assistant professor of political science, served as co-director of the project during its final two years and manager of the Policy Analysis Report. Co-directors of the study were Michael A. Chartock, associate professor of zoology and research fellow in S&PP; and R. Leon Leonard, associate professor of aerospace, mechanical, and nuclear engineering, and research fellow in S&PP (now a senior scientist with the Radian Corporation in Austin, Texas). Other S&PP team members were Edward J. Malecki, assistant professor of geography; Edward B. Rappaport, visiting assistant professor of economics (now in the Federal Trade Commission in Washington, D.C.); Frank J. Calzonetti, research associate (now assistant professor of geography in West Virginia University); Timothy A. Hall, research associate in political science (now senior scientist with the Radian Corporation in Austin, Texas); Gary D. Miller, graduate research assistant (now assistant professor of environmental sciences in the University of Oklahoma); Allyn R. Brosz, graduate research assistant (now assistant professor of political science in Virginia Polytechnic Institute and State University); Mark S. Eckert, graduate research assistant in geography; and Robert W. Rycroft, research associate (now assistant professor of political science in the University of Denver).

Don E. Kash, former director of S&PP and professor of political science and now serving as chief of the Conservation Division of the U.S. Geological Survey, assisted the team throughout the study. In many respects our approach to interdisciplinary research and technology assessment reflect the efforts of Don Kash and Jack White since the creation of the S&PP Program in 1970. Martha W. Gilliland, executive director of Energy Policy Studies, Inc., Omaha, Nebraska, played a major role at various stages in the project as author and reviewer. Other past and current members of S&PP who contributed to the study are Martin R. Cines, Rodney Freed, Phil Kabrich, Michael O'Hasson, Larry W. Parker, David Penn, and Cary Bloyd.

Radian Corporation was a major contributor to the impact analysis phase of this research. C. Patrick Bartosh, program manager, directed the Radian effort. Radian personnel who contributed to this part of the study were B. Russ Eppright, Thomas W. Grimshaw, Milton Owen, Ken Choffel, Timothy J. Wolterink, Jim Sherman, James L. Machin, Dennis D. Harver, David Cabe, Sam A. Gavande, W. F. Holland, Carl Heinz Michelis, and Michael W. Hooper.

Water Purification Associates (WPA) conducted studies of water requirements (including water conservation opportunities) for steam-electric power generation and synthetic fuel plants. WPA personnel who contributed directly to this study include Harris Gold, D. J. Goldstein, R. F. Probstein, J. S. Shen, and D. Yung.

The research reported here could not have been completed without the assistance of a dedicated research and administrative support staff. Members of the staff are an integral part of the interdisciplinary team approach employed by the S&PP. During the major part of the research project the administrative support staff was headed by Janice Whinery, assistant to the director, and Ellen Ladd, clerical supervisor. Mary Zimbelman is currently assistant to the director. Other administrative staff members during the typing of this manuscript were Amber Adams, Sue Bayliss, and Hazel Barnes; members of that staff that assisted in preparing earlier versions of this work include Cyndy Allison, Nancy Heinicke, Pam Odell, and Judy Williams.

The research support staff is headed by Martha Jordan, librarian, who played a crucial role in organizing research materials for the study and handling references for this book. Research team assistants were Mary Sutton, Diane Dean, Warren Dickson, and Lorna Caraway. Virginia Newman prepared most of the graphics in the final production of the book. Kate Douglas Torrey served as copy editor and has helped immensely in producing a consistent writing style from material originating from many authors.

Steven E. Plotkin, EPA project officer during the first three years of the study, provided continuing support and assistance in the conduct of the research. In addition to providing substantial review and comment, he was an invaluable source of information, and he assisted directly in the water policy analysis in chapters 3 and 4. Terry Thoem, EPA, Denver, also provided valuable assistance at various times during the project.

In addition, we wish to thank the members of our advisory committee who have assisted the team since the project was initiated in July 1975. The members of this committee and their affiliations at the time of this project were John Bermingham, attorney, Denver, Colorado; Thadis W. Box, Utah State University; Governor Jack Campbell, president, Federation of Rocky Mountain States; Bill Conine, Mobil Oil Corporation; Sharon Eads, Native American Rights Fund; Michael B. Enzi, mayor, Gillette, Wyoming; Lionel S. Johns, Office of Technology Assessment, U.S. Congress; Kenneth Kauffman, Bureau of Reclamation; S. P. Mathur, U.S. Department of Energy; Leonard Meeker, attorney, Center for Law and Social Policies; Richard Meyer, Abt Associates; Raphael Moure, Oil, Chemical, and Atomic Workers Union; Bruce Pasternack, Booz Allen Hamilton, Inc.; Robert Richards, Kaiser Engineers; H. Anthony Rickel, Sierra Club Legal Defense Fund; Warren Schmechel, Western Energy Company; and Vernon Valantine, Colorado River Board of California.

Finally, we wish to acknowledge the assistance of several persons who reviewed an early draft report which formed the basis for this book. This group, and their affiliation at the time of their assistance, includes Professor Larry Canter, Civil Engineering and Environmental Science, University of Oklahoma; John Cummings, housing specialist, State of Wyoming; William R. Frendeberg, Department of Sociology, Yale University; Edwin L. Hamilton, Central Nebraska Public Power and Irrigation District; Mary R. Hamilton, BDM Corporation; Charles O. Jones, Department of Political Science, University of Pittsburgh; W. W. Reedy, Bureau of Reclamation; John Reuss, National Conference of State Legislators; and Temple A. Reynolds, Glen Canyon National Recreation Area.

Although the project upon which this book is based was funded by the Environmental Protection Agency under Contract Number 68-01-1916, the opinions, findings, and conclusions or recommendations expressed in this book are those of the authors and do not necessarily reflect the views of the Environmental Protection Agency. Neither the University of Oklahoma nor any of the numerous organizations and individuals contributing to this project are responsible for the content. This book is the sole responsibility of the Science and Public Policy Program of the University of Oklahoma.

Contents

Illustrations

Tables

Acronyms and Abbreviations

AAR	Association of American Railroads
AC	alternating current
acre-ft/yr	acre-feet per year
BACT	Best Available Control Technology
bbl/day	barrels per day
BIA	Bureau of Indian Affairs
BLM	Bureau of Land Management
Btu	British thermal unit
BuRec	Bureau of Reclamation
CAA	Clean Air Act
CERT	Council of Energy Resource Tribes
CO	carbon monoxide
CO_2	carbon dioxide
COG's	councils of government
CONEG	Coalition of Northeastern Governors
CRB	Colorado River Basin
CWA	Clean Water Act
CWIP	construction work in progress
dBA	decibel(s) A-weighted
DC	direct current
DCF	discounted cash flow
DOC	Department of Commerce
DOE	Department of Energy
DOI	Department of the Interior
DOT	Department of Transportation
EDA	Economic Development Administration
EDF	Environmental Defense Fund
EIS	environmental impact statement(s)
EMARS	Energy Minerals Allocation Resource System
EPA	Environmental Protection Agency
EPRI	Electric Power Research Institute
ERA	Economic Regulatory Administration
EMS	emergency medical services
ESECA	Energy Supply and Environmental Coordination Act of 1974
FERC	Federal Energy Regulatory Commission
FGD	flue gas desulfurization

FHA	Federal Housing Administration
FHLMC or Freddie Mac	Federal Home Loan Mortgage Corporation
FmHA	Farmers Home Administration
FNMA or Fannie Mae	Federal National Mortgage Association
FPC	Federal Power Commission
FWPCA	Federal Water Pollution Control Act
FWS	Fish and Wildlife Service
GAO	General Accounting Office
GNMA or Ginnie Mae	Government National Mortgage Association
GNP	gross national product
HC	hydrocarbon(s)
HUD	Department of Housing and Urban Development
HVTL	high voltage transmission line(s)
H_2S	hydrogen sulfide
ICC	Interstate Commerce Commission
IRS	Internal Revenue Service
kW	kilowatt
kWh	kilowatt-hour
LNG	liquefied natural gas
mg/l	milligram(s) per liter
mtpd	thousand tons per day
mtpy	thousand tons per year
MM acre-ft/yr	million acre-feet per year
MMcfd	million cubic feet per day
MMgpd	million gallons per day
MMtpy	million tons per year
MWe	megawatt-electric
NAAQS	National Ambient Air Quality Standards
NAS	National Academy of Sciences
NEA	National Energy Act
NEPA	National Environmental Policy Act
NO_x	oxides of nitrogen
NPDES	National Pollutant Discharge Elimination System
NPRC	Northern Plains Resource Council
NSPS	New Source Performance Standards
OPEC	Organization of Petroleum Exporting Countries
ORV	off-road vehicle(s)
OSMRE	Office of Surface Mining, Reclamation, and Enforcement
ppm	parts per million
PSD	prevention of significant deterioration
psi	pounds per square inch
Q	quadrillion (10^{15}) Btu's
R&D	research and development
RARE	Roadless Area Review and Evaluation
RCRA	Resource Conservation and Recovery Act
RD&D	research, development and demonstration
S&PP	Science and Public Policy Program
SCS	Soil Conservation Service
SEAS	Strategic Environmental Assessment System

SEPA	state environmental policy act
SIP	state implementation plan(s)
SMCRA	Surface Mining Control and Reclamation Act
SNG	synthetic natural gas
SO_2	sulfur dioxide
SRI	Standford Research Institute
synfuels	synthetic fuels
TA	technology assessment
TDS	total dissolved solids
tpd	tons per day
tpy	tons per year
TSP	total suspended particulates
UCRB	Upper Colorado River Basin
UMRB	Upper Missouri River Basin
USDA	U.S. Department of Agriculture
USGS	U.S. Geological Survey
U_3O_8	uranium oxide and/or yellowcake
WESCO	Western Gasification Company
WESTPO	Western Governors' Policy Office
WISA	Wyoming Industrial Siting Administration
$\mu g/m^3$	micrograms per cubic meter

ENERGY FROM THE WEST

CHAPTER 1

An Introduction to Western Energy Development

If the U.S. is to decrease its dependence on foreign oil, then, in addition to using its available energy supplies more efficiently, the nation must increase domestic energy production. Given the substantial energy resources in the western U.S., this region is expected to be a major contributor to these increased domestic supplies. Table 1-1 shows reserve estimates for six major energy resources[1] in the eight western states of this study: Arizona, Colorado, Montana, New Mexico, North Dakota, South Dakota, Utah, and Wyoming.

Why Western Energy?

Coal is the region's most abundant resource, accounting for approximately 36 percent of the total 1975 U.S. coal reserves; western coal production in the eight-state study area has grown from 35.6 million tons in 1970 to approximately 148 million tons in 1978. Oil shale is also a vast resource, although none is currently being commercially produced; the oil shale reserves estimated in Table 1-1 translate into more than 81 billion barrels of oil. Virtually all of the nation's high-grade oil shale is located in the Green River formation in western Colorado, Utah, and Wyoming.

[1] Other energy resources in the area include hydropower, tar sands, solar radiation, and wind energy, but these are not included in this study.

Table 1-1: *Western Energy Reserve Estimates*

Resource	Reserves (Q's)	Percent of U.S. Total
Coal	3,430	36
Oil	12.2	6
Natural Gas	19.9	8
Oil Shale	464	approx. 100
Uranium	246	90
Geothermal[a]	650	22

Q = quad (equal to 10^{15} Btu. One Q equals approximately 175 million barrels of oil, 60 million tons of western coal, or one trillion cubic feet of natural gas.

[a]This figure includes reserves, submarginal resources, and paramarginal resources.

Similarly, uranium represents a tremendous resource, and most of the nation's high-grade uranium ore reserves are in this eight-state region. While the reserve estimates for oil and natural gas in the area are considerably smaller, these resources are more important than the others in supplying our short-term energy demands, and the estimates of undiscovered oil and natural gas are considerably larger than the reserve estimates. Finally, although estimates of geothermal resources and the future levels of its commercial production are highly uncertain, the long-term potential for this resource is significant. Some studies project as much as 50,000 megawatt-electric (MWe) of geothermal capacity for the U.S. by the year 2000.

3

Character of the Region

The large-scale development of all these energy resources will stimulate economic, environmental, social, and political changes in the region; such changes have already begun to create serious issues that promise to slow or even block development of some of the resources (see box). While many of these problems and issues would occur in any area of the country, a number of regional characteristics make the issues unique:

- Water Scarcity: As indicated in Figure 1-1, the area is largely semiarid. Precipitation levels range from less than 10 inches a year in the desert Southwest to 10 to 20 inches annually in the Northern Great Plains Region. Currently, available water supplies are largely used to support irrigated agriculture.

- Land and Resource Ownership: As shown in Table 1-2, the federal government and the Indians own almost 45 percent of the land in the study area. Together, the federal government and Indians own more than half of the land in Arizona, Utah, and Wyoming and more than a third of the land in Colorado, Montana, and New Mexico. Data on the ownership of resources are more difficult to obtain; however, it appears that the federal government owns about half the region's coal, geothermal, and uranium resources and about 80 percent of its oil shale resources. The 271 Indian reservations in the U.S. are estimated to contain 10 to 16 percent of the nation's coal reserves (Crittenden, 1978) and one-third of all lands considered promising for uranium exploration and development. Most of these resources are on a few of the approximately fifty Indian reservations in our study area (FTC, 1975).

- Sparse Population: Compared with much

Table 1-2: *Federal and Indian Lands in the Eight-State Study Area*

State	Federal and Indian Lands (thousands of acres)	Percent Federal	Percent Indian	Percent Federal and Indian
Arizona	72,688	43	29	72
Utah	52,697	66	4	70
Wyoming	62,343	48	3	51
New Mexico	77,766	34	8	42
Colorado	66,486	36	1	37
Montana	93,271	30	6	36
South Dakota	48,882	7	12	19
North Dakota	44,452	5	5	10
Total	518,585	35	9	44

Sources: U.S. DOI, BLM, 1977:10; and U.S. Dept. of Commerce, 1974.

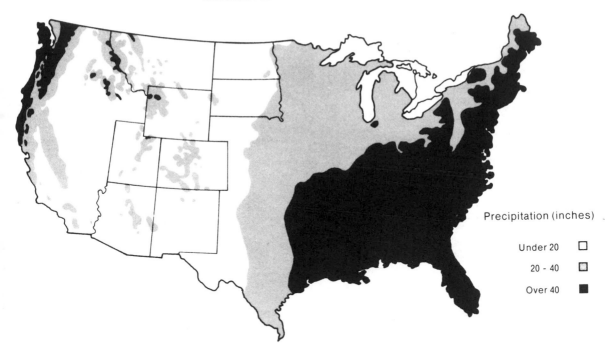

Figure 1-1: *Annual Precipitation*

of the rest of the nation, the states in our study area have small populations (see Table 1-3). All eight states have substantially fewer than 4.3 million people, the average population for the 50 states. Only three of the eight states have population densities of greater than 10 people per square mile; Wyoming, with fewer than 4 people per square mile, is the least densely populated state in the study area; the national average is approximately 60 people per square mile.

- Economic Activities: Agriculture, mining, government, and services are the major sectors of economic activity in the eight-state study area. Manufacturing is a major employer in only three of the eight states—Arizona, Colorado, and Utah.

Census data for 1970 indicate that the median family income in all eight of the states is below the national average of

Table 1-3: *Population Density for Eight-State Study Area*

State	Population (millions of people)	Population Density (per square mile)
Arizona	2.2	20
Utah	1.2	15
Wyoming	.4	4
N. Mexico	1.1	9
Colorado	2.5	24
Montana	.7	5
S. Dakota	.7	9
N. Dakota	.6	9
U.S.	214.5	60

Source: U.S. Dept. of Commerce, Bureau of the Census, 1976:46.

approximately $9,600 per year. Arizona, Colorado, and Utah, the three most populous states, come closest to the national average: Colorado (−$35); Arizona (−$403); and Utah (−$207). Median family income in Arizona would be above

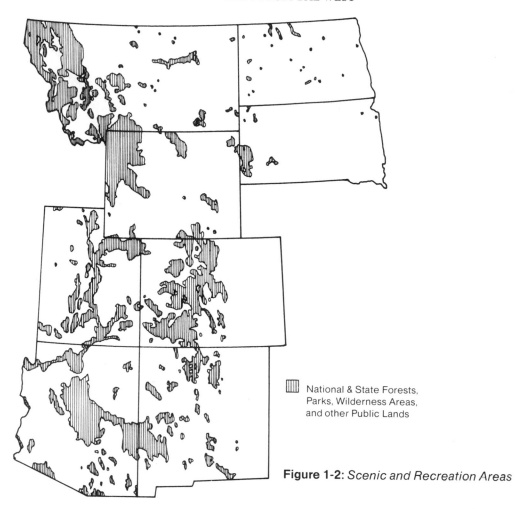

National & State Forests,
Parks, Wilderness Areas,
and other Public Lands

Figure 1-2: *Scenic and Recreation Areas*

the national average if the Navajo reservation were excluded, emphasizing the low income levels on the reservation.

• Public Values and Attitudes: Outside the five major metropolitan areas the region's life-style reflects the slower, more relaxed pace generally associated with small towns and rural areas. A sense of the frontier is still evident, and rugged individualism remains highly regarded.

An obviously strong feeling in the area is that "people ought to take care of their own" rather than depend on government.

If this feeling is not attributable to Mormonism, it appears to be greatly reinforced by it. Whatever the source, a strong strain of populism exists in the eight-state area. One manifestation of this is deeply felt opposition to most kinds of government intervention. There is also a very distinct antipathy toward planning, particularly any attempt to limit how owners use their real property.

• Scenic and Recreation Areas: The eight-state area includes some of the nation's most scenic and highly valued recreation land, including historic, wilderness, and recreation areas, parks, and forests (see Figure 1-2).

Westerners' View of Energy Development

It is difficult to generalize about western attitudes toward the development of energy resources. Assuring environmental quality, protecting scenic beauty, and maintaining present life-styles are widely valued goals, but so are higher personal incomes, greater job opportunities, more amenities, easier access to medical services, and various other benefits associated with energy resource development.

Several consistent themes do emerge, however. For example, outside the large metropolitan areas of the West, there is a general opposition to the intervention of outsiders whether the intervener is the federal government or the Sierra Club. Westerners generally seem to believe that they should be the ones to decide their futures; for example, public officials in southern Utah have strongly expressed the opinion that they should decide whether maintaining 75-mile visibility is more or less important than the anticipated economic and fiscal benefits of energy resource development.

There is little information about public opinion regarding the development of energy resources in the eight-state area. On the basis of contacts we established with public officials in local and state government during the course of this study, our impression is that westerners are more prodevelopment than is generally realized. This impression is supported by the findings of an opinion poll conducted in 1978 by *U.S. News and World Report* (reported in *Denver Post,* 1978). Compared with national averages, a higher percentage of residents of Colorado, Wyoming, and Utah considered the energy crisis serious (47 percent compared to 30 percent nationally) and resisted the idea that clean air and water are worth any price (53 percent to 41 percent nationally).

Of course, opinions are altered by events and vary considerably among residents of the eight states. People living in Utah and Wyoming seem to manifest the strongest prodevelopment position, and those in Montana, the weakest. In several of the eight states, it is not so much an antidevelopment attitude that is frequently expressed as a demand that the West not be treated as a national sacrifice area (for another discussion of western attitudes toward energy development, see Christiansen and Clack, 1976).

Westerners often ask, "Why should we develop our energy resources for the benefit of other regions of the country?" Gov. Jerry Apodoca of New Mexico put it explicitly: "We're not interested in being a colony for the rest of the country" (Congressional Quarterly, 1977:205). Colorado's Sen. Gary Hart has introduced legislation in the U.S. Senate calling for federal impact assistance to compensate the West and to help state and local governments mitigate the undesirable consequences of energy resource development (Inland Energy Development Impact Assistance Act, 1977). And several states have enacted high coal severance taxes; these are intended to ensure that consumers who enjoy the benefits of western energy compensate the West for the costs of development.

Policies Affecting Western Energy Development

The actual pattern and rate of the development of western energy resources will be influenced by a wide range of economic, social, and political factors. There is no such thing as a "western energy policy." For example, although energy development may be taken into account when air- or water-quality legislation is passed, the policy is set solely in terms of air or water quality. Another characteristic of the nation's political system is of course, that policies are made and implemented by various levels, branches, and agencies of government and by various interests in the private sector. These policymakers—at various levels of government and in different agencies of government, in Indian tribes, and in the private sector—all have different objectives and rank their interests and the values at stake differently.

For these reasons it is difficult to address western energy development comprehensively. It is clear that many of the decisions that will determine the consequences of development are and will probably continue to be made in the private sector. That is, energy developers —such as electric utilities and mining and energy companies—will decide such things as whether to "strip and ship" coal or to convert it at the mine mouth; whether to convert coal to electricity or synthetic fuels (synfuels); and whether to build a commercial-scale oil shale facility. Decisions made in the private sector, are, of course, affected by siting laws, leasing and land-use policies, air- and water-quality regulations, taxing and pricing policies, and other government programs and regulations. But to the extent that government attempts to shape the consequences of western energy development, it will primarily do so indirectly, by regulating facility siting, air and water quality, surface-mine reclamation, means of transportation, and other specific facets of development.

One of the key factors affecting the future level of western energy development is virtually beyond the control of the U.S.—namely, the world price of oil. Worldwide increases in oil consumption and the control of oil pricing by the Organization of Petroleum Exporting Countries (OPEC) will be crucial in influencing the commercial development of "new" energy options, especially synthetic fuels made from coal, oil shale, and the enhanced recovery of oil.

Another important factor in determining the level of development is the public's attitude toward the energy situation and its willingness to change behavior and life-styles. Opinions about whether an energy crisis actually exists, about the degree of its seriousness, and about who is to blame continue to

Table 1-4: *U.S. Energy Consumption, 1950-1978*

Year	Total Energy Consumption (10^{15} Btu's)[a]	Average Annual Growth Rate (percent)
1950	34.0	
1955	39.7	3.1
1960	44.6	2.3
1965	53.3	3.6
1970	67.1	4.8
1971	68.7	2.4
1972	71.6	4.2
1973	74.6	4.2
1974	72.3	-3.1
1975	70.7	-2.2
1976	74.2	5.0
1977[b]	76.5	3.1
1978	78.0	1.6

Sources: 1950-1971 from U.S. DOI, 1976; 1972-78 from U.S. DOE, Energy Information Administration, 1979.

[a]Total energy consumption is the sum of inputs into the economy of the primary fuels (petroleum, natural gas, and coal, including imports) or their derivatives, plus the generation of hydro and nuclear power, converted to equivalent energy inputs.

[b]Estimated.

shift (see *Public Opinion,* 1978). However, the ultimate measure of our attitudes is reflected in total energy consumption figures; as indicated in Table 1-4, consumption in 1978 was the highest it has ever been, even though the annual growth rate slowed considerably.

As this discussion suggests, many factors will influence the development of western energy, including private sector investment decisions, governmental regulations and programs, OPEC world oil pricing decisions, and individual attitudes. And although no comprehensive approach for making energy policy exists, there are certain areas of government policy concerning energy, the environment, and the economy that vitally affect the consequences of western energy development. The following discussion summarizes some of the most important aspects of how policy is made as it affects the development of western energy. The changing nature of the political and institutional system and current political conflicts are emphasized. Six topics are discussed: Energy, Environmental and Economic Policies; Public-Private Sector Relationships; Federal-State Relationships; Interstate Relationships; Intrastate Relationships; and Indian Tribes.

Energy, Environmental, and Economic Policies

At the national level, the development of western energy resources will be most directly affected by those policies that emphasize domestic energy development, promote the use of certain energy resources (particularly coal, uranium, and oil shale), and encourage the use of certain technologies (such as coal gasification and liquefaction). In addition, the economic and environmental elements of national and state energy policies, such as siting and water-use policies, will significantly affect how much, where, and in what manner western energy resources will be developed.

National energy policies since the Arab oil embargo of 1973 generally have encouraged the development of western energy resources by emphasizing domestic energy production. This trend continued with passage of the National Energy Act (NEA) of 1978.[2] The NEA encourages the direct burning of coal through policies that will gradually deregulate natural gas and prohibit the use of oil and natural gas in new industrial and utility boilers.

However, the NEA does not include a crude oil equalization tax or taxes on the use of oil and natural gas by electric utilities and industries, both of which had been proposed in President Carter's National Energy Plan. These taxes would have created further incentives for using coal. The NEA does not

subsidize coal liquefaction, coal gasification, or oil shale development, nor does it ensure private developers against the large financial risks that currently accompany these technologies.

In addition to the National Environmental Policy Act (1969), other environmental or environment-related legislation affects the development of western energy, including specific policies intended to protect water quality (the 1972 Federal Water Pollution Control Act [FWPCA] and the 1977 Clean Water Act [CWA]), maintain the natural character of wild and scenic streams (the 1968 Wild and Scenic Rivers Act), protect and preserve endangered species (the Endangered Species Preservation Act of 1973), control the handling and disposal of toxic substances (the Toxic Substances Control Act of 1976 and the Resource Conservation and Recovery Act [RCRA] of 1976), protect drinking water supplies (the Safe Drinking Water Act of 1974), restore surface-mined lands (the Surface Mining Control and Reclamation Act [SMCRA] of 1977), and develop comprehensive land-use plans (the Federal Land Policy and Management Act of 1976). In addition to the basic purpose of protecting environmental quality, all these policies will have the general effect of delaying or restricting the magnitude or location of development or of increasing the cost of domestically produced energy.

However, the environmental policies most likely to influence the development of western energy are those that deal with air quality. Energy development will have to meet either national ambient air quality standards (NAAQS) or more stringent state standards as well as the new source performance standards (NSPS), regulations governing the prevention of significant deterioration (PSD) increments, and requirements for using the best available control technologies (BACT) (Clean Air Act [CAA], 1970, 1977). Energy development in the region will be especially affected by the PSD regulations that are intended to protect air quality in areas where the air currently is better than that provided

---

[2] The NEA is a five-part package of legislation with provisions governing utility rate reform (Utility Regulatory Policies Act, 1978), energy taxes (Energy Tax Act, 1978), energy conservation (National Energy Conservation Policy Act, 1978), coal conversion (Powerplant and Industrial Fuel Use Act, 1978), and natural gas pricing (Natural Gas Policy Act, 1978).

for in the NAAQS, because there are many areas in the West that either already are or potentially are Class I PSD areas (such as national parks, national forests, and wilderness and recreation areas). One effect of the PSD requirements will be to limit the size and location of energy-conversion facilities (such as coal-fired steam-electric power and synfuels plants) in the West. The BACT requirements are also likely to be very important to energy development in the region, since they decrease the advantage low-sulfur western coal has over coal from other regions (see Chapter 5 for a discussion of this issue).

The economics of energy resource development in the region can also be influenced significantly by state severance taxes. Severance taxes on coal in the eight-state area range from 30 percent in Montana to 5 percent in New Mexico to no tax in Utah. While high taxes could have the effect of discouraging development in those states, there is no evidence that this is occurring. Severance taxes do, however, affect energy prices, and they have already contributed to regional economic conflicts.

In addition, many other national and state policies will influence energy development in the region. They include land-use and siting policies, water policies, and transportation policies such as those affecting railroads. Further discussion of the most important of these laws and policies can be found in Chapters 3 through 11.

Public-Private Sector Relationships

Relationships between the public sector and the private sector also influence the development of energy resources. Perhaps more than any other aspect of energy development, siting decisions provide the arena for the most important and most controversial aspects of these relationships. Siting decisions often involve government agencies, industry, environmental groups, landowners, local residents, chambers of commerce, labor unions, and others. Con-

flicts among these participants as a consequence of a siting decision can have important political and social impacts on the local community and surrounding area.

Changing relationships between the public and private sectors can also have consequences far beyond the immediate geographical area. As energy resource development decisions have become less a prerogative of the private sector, the number of participants, the range of interests to be accommodated, the time required, and the attendant uncertainty have also increased. While both positive and negative results can be linked to these trends, the general point is that they highlight the rapidly changing ways in which policies for energy development evolve. The trends also suggest that new mechanisms will need to be developed to accommodate the competing and conflicting goals among the various sectors of society.

Federal-State Relationships

Federal and state government relationships have also been changing rapidly, particularly as federal policies increasingly are developed in areas previously left to the states. Federal environmental policies during the past 10 years generally have had the effect of requiring states either to develop their own standards or to comply with federal regulations. Western states have objected to many forms of federal control, however, and several conflicts have surfaced during the past few years. These include the following: disputes over land use, particularly the extent of state authority over reclamation on federal lands; disputes over water-resource policy, particularly the cancellation of western reclamation projects and the proposals in the Federal Water Policy to increase federal responsibility for water-resource management; and disputes over air quality, particularly the implications for state economic development of BACT and PSD regulations.

Several factors suggest that these conflicts

are likely to continue, if not increase. On the one hand, the federal government will continue to be a prime actor in controlling energy development because of the extent of federal lands in the area, the inherently federal responsibility for energy policy that affects all 50 states, and various other factors. On the other hand, current trends also suggest that the states will continue to press for the authority to control development within their own borders. For some states this effort is reflected in attempts to resist "colonization" by such means as high severance taxes and strict environmental regulations. For other states, the effort is reflected in attempts to encourage development, for example, by keeping taxes low and reserving water resources specifically for energy uses. In both of these cases there is a common desire of western states to resist federal control and to establish or retain as much autonomy as possible.

Interstate Relationships

Just as common interests among the states contribute to a sense of regionalism and give them a common cause in conflicts with the federal government, divergent interests can and have led to conflicts among states. States often have different priorities, including the priority to be given to the development of energy resources. Utah, for example, seems

to rank energy development higher and environmental protection lower than does its neighbor Colorado. This raises the real possibility of conflicts between the two, especially concerning air quality. If Utah is successful in reclassifying large areas within its borders to Class III PSD areas, the resulting air-quality effect is likely to extend into Colorado, thereby threatening Colorado's attempt to preserve its air quality by instituting very strict standards.

Interstate conflicts are also developing in the competition for scarce resources, particularly water. Upper Colorado River Basin (UCRB) states seem to believe that Lower Basin states fail to take into account operational losses and uses of water in the Upper Basin, thus leading the Lower Basin to have unreasonable expectations of how much water the Colorado can supply. And some states are allocating their entire share to prevent losing it in the future to downstream users. In fact, some states, such as New Mexico, apparently have already allocated more than their share under the existing compact.

Energy development also affects the West's relations with other regions. Conflicts have arisen primarily over the economics of producing energy. For example, resentment of Montana's high coal-severance tax has led some nonwesterners to call Montanans "blue-eyed Arabs." In Michigan, where Detroit Edison has a 26-year contract for 193 million tons of Montana coal, one Detroit newspaper called the severance tax usurious, characterizing it as "fuel blackmail" (Richards, 1977). In the fall of 1977, Montanans voted to put 25 percent of their coal-tax revenues into a state trust fund, with deposits increasing to 50 percent by 1980. States whose residents are paying higher energy prices because of Montana's severance tax do not believe that Montana should be taxing to enlarge its general revenue fund; they believe that Montana's severance taxes should be set only at a level required to cover the impacts (*Electrical World,* 1978). Wisconsin Power and Light is challenging Montana's coal tax in the courts on the grounds that it is excessive and a burden on interstate commerce.

Interregional disputes are also emerging over the broader question of where the nation's major coal development will occur. In Congress, debate over an amendment called the "local coal use mandate" generated bitter regional rivalry, represented most vocally by Sens. Clifford Hansen of Wyoming and Howard Metzenbaum of Ohio. The purpose of the amendment is to protect the markets for interior and eastern coals from lower sulfur western coals. According to this policy, the governor of a state would be able to require industry and utilities to use local coal, in conjunction with appropriate emission-control technologies, if he determined first, that existing supply contracts would not be violated; second, that energy would not be wasted; and third, that significantly high costs to consumers would not result. President Carter supported this proposal and announced his intention to reassess "the appropriate contribution of western coal to the national energy budget" (Kirschten, 1977). When this amendment was voted on in June 1977, the Senate, dividing largely along regional lines, adopted it; it subsequently was incorporated into the 1977 CAA Amendments.

Intrastate Relationships

Energy development also creates conflicts among governmental units within states. However, these conflicts usually arise more in connection with controlling the impacts of large population fluctuations than with balancing energy and environmental concerns. Local governments in the West have traditionally paid little attention to the need for developing mechanisms to cope with energy-related growth problems—largely because this has not been much of a problem in the past. While local governments usually have ways to control development, such as zoning ordinances, building codes, and health and sanitation standards, these instruments also have serious

limitations. One is that energy development usually occurs near small or rural towns; these communities lack not only financial resources to carry out planning studies but also technical planning expertise and adequate information about development.

In the West local governments are often constrained by legal or constitutional restrictions imposed by the states. In some states debt ceilings, limits on interest rates, and existing tax structures limit the local capacity to finance community facilities. The Utah constitution, for example, prohibits the transfer of state revenues, including impact aid, to cities or counties. Some states restrict the freedom of local governments to require a developer to provide some kinds of public facilities, thus limiting growth-management alternatives (Kutak Rock Cohen Campbell Garfinkle & Woodward, 1974:77–78).

The combination of large, rapid population increases combined with existing institutional and legal constraints can create conflicts among the several governmental levels in a particular area. The city of Gillette, Wyoming, for example, has had housing developments halted by the state because of inadequate local water- and sewage-treatment facilities. In Montana officials of Miles City have expressed concern about not being able to obtain impact aid from the Montana Coal Board, both because they have been classified as only a secondary area of development and because they have not been able to finance an updated census needed to qualify for assistance (Kennedy, 1977).

Problems also emerge among city versus county and urban versus rural interests. The biggest city-county issue seems to be that the costs and benefits of developing energy are often distributed inequitably. Most school districts and county governments can expect to have a revenue surplus from property or severance tax receipts, since energy facilities are usually within county and school districts rather than municipal boundaries. However, it is generally the cities that experience the large population increases and that are expected to meet needs for housing, water, sewers, and other facilities and services.

Indian Tribes

Indians also have been affected by energy and environmental conflict, and relations between Indians and all levels of government have changed as a result. Indians have become an important group in formulating energy policies because, as noted earlier, substantial resources are on tribal reservations. Furthermore, Indian water rights could significantly influence the allocation of limited water supplies for the development of energy resources (and for other uses). Finally, Indians have become increasingly concerned about the impacts of development on the reservation.

Because of these factors, tribes have become more organized. They are represented by such interest groups as the Native American Rights Fund, Americans for Indian Opportunity, the National Congress of American Indians, the Council of Energy Resource Tribes (CERT), and the Native American Natural Resource Development Federation. The last organization represents the Northern Great Plains tribes for the purposes of defining and describing natural and cultural resources; developing programs to identify the impacts of energy resource development; and representing the tribes on federal and state land and water organizations. CERT, representing 22 tribes, was organized to form an energy combine that will maximize the benefits of developing tribal resources.

One example of the role Indians can play in energy development is the Northern Cheyennes' redesignation of their reservation to a Class I PSD status for air quality. As a result, only very small additional increments of particulates and sulfur dioxide (SO_2) are allowed. As discussed in detail in Chapter 5, this has had the effect of blocking construction of two coal-fired generating units near Colstrip, Montana. While there is still some uncertainty as to whether construction will eventually proceed, this is an indication of the potential power of Indian tribes.

This Study

The U.S. is under considerable pressure, both at home and abroad, to reduce its imports of foreign oil. To do so will require success both in limiting total energy consumption and in increasing the production of domestic energy. If domestic energy supplies are to be expanded, it is clear that a large part of that increase must come from the western U.S., which is rich in coal, oil shale, oil, natural gas, uranium, and other energy resources. As the preceding discussion indicates, however, the development of these resources will not only bring economic growth but also generate technical, environmental, economic, social, and political issues. The overall picture that emerges is that the eight-state region is large in land area but sparsely populated, energy rich but water poor, relatively unindustrialized, largely influenced by the decisions of the federal government and of Indian tribes who own large land areas, and strongly opposed to federal intervention, particularly when it affects westerners' rights concerning their personal property and lives.

The purposes of this study are to explore the consequences of the large-scale development of western energy resources and to analyze what can be done to realize the benefits while minimizing the costs of development. Chapter 2 briefly discusses the structure and approach of our study: energy-development technologies are described, and methodologies used for identifying impacts and analyzing policy alternatives are discussed. Chapters 3 through 11 discuss issues central to the question of developing western energy resources:

- Water Availability;
- Water Quality;
- Air Quality;
- Land Use and Reclamation;
- Housing;

- Growth Management;
- Capital Availability;
- Transportation; and
- Facility Siting.

Each of these chapters analyzes the problems and issues that result from energy develop-

ment and evaluates some policy alternatives for dealing with the problems and issues identified. Chapter 12 presents an overview and offers a conclusion by summarizing how various levels of government and the energy industries can address the issues most directly affecting them.

REFERENCES

Christiansen, Bill, and Theodore H. Clack, Jr. 1976. "A Western Perspective on Energy: A Plea for Rational Energy Planning." *Science* 194 (November 5):578–84.

Clean Air Act (CAA) Amendments of 1970, Pub. L. 91–604, 84 Stat. 1676.

Clean Air Act (CAA) Amendments of 1977, Publ. L. 95–95, 91 Stat. 685.

Clean Water Act (CWA) of 1977, Pub. L. 95–217, 91 Stat. 1566.

Congressional Quarterly, Inc. 1977. *Congress and the Nation. Vol. 4: 1973–1976.* Washington, D.C.: Congressional Quarterly.

Crittenden, Ann. 1978. "Coal: The Last Chance for the Crow." *New York Times,* January 8, Sect. 3, p. 1.

Denver Post, May 14, 1978, p. 3E.

Electrical World. 1978. "The Fuels Outlook." 189 (March 1):49.

Endangered Species Preservation Act of 1973, Pub. L. 93–205, 87 Stat. 884.

Energy Tax Act (1978), Pub. L. 95–618, 92 Stat. 3174.

Federal Land Policy and Management Act of 1976, Pub. L. 94–579, 90 Stat. 2743.

Federal Trade Commission (FTC), Bureau of Competition. 1975. *Report to the Federal Trade Commission on Mineral Leasing on Indian Lands.* Washington, D.C.: FTC.

Federal Water Pollution Control Act (FWPCA) Amendments of 1972, Pub. L. 92–500, 86 Stat. 816.

Inland Energy Development Impact Assistance Act of 1977, Senate Bill 1493.

Johnson, Haynes. 1975. "The Last Round-Up." *Washington Post,* August 3, p. C-5.

Kennedy, Barbara, City-County Planner, Miles City, Montana. June 1977. Personal communication.

Kirschten, J. Dicken. 1977. "Watch Out! The Great Coal Rush Has Started." *National Journal* 9 (October 29):1683–85.

Kutak Rock Cohen Campbell Garfinkle & Woodward. 1974. *A Legal Study Relating to Coal Development-Population Issues,* vol. 1, *Responding to Rapid Population Growth,* prepared for the Old West Regional Commission. Omaha, Neb.: Kutak Rock Cohen Campbell Garfinkle & Woodward.

National Energy Conservation Policy Act (1978), Pub. L. 95–619, 92 Stat. 3206.

National Environmental Policy Act (NEPA) of 1969, Pub. L. 91–190, 83 Stat. 852.

Natural Gas Policy Act (1978), Pub. L. 95–621, 92 Stat. 3350.

Powerplant and Industiral Fuel Use Act (1978), Pub. L. 95–620, 92 Stat. 3289.

Public Opinion. 1978. "The Energy Imbroglio." 1 (March/April):27.

Resource Conservation and Recovery Act (RCRA) of 1976, Pub. L. 94–580, 90 Stat. 2795.

Richards, Bill. 1977. "Boom in Strip Mining: Windfall for Montana." *Washington Post,* May 22.

Safe Drinking Water Act of 1974, Pub. L. 93–523, 88 Stat. 1660.

Surface Mining Control and Reclamation Act (SMCRA) of 1977, Pub. L. 95-87, 91 Stat. 445.

Toxic Substances Control Act of 1976, Pub. L. 94-469, 90 Stat. 2003.

U.S. Congress, Senate, Committee on Environment and Public Works. 1978. *Water*

Research and Development. Hearings before the Subcommittee on Water Resources, 95th Cong., 2d sess., April 7.

U.S. Department of Commerce. 1974. *Federal and State Indian Reservations and Indian Trust Areas.* Washington, D.C.: Government Printing Office.

U.S. Department of Commerce, Bureau of the Census. 1976. *Pocket Data Book, USA 1976.* Washington, D.C.: Government Printing Office.

U.S. Department of Energy (DOE), Energy Information Administration. 1979. "Part 1—Executive Summary." *Monthly Energy Review,* February, p. 4; April, p. 2.

U.S. Department of the Interior (DOI). 1976. *Energy Perspectives 2.* Washington, D.C.: Government Printing Office.

U.S. Department of the Interior (DOI), Bureau of Land Management (BLM). 1977. *Public Land Statistics 1976.* Washington, D.C.: Government Printing Office.

Utility Regulatory Policies Act (1978), Pub. L. 95-617, 92 Stat. 3117.

Wild and Scenic Rivers Act of 1968, Pub. L. 90-542, 82 Stat. 906.

CHAPTER 2

Scope and Structure of the Study

As indicated earlier, this study was undertaken to identify and analyze a broad range of consequences of the development of energy resources in the western U.S. and to evaluate and compare alternative courses of action for dealing with these consequences. The study approach used is one that has become known as "technology assessment." This chapter begins by describing our view of technology assessment as a kind of applied policy analysis; this is followed by discussions of the energy development technologies we considered, the scenarios we used to structure our analysis, and the organization of our study process.

Technology Assessment as Applied Policy Analysis

Technology assessments (TAs) have two kinds of objectives: first, to inform public and private policymakers as well as interested citizens about the likely consequences of developing and operating a particular technology; and second, to identify, evaluate, and compare alternative policies and implementation strategies for dealing with subsequent problems and issues—both those that are perceived as likely to arise and those that actually are likely to arise. To inform policymakers and citizens about the consequences of developing energy resources three questions must be answered:

- Are the anticipated consequences actually likely to occur?

- Are there also likely to be other consequences that have not been anticipated?
- If either the anticipated or previously unanticipated consequences do occur, how serious will they be?

To identify, evaluate, and compare alternatives, the answers to these questions must be related to the social and political context within which the particular technology will be developed and operated. The questions to be answered at this stage are:

- What alternative policies and strategies for implementation can reasonably be used both to maximize benefits and to minimize costs and risks when the technology is developed and put into operation?
- How will these alternatives distribute costs, risks, and benefits throughout society?

A diagram of the general conceptual framework used to structure our research effort is presented in Figure 2-1. This simplified diagram shows that when a technology is operated, it makes input demands (such as for capital, materials, and labor) and produces outputs (such as electricity, air emissions, and water effluents). When these inputs and outputs interact with conditions at the development site (such as the ambient air quality, the availability of water, and the availability of community services and facilities), a range of impacts occurs (such as changes in air and

17

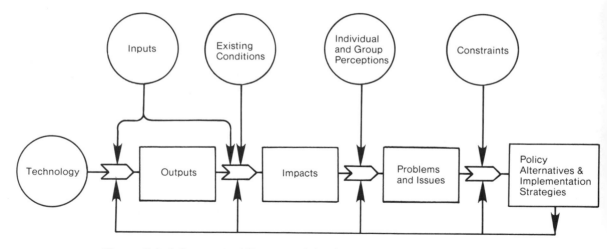

Figure 2-1: *A Conceptual Framework for Assessing Physical Technologies*

water quality and a need for more housing). Some of these impacts may cause problems and raise issues. When policymakers respond to these problems and issues, they operate under a variety of constraints (legal, technical, and political) which affect the feasibility and/ or acceptability of alternative policies and implementation strategies. The next two sections of this chapter describe the scope of our western energy study; the final section describes the process we followed to implement this conceptual framework.

Technologies for Developing Western Energy Resources

As indicated in Chapter 1, this study considers the development of six energy resources: coal, oil shale, uranium, natural gas, crude oil, and geothermal energy. We also considered a range of development technologies for each resource, as illustrated in Figure 2-2; the technologies selected are representative of those likely to be used to develop western energy resources during the next 25 years. A brief summary of these technologies follows. (More detailed descriptions can be found in White et al., 1979a.)

Coal Development Technologies

In this study, the technologies we assessed for producing and converting coal to other energy forms are coal mining, coal-fired steam-electric power plants, Lurgi and Synthane coal gasification, Synthoil coal liquefaction, and unit-train and slurry-pipeline transportation.

Coal Mining: Both surface and underground coal mining are considered in this study. Surface (or strip) mining begins with the removal and storage of topsoil from the area. (Topsoil is stored so that it can be replaced later during reclamation.) The overburden—that is the rock and soil material between the surface and the coal seam—is loosened by blasting and removed using a dragline. The dragline, which is usually electrically powered, lifts the overburden and places it on a spoils pile adjacent to the mining area (Figure 2-3). The exposed coal is then mined and loaded into large trucks or onto conveyor belts for transportation to either a conversion or a loading facility. Current regulations require surface-mined lands to be reclaimed. The present practice is for mining and reclamation to proceed simultaneously. The overburden is placed in the mined-out area, graded, and contoured; top-

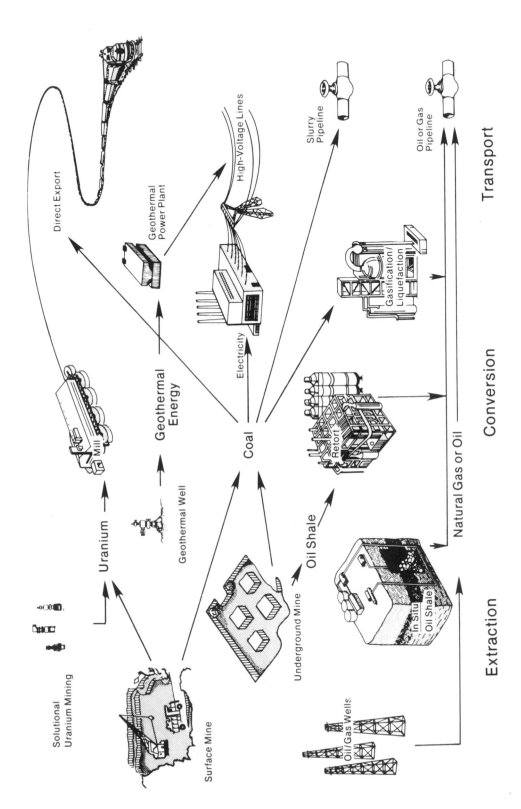

Figure 2-2: *Energy Resource Development*

Extraction   Conversion   Transport

Solutional Uranium Mining

Uranium → Mill

Surface Mine

Geothermal Well

Geothermal Energy

Direct Export

Geothermal Power Plant

High-Voltage Lines

Coal

Electricity

Underground Mine

Oil Shale

Retort

In Situ Oil Shale

Oil/Gas Wells

Natural Gas or Oil

Gasification/Liquefaction

Slurry Pipeline

Oil or Gas Pipeline

soil is then replaced and the area is revegetated.

Room-and-pillar mining is the predominant underground coal mining technique. As the coal is mined, pillars of it are left in place to support the mine's roof. (Roof supports are used in addition to the pillars.) Continuous mechanical miners are used to scrape the coal from the seam and load it directly onto a conveyor or into a rail car. Reclamation for underground mines involves permanently disposing of the spoils mined along with the coal and of the material removed to gain access to the seam. These waste materials are usually stabilized with lime and deposited in sealed landfills.

Coal-Fired Steam-Electric Power Plants: The power plants analyzed in this study are direct-fired boilers producing high-pressure, superheated steam, which is then expanded in a multistage turbine to produce mechanical energy. The mechanical energy is then converted to electricity in an electrical generator. We have not explored the details of boiler and turbine design, but we have assumed an overall efficiency for the plant of 34 percent, including environmental controls. These environmental controls include a wet limestone scrubber for flue-gas desulfurization (FGD) and an electrostatic precipitator for particulate removal. We have assumed there will be no cleanup technology for oxides of nitrogen (NO_x).

Because of thermodynamic limitations, almost two-thirds of the heat generated in a power-plant boiler must be dissipated. For the basic plant, we have assumed wet cooling towers will be used to dissipate the heat by evaporating water into the atmosphere. To examine the sensitivity of impacts to the cooling technology, we also examined wet-dry cooling system costs and water consumption.

High-Btu Coal Gasification: Two high-Btu coal gasification processes are considered. The Lurgi process was selected for study because it

Figure 2-3: *Dragline Used for Coal Surface Mining*

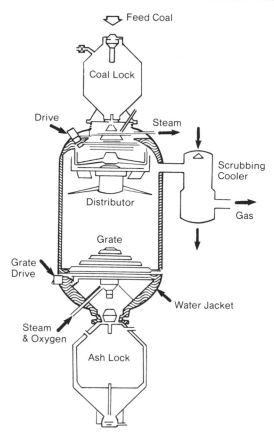

Figure 2-4: *Schematic of a Lurgi Gasifier*

a pressure of between 300 and 500 pounds per square inch (psi) atmosphere and at a temperature which varies between 1,800° F at the bottom of the gasifier and 1,000° F at the top (Figure 2-4). Because the rate at which coal is gasified in Lurgi units is low, between 25 and 30 reactors are needed for a plant producing 250 million cubic feet of synthetic natural gas per day. Lurgi units have a limitation because they will not process "caking coals" which are common in eastern U.S. regions. We assumed an overall energy conversion efficiency of 73 percent.

The Synthane process is a fluidized-bed gasifier involving three separate stages. The first stage pretreats the coal by partially burning it at 650 psi and 800° F so that it will not stick together during gasification. (The pretreatment is used only for caking coals.) From the surface treatment unit, the coal then enters the fluidized-bed unit at 1,750 to 1,850° F where the gasification occurs. Unconverted carbon (char) is discharged from the gasifier; it can be used to generate steam or process heat, but it can have a high sulfur content. Because it is a pressurized fluidized bed with a higher gasification rate, fewer commercial reactors will be needed than for the Lurgi process. We have assumed that the Synthane process operates at a conversion efficiency of 80 percent.

Following either the Lurgi or Synthane basic gasification process, solids and undesirable byproduct gases are removed before the methane is generated. Byproduct gases include carbon dioxide (CO_2) and hydrogen sulfide (H_2S). The H_2S is reduced to elemental sulfur and water in a Claus plant.

Water is used in these facilities both for the gasification process and for cooling. As with the steam-electric power plant, we assumed wet cooling towers would be used in the basic plant, but wet-dry and all-dry cooling are also examined.

Coal Liquefaction: Coal liquefaction processes are at an earlier stage of development than gasification, and therefore data on lique-

is a presently available commercial-scale technology. The Synthane process is selected as representative of a number of second-generation processes which could be commercially available by 1985 to 1990.

In high-Btu coal gasification, coal is transformed into gas by heating and then burning it with reduced oxygen and steam to produce carbon monoxide and hydrogen. This mixture of gases is then upgraded to create synthetic natural gas (primarily methane) in a separate reactor using a catalyst.

The Lurgi gasifier which is in commercial operation today in South Africa uses a moving grate and gasifies with steam and oxygen at

Oil Shale Country of Western Colorado

faction processes are somewhat limited and uncertain. The liquefaction process that we considered in this study is the Synthoil process developed by the Bureau of Mines.[1]

In the Synthoil process, crushed coal is slurried in (mixed with) oil, then the mixture is introduced (along with hydrogen) into a pressurized reactor (2,000 to 4,000 psi) containing a fixed bed of catalyst and heated to about 850° F. In the reaction that results, hydrogen is introduced into the coal molecules or their fragments, producing a liquid. The sulfur in the coal is converted to H_2S, and this is removed and converted to elemental sulfur in a Claus plant. Water is used both for cooling and as a source of hydrogen for the process.

[1]Since the time when these technologies were selected for study (late 1975), there has been considerable change in the technical progress on various processes. It now appears that Synthoil will not become commercialized; yet, its resource requirements and emissions are similar to other liquefaction processes based on hydrogenation. Therefore, the conclusions drawn do not depend greatly on the specific processes considered.

Coal Transportation: We have analyzed two options for transporting coal: unit trains and coal-slurry pipelines. For the purposes of this study, unit trains consist of 100 cars, each with a capacity of 100 tons. The capacity of a unit-train system is variable, depending upon the number of trains, their frequency of operation, and whether the system is a single track, double track, or a mixture of the two at various places along the route.

In a coal-slurry pipeline, the coal is pulverized, mixed with water, and pumped through a pipeline. Approximately equal parts (by weight) of coal and water are required. We have assumed that the carrying capacity of a coal-slurry pipeline is 25 million tons per year (MMtpy) of coal.

Oil Shale Development Technologies

The technologies we have considered for developing oil shale are underground oil shale mining and surface retorting using the TOSCO II process, and modified *in situ* recovery using the Occidental process.

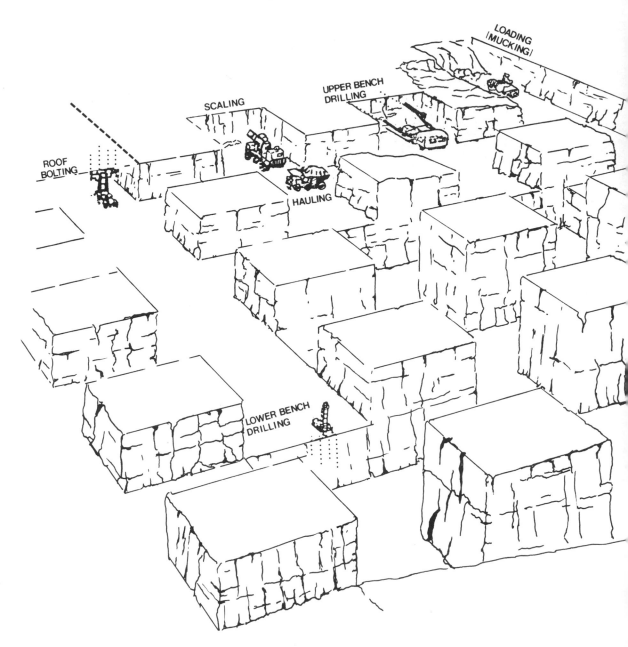

Figure 2-5: *Oil Shale Room-and-Pillar Mining*

Underground Oil Shale Mining: Conceptually, underground oil shale mining is similar to underground coal mining and most frequently uses the room-and-pillar method. Compared to coal mines, however, oil shale mines are very large, with roof heights of as great as 60 to 80 feet. These large rooms are mined in two zones. The top zone is mined with equipment extracting the shale from the wall (or face) of the resource, while the bottom zone is mined by extracting the shale from the floor (or bench). Large front-end loaders are used to load the mined shale into trucks which transport it to a sizing and crushing facility. Because of the enormous size of these mines, equipment more commonly seen in surface mines, such as large trucks and drill rigs, is used. A schematic drawing of an underground oil shale mine is shown in Figure 2-5.

Surface Oil Shale Retorting: We studied TOSCO II surface retorting. In the TOSCO II retort, one-half inch diameter ceramic balls are heated to about 900° F and then put into the retort with small pieces of raw shale. The retort vessel is then rotated so that the balls heat the shale by contact and, at the same time, crush it to a powder. This heating process is done in an inert gas atmosphere rather than in air, so that the released shale oil does not burn. The oil is collected, and the pulverized spent shale is separated from the balls with moving screens and carried away for disposal. A low-Btu gas also is generated; it is collected and used as a fuel in the heater for the ceramic balls.

Because the energy content of shale is relatively low (only about 25 to 35 gallons of shale oil can be extracted from a ton of ore), a large quantity of spent shale must be disposed of. The pulverized spent shale from the TOSCO II process can be set as a cement with about 13 percent water; this cement is stable enough for permanent disposal. We have assumed that a small canyon in the vicinity of the retort would be used for shale disposal and filled with the shale cement to a depth of several hundred feet. The disposal area would then be covered with topsoil and vegetated.

In Situ Oil Shale Retorting: We also analyzed the Occidental modified *in situ* retorting process as an alternative to surface retorting (see Figure 2-6). In the modified *in situ* process, oil shale is mined from a region below a high-quality ore deposit. The mined oil shale can either be discarded or processed in a surface oil shale retort. The high-grade ore is then broken into rubble with shaped-charge blasting to form large *in situ* retorts, measuring more than 100 feet on a side and nearly 300 feet high. Air and steam are circulated through the shale rubble, and the shale is heated until it ignites in a burning front which moves downward, releasing shale oil ahead of the combustion. The shale oil is collected in sumps, and the gases produced are treated to recover additional energy and to reduce emissions. *In situ* retorting requires the handling of a much smaller quantity of shale than surface retorting. As for disposal, the raw shale from *in situ* retorting is more stable than processed shale from surface retorting.

Uranium Development Technologies

The technologies for producing uranium include surface, underground, and solutional uranium mining; and acid-leach milling of the ore from surface and underground mines to produce "yellowcake" (uranium oxide [U_3O_8]).

Surface and Underground Uranium Mining: A surface uranium mine is more similar to pit mining than to the strip mining typically used to recover coal. Uranium is located in veins which vary in ore quality and are irregular in location. Mining is therefore carried out selectively, with lower-quality ore left as a spoil. As truckloads of ore leave the pit, radioactivity is measured to grade the ore for milling. The equipment used for mining uranium is similar to that used for mining coal

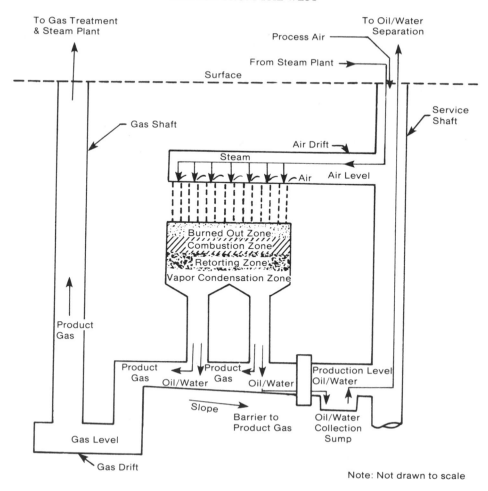

Figure 2-6: *Simplified Sketch of* In Situ *Oil Shale Retort*
(Source: Ashland and Occidental, 1977.)

but it usually is smaller in size because of the smaller quantities of ore involved.

Reclamation of uranium mines includes layering the overburden by composition to limit the potential for damage from trace elements in the overburden leaching into groundwater and to assure the stabilization of materials that may be radioactive.

Underground uranium mining is done by a room-and-pillar method. Ore is mined using continuous miners and removed by conveyors to the access shaft where self-dumping buckets are used to haul the ore out of the mine.

We have assumed that reclamation is necessary only for solid wastes from the mine. Thus

reclamation will be limited to stabilizing these solid wastes, placing topsoil over the landfill, and revegetating the area.

Solutional Uranium Mining: We included solutional uranium mining in this study to present a comparison with surface mining and also because the irregular seams, or veins, and the nature of the geology of uranium make this the most economical mining technique. In solutional mining an alkaline solution is injected into the ore formation through a group of injection wells. This solution dissolves the uranium compounds; the "pregnant" solution of uranium is pumped out of the formation in recovery wells; and the solution is then processed to recover yellowcake. Thus, the mining and milling processes are combined in solutional mining.

Reclamation after solutional mining involves cleansing the groundwater in the mined-out area to ensure that uranium and other heavy-metal compounds do not remain. This is accomplished by pumping clean water into the formation and recovering the remaining alkaline solution and dissolved compounds. The recovered solution is then either cleaned or disposed of in an evaporative pond.

Uranium Milling: Uranium milling is a process by which U_3O_8 is extracted from uranium ore and concentrated into a yellowcake which is about 90 percent U_3O_8. The milling process studied in this analysis is acid leaching. The ore is crushed, slurried, and pumped into heated, agitated tanks where sulfuric acid is added. The uranium ore dissolves in the acid, and the solution is then separated from the waste materials, or tailings, which are discarded in a tailings pond. The solution is concentrated with additional chemical processes, and finally the U_3O_8 is separated out of the solution and dried for shipment. We have assumed that the yellowcake will be transported by truck.

Oil Development Technologies

In this study, oil production is assumed to include rotary drilling wells in which water-based mud is used as a drilling fluid for lubrication and to remove cuttings. We have assumed that the wells are lined with steel-pipe casing cemented in place and that the oil is produced through tubing going from the bottom of the well in the oil reservoir to the surface. No wellhead processing of the oil is considered except for water removal, after which the oil is piped out of the region for refining elsewhere.

Enhanced Oil Recovery: The enhanced oil-recovery method we have analyzed is steam flooding. Two sets of wells are used, one to inject the steam and the second to recover the oil-and-water mixture. The steam increases the temperature of the oil so that its viscosity is reduced; thus, it flows more readily. The oil is then pushed by the steam pressure to the recovery wells. Most of the steam injected is recovered as water in a mixture with the oil. This water must be disposed of in evaporation ponds or cleaned and recycled to the steam generators.

Natural Gas Development Technologies

In this study, we have analyzed the conventional production of natural gas. As with oil, we assumed the use of rotary drilling using water-based mud as a drilling fluid; wells were assumed to be cased with a heavy pipe cemented in place; and the gas, produced through tubing (of a smaller diameter) from the production zones. The only processing assumed is drying using a glycol process.

Geothermal Development Technologies

Hot-water geothermal development is analyzed because it is one of the most likely geothermal resources to be developed in the eight-state study area. We have defined this resource as having a temperature above 300° F; we have also assumed that power is produced by either a flashed-steam or binary-cycle process.

In the flashed-steam process, the hot, high-pressure water is expanded to a low pressure so that it vaporizes. The vapor is then used to drive a low-pressure turbine, which in turn drives an electric generator. If the hot water is impure and contains salts or solid particulates, the flashed steam must be cleaned before entering the turbine.

In the binary-cycle process, the hot water is used to heat another fluid with a lower boiling point (such as propane) which is then vaporized and used to power a turbine. The binary cycle has fewer problems with impure water but is more complicated than the flashed-steam process. In both processes, cooling calls for wet cooling towers that use the geothermal condensate as a source of water.

The Energy-Development Scenarios

In order to analyze the various impacts that are likely to result from the development of western energy resources, we used two types of hypothetical energy-development scenarios: site-specific scenarios and regional scenarios. Six site-specific scenarios were constructed to combine the various development technologies with actual sites, representative of the

range of conditions that now exist in the study area. The components of these site-specific scenarios are summarized in Table 2-1, and a map of the Gillette, Wyoming, scenario is provided in Figure 2-7.

We also drew up two regional scenarios—one based on Low and one on Nominal national demands for energy between 1975 and 2000. These scenarios were created primarily using Stanford Research Institute's (SRI) inter-fuel competition model (Cazalet et al., 1975), but for this study, we made adjustments to include *in situ* oil shale and geothermal energy development. Table 2-2 summarizes the projected quantities of energy that will be produced from each resource for these two scenarios.

Table 2-1: *Standard size Facilities in the Six Site-Specific Scenarios*

	Navajo/ Farmington, New Mexico	Kaiparowits/ Escalante, Utah	Rifle, Colorado	Gillette, Wyoming	Colstrip, Montana	Beulah, North Dakota
Coal						
Mine/Export (25 MMtpy)						
Rail				1		
Slurry				1		
Power Plant (3,000 MWe)	1	2	1[a]	1	1	1
Gasification (250 MMcfd)						
Lurgi	1			1	1	2
Synthane	1			1	1	2
Synthoil (100,000 bbl/day)	1			1	1	
Oil Shale						
Surface Retorting (50,000 bbl/day)			1			
In Situ (57,000 bbl/day)			1			
Uranium						
Mine/Milling (1,000 mtpy)	1			1		
Solutional Mining (250 mtpy)				1		
Natural Gas Wells (3 MMcfd)				83		
Crude Oil Wells (125 bbl/day)			400			

MMcfd = million cubic feet per day. bbl/day = barrels per day. mtpy = metric tons per year.

[a]The Rifle power plant is only 1,000 MWe.

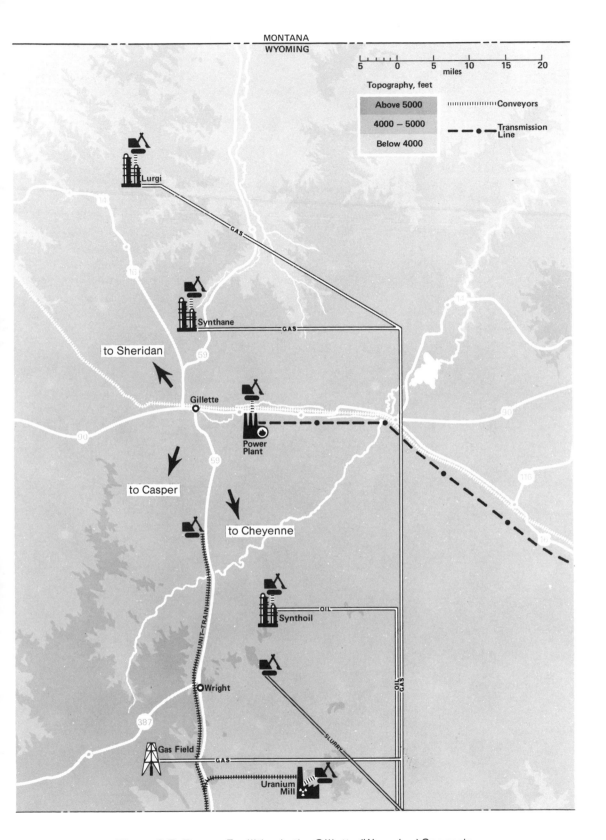

Figure 2-7: *Energy Facilities in the Gillette (Wyoming) Scenario*

Table 2-2: *U.S. and Regional Projected Energy Development*

Resource and Levels of Development	Total U.S. Production (Q's)			Production in the Eight-State Study Area					
				1980		1990		2000	
	1980	1990	2000	Q's	Percent U.S. Total	Q's	Percent U.S. Total	Q's	Percent U.S. Total
Coal									
Nominal Demand	15.12	25.12	50.99	5.03	33.3	12.61	50.2	29.93	58.7
Low Demand	13.36	20.24	38.65	4.28	32.0	9.80	48.4	22.11	57.2
Oil Shale									
Nominal Demand	.001	.95	4.76	.001	100.0	0.95	100.0	4.76	100.0
Low Demand	.001	0.19	1.91	.001	100.0	0.19	100.0	1.91	100.0
Uranium Fuel									
Nominal Demand	5.34	13.90	26.10	4.77	89.3	13.64	91.1	23.75	91.0
Low Demand	4.56	10.40	18.80	4.15	91.0	9.46	91.0	17.11	91.0
Gas (Methane)									
Nominal Demand	23.73	26.02	18.34	1.97	8.3	2.08	8.0	1.06	5.8
Low Demand	23.12	24.61	17.69	1.99	8.6	1.89	7.7	1.19	6.7
Domestic Crude Oil									
Nominal Demand	21.10	25.96	22.79	1.69	8.0	1.32	5.1	1.03	4.5
Low Demand	21.16	25.87	22.62	1.74	8.2	1.34	5.3	.90	4.0
Geothermal									
Nominal Demand		1.49						0.15	10.0
Low Demand		0.21						0.02	10.0

Q = 10^{15} British thermal units. One Q equals approximately 172 million tons of western coal or one trillion cubic feet of natural gas.

In the SRI model, the geographical distribution of development was carried out by dividing the western states into only two subregions, the Powder River and the Rocky Mountain areas. For some of our impact analyses, it was necessary to disaggregate further, to an individual state. The number of facilities by state that we have projected and used in our analysis is given in Table 2-3. For the near future (up to 1985), disaggregation was done on the basis of the locations of energy facilities that have already been announced (Denver Federal Executive Board and Mountain Plains Federal Regional Council, 1975). Thereafter (for 1990 and 2000), development was assumed to be proportional to the proven reserves in each state. Disaggregation based on resource levels was done only to provide a basis for impact analysis. As discussed in Chapter 11, the actual siting of facilities depends upon a number of other factors, including site characteristics, the availability of land, and several legal and institutional factors.

Organization of the Study Process

In order to achieve the broad goals of technology assessment as described in the beginning section of this chapter, the analytical skills and perspectives of many disciplines are required. We used an interdisciplinary team approach to ensure that a range of skills, perspectives, and procedures were available to review extensively the results of our research.

Table 2-3: *Number of Standard-Size Facilities[a] by State in the Low- and Nominal-Demand Scenarios*

State	1975	1980 Low	1980 Nominal	1990 Low	1990 Nominal	2000 Low	2000 Nominal
Colorado							
Power plants		1	1	1	2	1	2
Modified *in situ*		0	0	1	3	3	13
Uranium		0	0	1	2	2	3
Natural gas		0	0	0	8	4	4
TOSCO II		0	0	0	2	2	10
New Mexico							
Natural gas	19	22	22	21	15	9	8
Crude oil	7	8	8	6	6	5	5
Uranium	4	9	10	19	22	31	42
Power plants		0	1	1	1	1	1
Geothermal		0	0	1	10	10	71
Gasification		0	0	0	0	1	2
Utah							
Power plants	1	1	1	1	2	1	2
Uranium		0	0	0	2	2	3
TOSCO II oil shale		0	0	0	0	0	2
Montana							
Power plants		1	2	3	5	5	6
Gasification		0	0	0	1	9	15
Liquefaction		0	0	0	0	1	1
Wyoming							
Power plants	3	1	1	3	2	3	3
Uranium		5	6	12	16	22	31
Gasification		0	0	0	0	5	9
Liquefaction		0	0	0	0	1	1
North Dakota							
Power plants		2	2	4	6	6	9
Gasification		0	0	2	2	13	21

[a]The size of facility is indicated in Table 2-1.

32

Interdisciplinary Research Approach

As the research progressed and draft papers were prepared by members of the research team, each draft was subjected to an intensive critical review by the entire team and then redrafted. The result of this internal review process was still considered a draft and was critically reviewed by an external advisory committee, subcontractors, consultants, and a broad range of interested parties.

The internal reviews were intended to bring the team's collective knowledge, analytical skills, and perspectives to bear on the TA topic and to produce a well-informed, technically correct, reasonably comprehensive, and —to the extent possible—unbiased final report. The external reviews had two major purposes: first, to involve potential users of our research, thus increasing the likelihood that the results of the TA would be useful and used; and second, to provide an additional guarantee that the research and conclusions would be as well informed, technically correct, adequately comprehensive, and unbiased as possible (see White et al., 1976; Kash and White, 1971; and White, 1975).

In addition to providing a variety of expertise and perspectives, the interdisciplinary team approach also helps to temper what could become an excessive analytical structure. The procedural steps for carrying out a TA are not nearly as sequential and formal as they might seem from the conceptual framework depicted in Figure 2-1. We purposely employed the creative tensions generated by the interdisciplinary team and by the extensive internal and external review processes to prevent this research effort from becoming a mechanical exercise.

The Phases of a Technology Assessment

The tasks that must be completed when the general conceptual framework (Figure 2-1) is put into operation can be divided into three categories: descriptive, interactive, and inte-

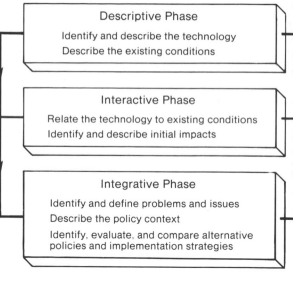

Figure 2-8: *The Phases of a Technology Assessment*

grative (see Figure 2-8). Each of these phases is described briefly below.

The Descriptive Phase: Results of the descriptive phase characterize both the technologies used to develop western energy resources and the conditions that exist in the vicinity of the sites where these technologies might be operated. Technologies for the development alternatives are described in more detail in White et al. (1979a). The resource base, technological alternatives, input requirements, product and residual outputs, economic costs, energy efficiencies, and resource-specific laws and regulations for each of the six resources are described.

The descriptions of existing conditions at the six scenario sites and for the entire eight-state area more generally are presented in White et al. (1979b). The six scenario sites were chosen because they seem to be representative of the sites where energy development either is taking place or is likely to take place in the eight-state area in the future.

The Interactive Phase: Technology and site descriptions are the primary data base for analyzing the impacts of deploying a particular technology or some combination of technologies at a specific energy development site or of developing specified quantities of energy within the eight-state region. Four major categories of site-specific impacts have been analyzed:

- Air Quality;
- Water Quality and Availability;
- Quality of Life; and
- Ecology.

In addition, we have analyzed certain local impacts which are not particularly site-specific or which are highly speculative because of inadequate knowledge, data, and/or analytical tools. Thus, in addition to those categories listed above, the regional impact analyses cover the following topics:

- Aesthetics;
- Health;
- Transportation; and
- Noise.

The results of this interactive phase are described in detail in White et al. (1979b). The findings of our impact analysis are briefly summarized at the beginning of each of the chapters that follow.

The Integrative Phase: The descriptive and interactive phases are intended to provide results which make it possible to evaluate and compare technologies and potential energy developments on the basis of scientific and technical criteria. Together, the descriptions and the analyses of impacts can be used to inform policymakers about the costs, risks, and benefits of various technological and siting options for the development of western energy resources. For example, on the basis of these technical analyses, policymakers can make better-informed choices about which combinations of technology and location to encourage or discourage or permit or prohibit. But the results of technical analyses such as these are almost always incomplete, largely

because of the limited explanatory power of existing theories and because necessary data and analytical tools are either inadequate or unavailable. But even if it were possible to overcome these limitations, the results of these technical analyses alone would not be an adequate basis for policymaking; policymakers need to know more than the results of these analyses tell them. They need to know how the costs, risks, and benefits identified by the technical analyses will be distributed, which of society's interests and values will be promoted at the expense of which others, how to advance the desired interest and values, and how to avoid unwanted costs and risks. Providing policymakers with this kind of information is the objective of the integrative phase of a TA. As indicated in Figure 2-8, this phase consists of three components, each of which is described below.

The Identification and Definition of Problems and Issues[2]*:* Some of the problems and issues associated with the development and operation of a specific energy technology in the western U.S. were identified when the existing conditions and the location of deployment were being described. That is, some problems and issues were anticipated independently of the formal impact analyses, often on the basis of past experience, analogy, and informed speculation. Some problems and issues, however, were not anticipated; consequently, a beginning step in the integrative phase was for the interdisciplinary research team to review systematically the results of the technical analyses to guard against any otherwise unanticipated consequences being overlooked.

---

[2] The terms "problems" and "issues" are not synonyms. Problems such as those resulting from the labor and capital intensity of a technology, may or may not lead to an issue being raised. The key distinction is that issues involve conflict among competing interests and values; not all problems produce conflicts. Consequently, both terms often must be used.

The Description of the Policy Context: Problems and issues must be related to the social and political context within which the development and operation of the technology is expected to take place. This requires that the relevant political systems be identified and described in substantive terms. The interests and values at stake, the relevant institutional arrangements, the applicable laws and regulations, the governmental and nongovernmental participants, and the intensity of involvement by various participants all can vary with the substance of the problem.

For an issue that the political system has dealt with in the past, the first step is to examine key elements in the historical development of the issue:

- When did the issue first arise?
- Which groups perceived it as an issue, and what interests did they represent?
- When did government respond?
- How did government respond, and what policies were enacted?
- Who administers these policies?
- How have these policies affected the issue?

This step also includes a more detailed identification and description of the existing system for dealing with the issue:

- What are the relevant existing institutional arrangements—public and private, formal and informal?
- What interests and values are at stake?
- Who represents these interests, and what strategies and tactics are they employing?
- Are there social or physical/environmental conditions that affect or potentially could affect whether and how the issue is dealt with?

One essential task in this step in policy analysis is to identify the interests and values that are at stake. This is a difficult task to perform systematically and comprehensively because of the range of interests and values

potentially at stake and the numerous ways they can be categorized. Therefore, we compiled a checklist to encourage the researcher to search broadly when taking an inventory of interests and values.

The Identification, Evaluation, and Comparison of Policy Options: To structure the identification, evaluation, and comparison of alternative policies and strategies for implementation, we considered:

- The policy objectives of each problem or issue category;
- General categories or kinds of alternatives to achieve each objective;
- Specific alternatives within each general category;
- Implementation of each specific alternative; and
- The implications of choosing and implementing each alternative and strategy.

For example, in the question of water availability, if policymakers choose the goal of meeting the expanded needs of water users, one general way to accomplish this objective is to increase the efficiency of using existing supplies. Specific options within this general alternative include conservation by energy industries, by irrigated agriculture, or by municipalities. Each of these options has a variety of specific implementation strategies which can be carried out by several levels of government; and each strategy has various implications for environmental quality, cost of energy products, etc. Focusing on this sequence of questions helps to make the analysis of policy alternatives systematic. Of course, at each step considerably more detail is added.

Criteria for Evaluating and Comparing Alternatives: In general, five basic criteria—effectiveness, efficiency, equity, flexibility, and implementability—were used in the following chapters to evaluate and compare alternative policies and implementation strategies (Table 2-4). Of course, each criterion has to be de-

Table 2-4: *Evaluation Criteria*

Criterion	What Does It Evaluate?
Effectiveness	Achievement of Objective Does it avoid or mitigate the problem or issue? Is it a short- or long-term resolution or solution? Is it dependent on state-of-society assumptions?
Efficiency	Costs, Risks, and Benefits What are economic costs, risks, benefits? What are social costs, risks, benefits? What are environmental costs, risks, benefits? Is it reversible/irreversible, short- or long-term?
Equity	Distribution of Costs, Risks, and Benefits Who will benefit? experience costs? assume risks?
Flexibility	Applicability/Adaptability Are local and regional differences accommodated? Are differences among social groups and economic sectors taken into account? How difficult will it be to administer? How difficult will it be to change?
Implementability	Adoptability/Acceptability Can it be implemented within existing laws, regulations, and programs? Can it be implemented by a single agency or level of government? Is it compatible with existing societal values? Is it likely to generate significant opposition?

fined specifically shaped to the kinds of problems and issues being evaluated and compared; and appropriate qualitative and quantitative measures also must be specified for each. For example, as applied to water availability, the effectiveness of an alternative is defined in terms of how much water would be saved (or added to the overall supply), whether the alternative would avoid or mitigate the supply problem, and whether the alternative would offer a long- or short-term solution. Among the quantitative measures used in this case are gallons or acre-feet per year saved (or added), and the percentage of increase in overall supply. In this instance, then, the anticipated degree to which the supply problem is avoided or mitigated is an example of a qualitative measure.

The basic point about measures is that while many policymakers may want to be presented with a "bottom line," no single measure or evaluative criterion can provide an adequate summary of the costs, risks, and benefits of alternative policies and the strategies for implementing them. The combination of measures and criteria we have used was determined both by what was being evaluated and by the interests and values at stake. Although we used economic measures and criteria most frequently, they were not always applicable and did not always provide an adequate basis for evaluation. For example, dollars are not an adequate measure of aesthetic values, nor do dollars always provide the best indication of how equitably an alternative may distribute the costs, risks, and benefits. And while it is possible to determine the dollar cost of environmental controls, the associated social costs often cannot be determined. By themselves, economic measures and criteria can be used to evaluate only one component of overall costs, risks, and benefits.

REFERENCES

Ashland Oil, Inc., and Occidental Oil Shale, Inc. 1977. *Supplemental Material to Modified Detailed Development Plan for Oil Shale Tract C-b,* prepared for Area Oil Shale Supervisor. June.

Cazalet, Edward, et al. 1976. *A Western Regional Energy Development Study: Economics,* Final Report, 2 vols. Menlo Park, Calif.: Stanford Research Institute.

Colony Development Operation. 1974. *An Environmental Impact Analysis for a Shale Oil Complex at Parachute Creek, Colorado.* N.p.

Denver Federal Executive Board, Committee on Energy and Environment, Subcommittee to Expedite Energy Development; and Mountain Plains Federal Regional Council, Socioeconomic Impacts of Natural Resource Development Committee. 1975. *A Listing of Proposed, Planned or Under Construction Energy Projects in Federal Region VIII.* N.p.

Kash, Don E., and Irvin L. White. 1971. "Technology Assessment: Harnessing *Chemical and Engineering News* 49 (November 20):36–41.

White, Irvin L. 1975. "Interdisciplinarity." In *Perspectives on Technology Assessment,* edited by Sherry Arnstein and Alexander N. Christakis, pp. 87–96. Jerusalem: Science and Technology Publishers.

White, Irvin L., et al. 1976. *First Year Work Plan for a Technology Assessment of Western Energy Resource Development.* Washington, D.C.: U.S. Environmental Protection Agency.

White, Irvin L., et al. 1979a. *Energy From the West: Energy Resource Development Systems Report,* 6 vols. Washington, D.C.: U.S. Environmental Protection Agency.

White, Irvin L., et al. 1979b. *Energy From the West: Impact Analysis Report,* 2 vols. Washington, D.C.: U.S. Environmental Protection Agency.

CHAPTER 3

Water Availability

Problems of water availability are among the most critical associated with the development of western energy resources. Although water has always been scarce in the West, in general enough has existed to ensure users—primarily irrigated agriculture and municipalities—of adequate supplies. However, the question that is central to the future development of the region is whether enough water exists to support both the demands of traditional users and the growing demands of energy developers, environmental interests, Indians, and others.

There is no single answer to this question, in part because data on water availability, current uses, and future demands are incomplete. In addition, there is a complex system of appropriated rights, interstate compacts, court decisions, and international treaties, as well as a multiplicity of political jurisdictions, that regulate water use. However, it seems clear that as existing supplies are increasingly depleted, perhaps to their physical limits, conflicts over how water should be used will also increase. In fact, such conflicts are already serious, involve energy, environmental, agricultural, Indian, and municipal users, and affect all levels of the public and private sectors. Thus, finding appropriate ways to deal with the problems of water availability and to resolve political conflicts will be critical not only to the future development of energy resources but also to the overall growth and character of the region.

Consequences of Energy Development

Figure 3-1 shows the major surface waters in our eight-state study area that are likely to be affected by the development of the energy resources.

Impacted Surface Waters

The Upper Missouri, especially the Yellowstone, Belle Fourche, and the Little Missouri tributaries will be affected by developments in Montana, Wyoming, and North Dakota. Upper Colorado tributaries, especially those in the Colorado mainstem and the San Juan system, will be affected by developments in Colorado, Utah, and New Mexico. Although these will be the surface waters that are most affected, the withdrawal of water from these streams will also have effects downstream.

Water Requirements for Energy

As shown in Figure 3-2, the water requirements of energy facilities vary considerably. To produce an equivalent amount of energy, coal-fired electric power generation requires more water than any of the synthetic fuel technologies considered and more water than slurry pipelines. Based on the best available data, high-Btu coal gasification consumes more water than either coal liquefaction or *in situ* oil shale production, but less than oil shale

37

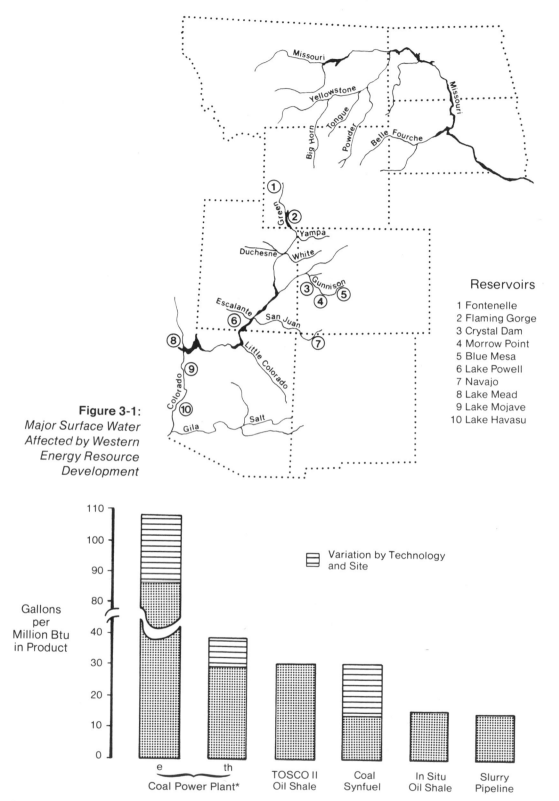

Figure 3-1:
Major Surface Water Affected by Western Energy Resource Development

Reservoirs

1 Fontenelle
2 Flaming Gorge
3 Crystal Dam
4 Morrow Point
5 Blue Mesa
6 Lake Powell
7 Navajo
8 Lake Mead
9 Lake Mojave
10 Lake Havasu

*To provide some basis for comparing power plants with synthetic fuel plants, emissions are expressed on both an electrical output basis and a Btu thermal input to the plant basis (assuming a 35 percent power plant efficiency). However, neither measure is completely satisfactory since electric energy has high quality uses not possible with oil and gas (such as electric home appliances and lighting). A Btu of electricity is more valuable than a Btu of oil or gas, but a Btu input to the power plant is less valuable than a Btu of oil or gas produced from coal.

Figure 3-2: *Water Requirements for Energy Facilities*

Wet Cooling Towers for a Coal-Fired Power Plant

conversion by TOSCO II surface retorting. Water requirements for shipping coal by slurry pipelines are less than for any conversion facility considered in this study.

Water requirements for processing are generally much smaller than are requirements for cooling. The single exception is TOSCO II, the most water-intensive of the surface oil-shale retorting processes currently being developed. When wet cooling is used, the cooling-water requirements account for 53 to 96 percent of the total water needed for electric

power plants and synthetic fuel facilities.[1]

Water requirements for these facilities will be affected by location, primarily because of differences in coal characteristics and in climatic conditions. For example, water requirements for a Lurgi facility in the Four Corners area would be about twice that required for the same facility in the Northern Great Plains. This is because of the low moisture content of the coal in New Mexico (the Lurgi process uses the water in the coal during the gasification process), the high ash content of New Mexico coal (which means more water is needed for disposing of the ash), and the need for supplemental irrigation in the Southwest to reclaim the land.

Water requirements for mining, for waste disposal, and for the increased population

[1] Water requirements can be decreased considerably by using some combination of wet and dry cooling, by using dry cooling, and by incorporating process design changes. These alternatives are discussed below in the section on policy alternatives.

related to energy development represent at most 10 percent of the water required for conversion facilities. Even so, these demands may still lead to water problems, especially if treatment and distribution systems are needed to supply the water. Water requirements for most small communities are satisfied by groundwater. But with withdrawals for mine dewatering and agricultural purposes, aquifer depletion is a potential problem.

Assuming wet cooling, our estimates of the annual water requirements for the development of energy resources in our eight-state study area by the year 2000 are 1.4 million acre-feet per year (acre-ft/yr) for the Low Demand scenario and 2.3 million acre-ft/yr for the Nominal Demand scenario. These estimates call for between 350,000 and 1,117,000 acre-ft/yr to be used in the Upper Colorado Basin and between 990,000 and 1,225,000 acre-ft/yr in the Upper Missouri Basin.

Water Availability

One approach to discussing future water problems is to compare estimates of water availability and water requirements. However, such estimates vary widely, especially for the Colorado River, and very little agreement exists about which estimate is most accurate. Furthermore, a strict reliance on any set of estimates to predict future problems ignores the substantial complexity and uncertainty of the current political system, including conflicts over Indian water rights, instream water requirements to protect the ecosystem, and salinity controls. The resolution of any one of these, or other conflicts, could substantially reduce the amount of water that would be available for other uses, including the development of energy resources. Thus, the following discussion should be considered as only part of the water-availability picture.

Under the provisions of the Colorado River Compact (1922), Upper Basin states guaran-

tee Lower Basin states 75 million acre-ft/yr over each consecutive ten-year period, or an average of 7.5 million acre-ft/yr. In the Mexican Water Treaty (1944), the U.S. agreed to guarantee Mexico 1.5 million acre-ft/yr from the Colorado River. Assuming that the Upper Basin states are responsible for supplying half this amount, then on the average 8.25 million acre-feet must flow into the Lower Basin each year. The Upper Basin states apparently do not make this assumption, however, and this adds further uncertainty to estimates of the share of water for each basin and each state.

Three of the most frequently cited annual virgin-flow estimates for the Colorado are shown in Figure 3-3. If it is assumed that 8.25 million acre-feet will be delivered to the Lower Basin each year, an estimated 5.25 to 7.95 million acre-feet would be available on the average to the Upper Basin.

These estimates are based on average flows, around which considerable variation occurs from year to year. For example, one estimate puts the Colorado's average yearly virgin flow at 13.5 million acre-feet with a standard deviation of 3.4 million acre-feet. This means that in 67 percent of the years, the flow would be between 10.1 and 16.9 million acre-feet; in drought years, the flow could be much less. In fact, 1977 was one of the driest years on record, with a virgin flow at Lees Ferry estimated at only 5.3 million acre-feet (Stockton and Jacoby, 1976; Colorado River Basin Salinity Control Forum, 1978:3).

Protection against dry years is provided by the Interior Department's Bureau of Reclamation (BuRec) storage reservoirs (see Figure 3-1). The maximum storage capacity of the ten reservoirs is estimated to be about 61 million acre-feet. As shown in Table 3-1, from October 1976 through September 1977 storage ranged from about 24 to 49 million acre-feet. Technically, storage capacity is about the same for the Upper and Lower Colorado basins. However, about 90 percent of the

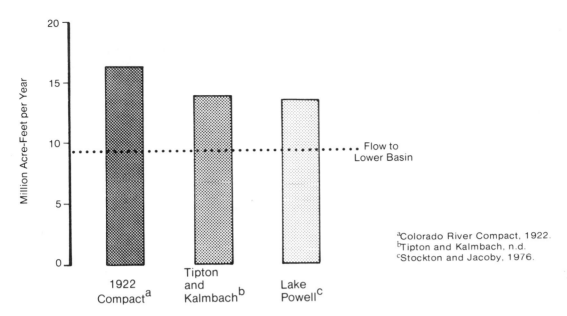

Figure 3-3: *Estimates of Average Flow in Colorado River*

Table 3-1: *Reservoirs of the Colorado River Basin*

Reservoir	Maximum Storage[a]	Actual Storage 1976 – 1977[b]	Releases
	Characteristics (1,000 acre-feet)		
Upper Basin			
Fontenelle	345	220–300	. . . .
Flaming Gorge	3,749	2,000–2,500	. . . .
Blue Mesa	830	250–700	. . . .
Morrow Point	117	90–110	. . . .
Crystal Dam	18	0–14	. . . .
Navajo	1,696	1,000–1,250	. . . .
Glen Canyon/Lake Powell	25,002	1,500–19,650	8,230
Total	31,757	5,060–24,524	8,230
Lower Basin			
Hoover Dam/Lake Mead	23,377	17,250–22,000	7,560
Davis Dam/Lake Mojave	1,810	1,450–1,790	8,050
Parker Dam/Lake Havasu	619	560–600	6,710
Total	29,806	19,260–24,390	22,320
Basin Totals	61,563	24,320–48,914	30,550

Source: U.S. DOI, BuRec, 1978.
[a]Does not include dead storage (not recoverable), approximately 4.5 million acre-feet.

[b]Water year is from October 1976 through September 1977.

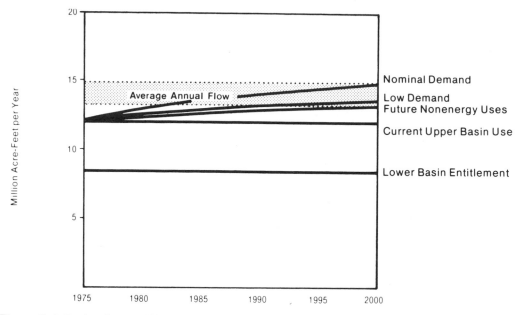

Figure 3-4: *Projections of Water Requirements and Availability in Colorado River, Year 2000*

storage capacity of the Upper Basin is contained in the Glen Canyon/Lake Powell reservoir located on the southern border of the Upper Basin (U.S. DOI, BuRec, 1978). Thus, most of the total storage of the Colorado Basin is available to the Lower Basin.

The current level of water consumption in the Upper Basin is estimated to be between 3.2 million and 3.7 million acre-ft/yr (U.S. DOI, BuRec, 1975). One estimate of increased consumption by nonenergy users is 1.5 million to 1.6 million acre-ft/yr by the year 2000 (U.S. DOI, Water for Energy Management Team, 1974). As shown in Figure 3-4, if the maximum estimates of current consumption and the projections for energy and nonenergy water use are combined, by the year 2000 water requirements in the Upper Colorado River could exceed average flow by over 1 million acre-ft/yr.[2] If the minimum estimates of water requirements are used, a large per-

centage (approximately 95 percent) of the current annual supply could be consumed.

Water availability does not appear to be as acute a problem in the Upper Missouri Basin as a whole since approximately 12 million acre-feet of water is currently unallocated. However, demands from energy, environmental, agricultural, and other uses are increasing. In the energy-rich Yellowstone River Basin, for example, current demands exceed supplies (U.S. DOI, BuRec, 1975: 298–300).

Groundwater is also available in both the Colorado and Upper Missouri River Basins and, in some cases, it might provide an alternative water source for energy development. However, relying on groundwater can also produce problems, particularly if it is consumed faster than the aquifer from which it is being withdrawn is recharged—a phenomenon called groundwater mining—or if using groundwater affects the surface stream flow. An estimated 115 million acre-feet is contained in aquifers in the UCRB at a depth of less than 100 feet (Price and Arnow, 1974); it is estimated that these aquifers are recharged at a rate of 4 million acre-ft/yr. In addition,

[2] The most commonly used average flow estimates are from 13.5 to 14.2 million acre/ft/yr. The estimate of 16.2 million acre/ft/yr from the 1922 compact is generally accepted to be in error.

substantially greater quantities of water are estimated to occur in deeper reservoirs. In the Upper Missouri River Basin (UMRB) a total of 860 million acre-feet is estimated to be stored in the upper 1,000 feet of rock (Missouri Basin Inter-Agency Committee, 1971). Alluvial aquifers and aquifers in the Fort Union formation occur in the UMRB and are presently used for municipal and agricultural needs. Most of the aquifers, however, are not highly productive and will probably be used mainly to supply water for population growth. The Madison aquifer in northern Wyoming, southern Montana, and western North Dakota is currently being studied as a possible source of water for energy development (U.S. DOI, Geological Survey, 1975). But its use is limited both by insufficient knowledge of its location and properties (Swenson, 1974) and by its great depth in most areas of the basin. For example, at Colstrip, Montana, the Madison aquifer is located at about a 7,500 foot depth.

The use of aquifers, especially in the UCRB, can affect nearby streams which depend on groundwater to maintain their flow. Thus, using aquifers can result in conflicts with those who have surface water rights. Until legal procedures are well defined for resolving conflicts over the interrelationship between groundwater and surface water flow, the use of groundwater for energy development is likely to be restricted.

Problems and Issues

As suggested above, the quantity of water that is available and its ultimate use can depend on the manner in which several problems and issues are resolved. In the following sections, we will describe four of the most seriuos issues and suggest their potential effects on the availability of water. These issues are: the uncertainty and complexity of water policies; new demands for water use; reserved water rights, and jurisdictional disputes.

Uncertainty and Complexity of Current Water Policies

A complex combination of state water laws, federal water policies, court cases, interstate agreements, and international treaties combine to form current water policies. Although this system has dealt successfully with many difficult water problems, it also creates barriers to change, discourages a diversity of water uses, and provides few incentives for the efficient management and conservation of water.

Because water has always been a vital but scarce resource in the West, the history of legal and physical controls on its use is complex and lengthy. In general, such controls have promoted two primary interests. The first is geographic equity—that is, diversions, impoundments, legal doctrines, and interstate compacts protect downstream users within a state and provide at least minimal water supplies to states of the Colorado and Missouri river systems. The second interest is irrigated agriculture, a basic industry in each of the states, which often depends on water from distant sources. Irrigation generally accounts for 80 to 85 percent of the total consumption of water in the West (Glenn and Kaufman, 1977).

Several intergovernmental arrangements regulate western water resources. As shown in Table 3-2, waters of both the Colorado and Missouri rivers are divided by interstate compacts or agreements. One of the most important of these is the 1922 Colorado River Compact (discussed previously), because of the requirements it establishes for water distribution between the Upper and Lower basins. The amount of water available to each state of the Colorado River Basin (CRB) is affected as well by the 1944 Mexican Water Treaty (also discussed above).

Within these arrangements, states have the primary authority for allocating water. All eight states in our study area operate under an appropriation system. The cornerstone of

Table 3-2: Interstate Compacts and Agreements

Compact or Agreement	Lower Colorado River Basin			Upper Colorado River Basin				Upper Missouri River Basin		
	Arizona	California	Nevada	Colorado	New Mexico	Utah	Wyoming[a]	Montana	North Dakota	South Dakota
Colorado River Compact, 1922	Guarantees 7.5 MM acre-ft/yr to the Lower Basin									
Boulder Canyon Project Act, 1928, and Arizona v. California, 1963[b]	2.8 MM acre-ft/yr	4.4 MM acre-ft/yr	0.3 MM acre-ft/yr							
Upper Colorado River Basin Compact, 1948	50,000 acre-ft/yr			51.75% of flow remaining after Lower Basin and Arizona	11.25% of remainder	23% of remainder	14% of remainder			
Yellowstone Compact, 1950[c]										
Clark's Fork River							60%	40%		
Big Horn River							80%	20%		
Tongue River							40%	60%		
Powder River							42%	58%		
Belle Fourche River Compact, 1943[d]							10%			90%

MM acre-ft/yr = million acre-feet per year.

[a] Surface waters in Wyoming are part of both the UCRB and the UMRB.

[b] In *Arizona v. California*, the U.S. Supreme Court held that the compact-apportioned water available to the Lower Basin states was to be divided among them as follows: California, 4.4 MM acre-ft/yr; Arizona, 2.8 MM acre-ft/yr; and Nevada, 0.3 MM acre-ft/yr. Although the division was not agreed to by the states, the terms of the division were made final by the Supreme Court's decree.

[c] Recognized existing appropriations for beneficial uses and divided the remaining waters as shown.

[d] Recognized existing water-right priorities of South Dakota and Wyoming; remaining unappropriated water allocated as shown.

this system is that senior right-holders are granted priority rights, based on the date of the water-right application. Junior users, whether upstream or downstream, are entitled only to whatever remains after senior users have met their needs. In drought years or in areas of limited supply, water diversions are shut off in reverse order of the date of the right. An appropriation of water generally is quantified by the state and sustained only by actual and continuous beneficial use. "Beneficial use" is defined differently by each of the eight states and is a critical element of water policy, in part because for the use to be considered beneficial, water usually must be withdrawn from a stream.

The availability of water in the West is also affected by federal policies, principally the BuRec reservoir systems (mentioned above) and federal reserved water rights. The Reclamation Acts of 1902 and 1939 gave BuRec a general mandate to make arid lands available for agricultural development and to provide water for power generation and other uses such as recreation. A critical element of the bureau's responsibility is to establish prices for the sale of water from federal projects. These projects and the corresponding federal pricing policies have subsidized irrigated agriculture by selling water at well below the actual cost of providing it.[3] However, most of the West's water resources have been developed privately: approximately 75 percent of all water used for irrigated agriculture in the eight states is based on private or nonfederal development (National Water Commission, 1973:127).

The term "reserved water rights" means that when the United States establishes a federal reservation, such as a national park, a military installation, or an Indian reservation, a sufficient quantity of water is reserved to accomplish the purposes for which land was set aside. The reservation doctrine has been affirmed in the courts, guaranteeing that federal reserved rights are not subject to state appropriation laws and that federal water rights are not lost if they are not used.[4] These doctrines are significant because the federal government owns large amounts of western land—for example, about 70 percent of the land in the CRB—and because Indians have claimed rights to as much water as is needed along many segments of the Colorado. (This issue is elaborated below.)

The policy system which has developed around these basic elements has successfully dealt with many water problems, in part because even though water has been scarce, shortages have been infrequent. However, increasing demands for water already are challenging the responsiveness of the system: the current system creates barriers to change, discourages a diversity of water uses, and provides few incentives for the efficient management and conservation of water.

The "beneficial use" aspect of water appropriation and a series of related doctrines combine to create much of the present inertia. Because states traditionally have favored agricultural water uses, it has been difficult for other interests, particularly environmental interests, to achieve legal recognition. For example, since a beneficial use generally requires withdrawal of water from the stream, "instream uses," such as scenic beauty, are not usually considered beneficial. As a result, many environmental groups have pursued their interests in the courts. But because of the time required for litigation and because court rulings typically address only particular segments of an issue, the appropriation of water for environmental needs remains uncertain.

Another aspect of the appropriation system which generally favors agricultural interests

---

[3]For a comprehensive overview of the traditional federal role in water-resource management, see Ingram and McCain, 1977:448–55.

[4]Specific extensions of the reservation doctrine are the product of more than fifty years of refinement of the *Winters* ruling (*Winters* v. *U.S.,* 1908); this doctrine also has been confirmed in *Arizona* v. *California* (1963, Decree 1964).

is preference clauses, which can be found in six states in our study area (Arizona, Colorado, North and South Dakota, and Wyoming). In situations of drought or in other circumstances where there is insufficient water for a particular use, these clauses allow the condemnation of water rights of less-preferred uses, even if the preferred use has a junior right. The order of preferred uses is municipal, agricultural, and industrial. Since most energy water rights are junior, the effect of these preference clauses on energy facilities is minimal unless industry attempts to buy the more senior, agricultural water rights.

State appropriation systems also regulate how an individual may use a water right. Consumptive use, nonimpairment, historical use, "use it or lose it" policies, and a variety of other doctrines have been developed to protect the rights of water users (White, 1975).[5] These doctrines have also built rigidity in the appropriation system. For example, if a user has been diverting only a portion of his total water right for several years, it is not likely that he would be able now to begin diverting all of his allocation if downstream users had in the meantime become dependent on his unused portion. This aspect of water appropriation, sometimes referred to as "historical use," creates an incentive for users to "use or lose" all of their water rights. Since BuRec operates within state water laws, BuRec contracts for water from federal reservoirs also operate under a "use or lose" policy.

The rigidity of the system for appropriating water is one of the biggest barriers to conservation: farmers risk losing the future right to any water which they might save by using, for example, more efficient conveyance or irrigation systems. This occurred in the case of the *Salt River Valley Water Users Association* v. *Kovacovich* (1966) in which the Arizona Supreme Court denied the right of a water user to apply water that had been "salvaged" by conservation to lands other than those to

Strip Mine in Farmlands

which the water was originally appurtenant.

The state appropriation systems also create barriers to water management. This is because the systems are designed primarily to identify the legality of water rights and water uses rather than to manage the precious resource. For example, although states record the holder, date, priority, and use to which a water right is granted, little if any information is recorded about the actual patterns of use. In part, this results from some uses not being "perfected"—that is, even though a right has been granted, the user may have several years to develop the right to its full use. Furthermore, little data seems to exist on actual runoff patterns; typically, the gathering of such data begins only when one user claims his right has been impaired by another user. Hence, it is difficult to effectively administer the transfer of water, as well as to enforce the beneficial use and nonwaste provisions of the law.

---

[5] For a comprehensive overview of western water law and irrigation, see Radosevich, 1978.

of Northern Great Plains

New Demands for Water Use

Although agriculture still enjoys a preferred status in western water law, demands for other uses of water are rapidly increasing and may create water shortages in many areas. At a minimum, these demands will continue to create conflicts among existing and new users.

The preferred status of agricultural water use has come under increasing criticism, including charges that a great deal of irrigated agriculture wastes water, that it is uneconomical (in the sense that the crops don't return the price of the water used to produce them), and that it is environmentally damaging. While these criticisms raise important questions, the fundamental change occurring in the availability of western water is that a diversity of users are requesting a larger portion of the scarce resource. The largest requests come from the developers of energy resources, from environmental interests, and from Indians. The conflicts among agricultural, energy, and environmental users are discussed in the next section; Indian water needs are discussed in the succeeding section.

Energy Needs: Water requirements for the development of energy resources have been discussed above. The biggest problems are likely to occur if water-intensive coal conversion facilities are sited in locations where little water is available. In our study area, the most troublesome locations appear to be in the Yellowstone River Basin, in western Colorado along the Colorado main stem, and in the San Juan River Basin. However, energy development already is (and will likely continue) contributing to conflicts over water in many other areas where other users' interests are threatened.

One of the important questions in this regard is whether energy industries will be able to acquire enough water and, if so, whether that water will be at the expense of irrigated agriculture. In some cases, state governments have tried to ensure adequate supplies for industrial growth. Utah appears to favor guaranteeing a water supply for the development of the Uinta Basin; the state water code was revised in 1976 to allow the state engineer to approve water appropriations for industrial, power, mining, or manufacturing purposes (Laws of Utah, 1976). Other states, particularly Colorado, Wyoming, and Montana, are apparently much less willing to provide water for energy development as a state policy. Thus, in these states it may be more likely for energy developers to buy water rights from current users, primarily irrigated agriculture.

How often this will occur, whether such sales will provide enough water for energy, and the consequences for the West are unclear. In general, energy resource developers can pay substantially more for water than can irrigated agriculture. It is estimated that water for agriculture is economic up to about $25 per acre-foot, while water for energy can be economic at $200 per acre-foot and above (Andersen and Keith, 1977:161). This has led some observers to suggest that water for energy will not be a problem—the energy industries will simply buy up agricultural water rights if other sources do not exist.

However, several other considerations suggest that the transfer of water from agriculture to energy development will not be so simple. First, some coal conversion facilities have sizeable water requirements: a 3,000 MWe wet-cooled power plant requires at least 24,000 to 30,000 acre-ft/yr; and a 100,000 bbl/day wet-cooled synthoil liquefaction plant requires about 9,000 to 12,000 acre-ft/yr. In order to meet these requirements, the water rights of several farms in the vicinity may be needed. In some cases, these requirements may even exceed the total amount of water consumed in an entire irrigation district or reclamation project. For example, the Collbran reclamation project irrigates nearly 20,000 acres of land along the Colorado main stem in western Colorado. It is estimated that farms in the project consume about 20,000 acre-ft/yr, substantially less than the water requirements of a single 3,000 MWe wet-cooled power plant (U.S. DOI, BuRec and BIA, 1978).

Even if enough water for energy development exists in an area, farmers must be willing to give up their water rights and/or their farms. In some cases, farms have been worked by a family for generations. And, throughout the West, farmers' groups are organizing to protect their interests. For example, the Northern Plains Resource Council (NPRC) represents farmers, ranchers, and citizens concerned about coal development in the northern plains. In addition to opposing the appropriations of water for power plants in Montana, NPRC supports agricultural expansion in the Yellowstone Basin (*The Plains Truth,* 1979: 5). John Stencel, president of the Rocky Mountain Farmers Union, expresses a similar view of the competition for water between the development of energy and agriculture:

We urge adoption of legislation . . . to prevent future power and energy plants from consuming water to the detriment of agriculture. . . . We also feel that the sale of adjudicated irrigation water rights should be limited to agricultural uses [U.S. Congress, Senate Committee on Energy and Natural Resources, 1978:187].

While it is apparent that some farmers will be willing to sell their water rights to energy developers, it is also clear that the prospect of a large-scale transfer of water from agricultural to energy uses, representing a significant change in the economy of the West, is troublesome to many westerners.

The appropriation system itself will also constrain these transfers. In all eight states, a water right granted for irrigation becomes appurtenant to the lands described in the permit and cannot be transferred to other lands without state approval (Radosevich, 1978:63). In Colorado, the water right does not necessarily transfer with the sale of the land (*James* v. *Barker,* 1973). In 1975, Montana prohibited the change in use of an agricultural water right for 30 or more acre-feet to an industrial use (R.C.M., 1975).

Perhaps the most serious constraint is the requirement that transfers—either in their purpose, use, or place of diversion—do not impair the right of downstream users to the return flow (runoff). This is the nonimpairment doctrine. Since water used for energy facilities will have no return flow, the probability of impairing others' rights is high, particularly if water from several farms or an entire irrigation district is transferred to an energy user. In Wyoming, a transfer is allowed only if the quantity of water transferred does not exceed the historic consumptive use, reduce the return flow, or injure any lawful appropriations (W.S.A., n.d.). Thus, the possibility exists for one farmer, dependent on return flow, to prevent a transfer of the total water diversion to an energy plant. If only the water that is consumptively used—for example, in irrigation—is transferred, a large number of such transfers would be needed to meet the total water requirements of most energy facilities.

Energy conversion facilities are not the only energy developments creating water problems. Although coal slurries require less water per unit of energy than any energy conversion process studied, they are an increasing concern in the West, particularly in Colorado and Wyoming. A proposed slurry pipeline from Walsenburg, Colorado, to Texas would carry 15 million tons of Colorado and Wyoming coal per year and would use about 10,000 acre-feet of water annually. However, Colorado farmers, ranchers, legislators, environmentalists, and railroad interests are opposing the development of this line by the Houston Natural Gas Company and the Rio Grande Industries Corporation of Colorado (Davidson, 1977). Railroad interests, who would stand to lose a good deal of coal-hauling business and who say they currently operate at only 30 percent of capacity, have aligned with environmentalists to prevent the energy companies from acquiring rights of eminent domain.

Water requirements for this slurry pipeline also concern Colorado legislators, some of whom have proposed legislation that would allow the line to run through the state but would prohibit the use of Colorado water for exportation. This plan would produce between $2 million and $3 million in taxes annually for Colorado. The issue is also being addressed in the courts, where the pipeline companies must prove that their plan to pump ground-water will not injure the water supplies of present users. The possibility of reducing the water table, coupled with the fact that the pipeline companies have already purchased water rights from some area farmers, apparently has infuriated other farmers (Strain, 1977).

Environmental Needs: As discussed above, environmental demands for water contribute to the increasing complexity of water policies. In some states, such as Montana, instream environmental needs have been legally recognized as beneficial uses. Subsequent requests for instream appropriations, combined with requests from industry, led to a state-imposed moratorium on new appropriations of water from the Yellowstone River (see box).

In addition, suits have been filed in each of the eight states to preserve the scenic beauty, natural habitat, and recreational uses of rivers. For example, the Colorado Water Conservation Board, representing the Colorado Division of Wildlife, is currently suing the Colorado River Conservation District, which represents those who hold water rights. The issue is the preservation of minimum stream flows in the Crystal River near Carbondale, Colorado, in order to protect the natural environment, including ensuring enough water for trout (Saile, 1977). This case is indicative of the challenges being faced in many streams of the West which, like the Crystal River, are already overallocated. The uncertainties inherent in the evolv-

ing regulatory system may become too great
for some potential users of large quantities
of water, such as capital-intensive energy
industries (see box).

The environmental concerns articulated in
the Wild and Scenic Rivers Act (1968) have
also made new demands on water resources.
In our eight-state study area, portions of
twelve rivers in the Colorado Basin have been
considered for inclusion in the wild and scenic
rivers system. Those most likely to affect energy development are the Escalante, White,
and Green rivers and the Colorado mainstream
between its confluence with the Gunnison and
its confluence with the Dolores. The significance of the act is that it may be used to preserve minimum stream flows in order to protect instream uses. This could dramatically
affect the amount of water that is available
for energy development and other uses and
could prevent impoundments or any water
project which alters the nature of flow in a
wild or scenic river. However, state water
administrators generally do not believe that
this act will reduce the amount of water that
is available for appropriation.

Concerns about water quality, especially
salinity, also are related to patterns of water
use. The most important question concerns
irrigation practices; most agricultural lands
in the West are irrigated simply by flooding,
which requires a large diversion of water and
produces a high percentage of runoff. One
estimate suggests that at least twice as much
water is typically diverted as is needed for
maximum plant growth (Utah State Univ.,
Utah Water Research Lab., 1975:60). Thus,
many environmentalists prefer more efficient
irrigation systems, which would leave more
water instream and would reduce the salinity
caused by irrigation runoff.[6] However, irriga-

---

[6] Trickle systems, using pipes to drip water to the
soil through outlets near each plant, and sprinkler
systems are more efficient methods. The costs and
benefits of these methods are elaborated in the discussion of policy alternatives below. For an elaboration of salinity issues, see Chapter 4.

EDF Sues for Salinity Control

*The Environmental Defense Fund (EDF)
is suing EPA to require stricter control in
the Colorado River Basin. An EDF spokesman, calling salinity the most serious water
quality problem in the basin, estimated the
salinity cost for agricultural, municipal,
and industrial users to be over $50 million
annually. EDF will try to get EPA to enforce existing laws and force states to meet
control deadlines. Another EDF
spokesman called for the creation of a
system of "salinity rights" that could be
treated like current water rights in order
to regulate the quality of water returned
to surface waters.* —Gill, 1977; Denver
Post, 1977a.

---

tion runoff has some positive environmental
effects. Since runoff creates marshes and wetlands, which can be scenic and can provide a
habitat for wildlife, many environmentalists
are reluctant to push for more efficient irrigation systems.

Salinity issues may also affect energy development. Since withdrawals of water for energy-
related uses generally cause the levels of
salinity concentration to increase downstream,
stricter standards could restrict future consumptive uses of water. In fact, salinity problems have already created political conflicts
(see box).

Reserved Water Rights:

*The reserved-rights doctrine may entitle
Indians and the federal government to large
quantities of both surface water and groundwater. Indian water rights in particular
have already brought Indians into direct
political conflicts with state governments;
the possible magnitude of Indian water
claims could restrict other uses in the
future, including the development of energy
resources. If Indian water rights are denied, the development of Indian lands may
be restricted and damage to Indian values
and cultures may occur.*

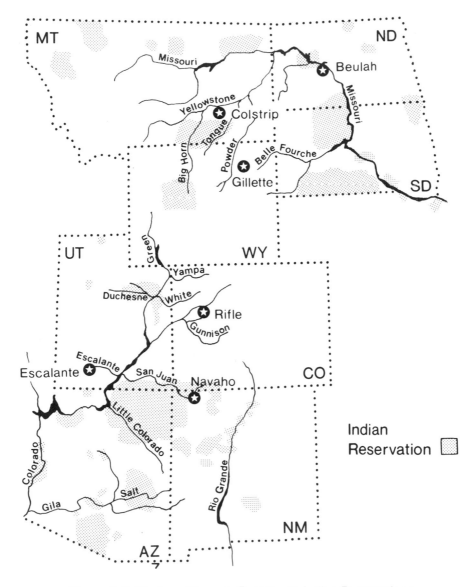

Figure 3-5: *Western Rivers in Relation to Indian Reservations*

Because of the potentially large amount of water at stake, one of the most important conflicts is the question of quantification of Indian reserved rights. Under the reservation doctrine, some Indian tribes have claimed rights to large quantities of water, threatening supplies to many current users who hold appropriated rights. These conflicts generally occur between Indian tribes and state governments. To defend or claim water rights, In-dian tribes usually have to file suit initially in state courts.[7] Many Indians believe that the state courts are inherently more sympathetic to state interests; they also observe that federal courts generally are not inclined to

[7] This is based on the McCarren Amendment, as affirmed in *U.S.* v. *District Court in and for County of Eagle* (1971); and *U.S.* v. *District Court in and for Water Division No. 5* (1971).

review findings of fact from lower courts[8] and conclude, therefore, that the system for judging water rights is biased against them.

As shown in Figure 3-5, Indian tribes potentially control large quantities of western surface waters, many of which are or could be prime sources of water for energy development. If Indians are ultimately found to hold prior and paramount water rights, the existing allocations and appropriations among and within western states could be seriously affected. In one case, the Supreme Court has held that Indian water rights were included in interstate compacts (*Arizona* v. *California*, 1963, Decree 1964). If this interpretation is applied generally, states would have to absorb any changes resulting from the quantification of Indian water rights.

Federal reserved rights may be as important as Indian reserved rights because the amount of water necessary to fulfill the purposes for which federal reservations were created has not been quantified. This lack of quantification is a problem because it leaves other users uncertain as to how much water will remain to fulfill their rights. This uncertainty is especially great with regard to state water laws. Moreover, the legal system developed to deal with federal and state water rights grants state and federal courts concurrent jurisdiction. So a determination of federal rights in state courts, and vice versa, adds to the complexity by creating additional arenas in which quantification issues are addressed but are not finally and definitely resolved.

Jurisdictional Disputes:

The uncertainty and complexity of water policies along with the growth in water demands, have created conflicts over political autonomy, authority, and responsibility among governmental units and between the public and private sectors.

Conflicts over water rights, beneficial uses, and water conservation have raised questions about the roles and responsibilities which various governmental units should play. As indicated above, the issues surrounding water policies involve many conflicts in intergovernmental relations, including both intrastate problems, already occurring between west- and east-slope interests over the diversion of Colorado River water to Denver, and interstate problems—for example, between the Colorado's Upper and Lower basins over which states are responsible for providing 1.5 million acre-ft/yr to Mexico. The most important of these conflicts with respect to water policy in the West appears to be the tension between individual state responsibility and pressures for a centralized federal water policy.

Several proposals of the Carter Administration have threatened the western states, beginning with the proposals to limit or stop funding for several water-development projects. Citing a general need for wise management and conservation, President Carter identified specific western projects as being inefficient and damaging to water quality and to the environment (Kirschten, 1977a). Of the nine water projects halted completely by Congress, five were in our study area (*CQ Weekly,* 1977).

This "hit list" was followed by an administration porposal to enforce a 1902 law limiting to 160 acres the amount of land a farmer is allowed to irrigate with water from federal reclamation projects. This proposal brought sharp reactions from western politicians in several public hearings. Wyoming's Governor, Ed Herschler, for example, said the proposal would "upset the entire western agriculture" and called for equivalency standards which would allow ranchers in arid regions to irrigate more land than those in humid areas (*Denver Post,* 1977b; Kirschten, 1978).

Probably the most threatening federal ac-

---

[8]For example, state findings concerning questions such as "practicable irrigable acreage" could be combined with acceptable legal principles to minimize water available to tribes; see MacMeekin, 1971.

tion, however, has been the Carter Administration's federal water policy, originally proposed in July 1977. Recommendations of the draft water-policy proposal closely parallel the conflicts outlined above: more efficient irrigation practices, new water-pricing strategies, better management of groundwater supplies, and water-resource management that would protect the environment. According to Interior Secretary Cecil Andrus:

The day of considering money to be the only solution to water problems is over. We want results not in the form of more dams and canals and the like, but in the form of more rational use of this very precious resource [Kirschten, 1977b].

The draft policy originally included provisions for a comprehensive national water-resources policy that would fundamentally change state water laws. A formal response to the proposals (see box), submitted on behalf of several Colorado water users' associations and conservation districts, charged that the draft's premise—that state policies are inept—and its conclusion—that policies developed by the federal government for the West could be an improvement—"surpasses credulity" (Fischer, 1977).

Although there is still uncertainty about how these relationships will unfold, President Carter stepped back from many of his original proposals in announcing the revised national water policy on June 7, 1978. Although this plan retains proposals to encourage conservation programs and to require states to contribute a percentage of the cost of new water projects, it explicitly recognizes state responsibility for water management. In his announcement of the policy, President Carter said, "These water policy reforms will not preempt state or local water responsibilities" (Strain and Larsen, 1978). Colorado's Governor Lamm responded, "The West can definitely claim a limited victory" (Strain and Larsen, 1978).

The extent to which intergovernmental disputes will influence energy development in the West is unclear. However, these disputes are indicative of the increasingly serious problem of water availability. The management of water resources has always been essentially a state prerogative, yet the implications of various water uses create interstate and national issues. Among the most important of these is whether western water will be used to meet national energy needs. Thus, these intergovernmental disputes suggest that new institutional mechanisms may be required to cope with the variety of demands being made on scarce western water.

Summary of Problems and Issues

The importance of water has always been appreciated in the West, but new and larger demands on the region's existing resources increase the visibility and significance of water policies. In addition to raising questions about the adequacy of resources and the distribution of limited supplies, this situation also raises questions about the adequacy of existing political institutions to cope with new demands both for resources and for participation in the policymaking processes. The courts have played a major policymaking role; yet they characteristically operate very slowly and provide piecemeal, localized, and short-term resolutions to problems. While the judicial and the legislative processes have been successful in some senses, they may not be adequate in the future as problems become regional and begin to influence national goals, such as a decreased dependence on foreign energy sources.

The fragmented way in which water resources are managed also raises questions, particularly about the inadequacy of water-resource and water-use data and about resistance to institutional innovations. Hence, the mechanisms for resolving the problems of water availability and the issues associated with increased energy development will need to consider more comprehensive approaches to the management of water resources, will need to accommodate diverse interests and values, and will need to facilitate compromises before, rather than after, the large-scale impact is felt.

Table 3-3: *Water Availability Alternatives*

General Alternative	Specific Alternative

Increase supply (augmentation)	Surface water diversion, transfer, and storage
	    Interbasin
	    Intrabasin
	Groundwater storage
	Groundwater use
	Weather modification
	Vegetation management
	    Runoff modification
	    Removal of nonproductive vegetation

Make more efficient use of existing supply (conservation)	Conservation for the development of energy resources
	    Choice of technology
	    Cooling and process designs
	    Municipal wastewater for cooling
	Conservation for agriculture
	    Irrigation efficiency improvements
	    Crop selection
	Conservation for municipalities

Policy Alternatives for Water Availability

As suggested above, conflicts among water users will almost certainly increase as water shortages become more likely. If policymakers try to deal with these conflicts by meeting the expanded needs of water users, two general approaches can be used: to increase water supply (augmentation); and to increase the efficient use of existing supplies (conservation). As suggested in Table 3-3, several specific options exist for both of these categories.

Increasing the supply of water has been the traditional response to water availability problems in the West. This approach could be continued by adopting specific augmentation alternatives, such as intrabasin diversions and transfers and increased storage in surface impoundments. Existing water supplies can also be augmented by other alternatives such as interbasin transfers, groundwater storage, weather modification, and controlling or managing vegetation. In order to meet the expanded needs of water users, policymakers may also choose to improve the use of exist-

ing supplies through conservation in energy resource development, in irrigated agriculture, and in municipal and industrial uses. These alternatives are described and evaluated below. Then, in a summary section the alternatives are compared, based on the five evaluative criteria identified in Chapter 2.

Policymakers may also choose a more radical approach by allocating available supplies to those uses most valued in specific areas. In fact, as discussed above, some states have already begun to do this by adopting policies which explicitly favor energy, agricultural, or other uses. This policy approach also could be adopted by the federal government; for example, by increasing the price of water from reclamation projects, energy development would be favored over irrigated agriculture. A comprehensive evaluation of strategies to decrease or eliminate some of the uses will not be considered here. However, it is likely that basic choices between various uses are likely to become more focused as greater demands are placed on water resources.

Augmentation Alternatives

Surface Water Diversion, Transfer, and Storage: In the past, diversion, transfer, and storage of surface water has been the traditional response to the management of water resources in the West. In fact, the Colorado River is generally considered to be the most regulated river in the world; as discussed above, reservoirs and storage projects, which can store up to about 60 million acre-feet of water, play a critical role, particularly in maintaining stream flows and supporting irrigated agriculture.

However, there is considerable uncertainty about just how much water can be added to existing supplies by diversion, transfer, and storage. For intrabasin transfers, this is largely because so many impoundments and diversions already exist that few acceptable locations may remain (National Water Commission, 1973:319–33). Since intrabasin transfers and impoundments redistribute water from one area to another, rather than adding to total supplies in a basin, the suitability of locations is obviously a critical consideration. Intrabasin transfers have been considered in southern and central Utah, western Colorado, and along the Yellowstone River in Wyoming and Montana (U.S. DOI, BuRec, 1975:143–49; NPC, 1972:245–48). Interbasin transfers potentially could provide large amounts of water, particularly if water from the Columbia and Upper Missouri river basins are diverted to the Upper Colorado. One study estimated that from 2.5 to 15 million acre-feet could be transferred annually depending on the specific location (National Water Commission, 1973:317). However, some transfers are currently prohibited; the CRB Project Act (1968) prohibits the federal government from even studying the transfer of water to the CRB from other river basins, and the Yellowstone River Compact (1950) requires approval of Wyoming and Montana for interbasin transfers.

Several concerns have been raised about surface water diversions and transfers. First, the financial cost of interbasin and intrabasin transfers is likely to be high. Estimates depend on interest rates, the rate of flow through pipelines, and construction and operation costs that vary with location. For pipelines transporting about 30 acre-feet per day, the cost will range from $3 to $7 per acre-foot per mile (Gold and Goldstein, 1979). Larger federal projects, transporting about 1,000 acre-feet per day, cost from $.60 to $1.10 per acre-foot per mile (WPA, 1978). For example, BuRec has studied the feasibility of diverting water from the Green River in the Upper Colorado to Gillette, Wyoming, 225 miles away, via the North Platte River and a pipeline. Assuming existing reservoirs are used, the capital costs for pipes and pumping stations would be about $430 million, and operating costs would be about $2.6 million per year. The total project would cost about $132 per acre-foot (U.S. DOI, BuRec, 1972).

A second major concern with this option is environmental. Surface water diversions and storage projects have been severely criticized because of damage to scenic and wilderness areas and because stream flows decrease at the source of the diversion. Furthermore, surface impoundments lose water through evaporation; in the UCRB, approximately 14 percent of the total withdrawals are lost to evaporation from surface impoundments (U.S. DOI, Water for Energy Management Team, 1974). On the other hand, surface impoundments do provide a direct environmental benefit by helping to maintain minimum flows in drought years.

Questions have also been raised regarding how equitably diversions, transfers, and storage of water will distribute the costs, risks, and benefits. It is difficult—if not impossible—to calculate in economic terms the various issues involved in transferring water from one basin or subbasin to another. For example, the value of water for economic growth, hydroelectric production, recreation, commercial shipping, aquatic life, and scenic and aesthetic amenities may need to be compared to the value of water for the development of energy resources (National Water Commission, 1973:319–31). Even if these values could be expressed in economic terms, considerable uncertainty would still exist regarding the long-term projections of costs, including construction costs and the cost-recovery aspect of these projects.

Nevertheless, some general issues regarding who will pay and who will benefit can be addressed. Transferring and storing water will ultimately mean that the taxpayer pays, since the capital and operating costs of these alternatives will probably necessitate federal authorizations. This setup would be equitable if the water is being transferred to meet national energy goals. Since areas receiving the water would gain other benefits—for example, those associated with economic growth—they can be expected to also pay increased costs. Hence, recent proposals for states to share the cost of such projects with the federal government could redistribute some of the expense to the prime beneficiaries. If the ultimate users of water are asked to pay for water augmentation, energy industries could probably best afford these water prices, while agricultural interests generally would not be able to afford them.

Interbasin transfers also face implementation barriers. Under current law as mentioned above, interbasin transfers to the CRB cannot be subject to detailed study by the Department of the Interior (DOI), and transfers from the Yellowstone would require agreement among the signatories to the Yellowstone River Basin Compact. Furthermore, impoundments in many areas of the CRB and a few areas of the UMRB are precluded by the Wild and Scenic Rivers Act (1968). Water transfers and impoundments are also constrained by the current political system, including general opposition by the Carter Administration and widespread opposition among environmental groups.

Groundwater Storage and Use: Present technologies for the development of groundwater include withdrawal and the use of

aquifers for storage. The withdrawal of groundwater is already common, particularly to provide municipal supplies. Because of the shortage of surface water and the conflicts associated with its use, groundwater is frequently mentioned as an alternative source for energy development. As discussed previously, there are considerable groundwater resources in our eight-state area.

The use of groundwater can apparently increase existing supplies of water by millions of acre-ft/yr (National Water Commission, 1973:232–45). However, since groundwater mining (that is, consumption exceeding aquifer recharge) will deplete supplies and reduce surface water recharge, it is reasonable to expect that groundwater would be mined for a very short time and consumed thereafter only at the recharge rate.

Groundwater storage would be used primarily to reduce the amount of water lost through evaporation in surface impoundments. Such storage, however, is geographically limited by the location and accessibility of large aquifers. In the CRB, it is estimated that 2 million acre-feet of water currently evaporates annually from major storage reservoirs and lakes (U.S. DOI, BuRec, and USDA, Soil Conservation Service, 1977:II–30); underground storage could greatly reduce these losses.

Although less expensive than surface water transfers, groundwater storage and mining would require a fairly large investment for pumping systems to store and/or retrieve the water. Estimates of pumping costs for groundwater mining in Montana's Madison aquifer range from $30 to $50 per acre-foot (Montana DNRC, Water Resources Division, 1976). Costs of groundwater storage can be twice this high, depending on how water is introduced into the aquifer (U.S. Army, Corps of Engineers, 1978). The pumping systems which may be required to inject water into deep aquifers may increase the cost to as much as $100 per acre-foot. If aquifers are shallow, large wells or percolation ponds can be used to reduce costs. Of course, the cost of groundwater storage and mining will increase if water from the aquifers has to be transported to where it will be used.

Although groundwater storage does not appear to be environmentally damaging, groundwater mining raises environmental concerns because it probably would reduce the rate at which surface streams would be recharged. Furthermore, if groundwater is continuously used beyond recharge rates, the resource eventually will be exhausted; thus, future water users will almost certainly pay the costs of present use. Finally, groundwater use may be constrained by legal systems. In Arizona, groundwater is a private resource, conveyed with the land (excluding underground streams with defined beds). Thus, surface water, recharged from groundwater become the property of landowners and are outside governmental control.

Weather Modification: Weather modification by cloud seeding is already being used in the West, particularly during drought conditions. Although the technologies are still developing, experiments in Colorado, Utah, and New Mexico suggest that these techniques can produce detectable increases in snow accumulation and subsequent runoff (NAS/NRC, 1973:80). Weather modification appears to be applicable to many locations, although the most appropriate climates and topography occur along the Rocky Mountains.

Just how much supplies could be augmented in this way is unclear. Some estimates project that weather modification can offer a considerable potential for augmenting natural supplies of water at a low cost:

> Precipitation management technology appears to be sufficiently advanced to offer an important new source of water supply to water and potential water short areas of the West at an estimated $5 per acre-foot [U.S. DOI, BuRec, 1975:442].

However, other estimates are considerably less optimistic, suggesting that success varies considerably with location and yearly conditions (MacDonald et al., n.d.). The economic costs of weather modification appear to be low compared to other augmentation choices. In fact, the estimate of $5 per acre-foot of the study cited above is about twice as high as other estimates of $2 to $3 per acre-foot (Weisbecker, 1974:ix).

In light of relatively limited experience in actual weather modification, the environmental costs and risks are important, unanswered questions. Although weather modification may contribute to hail damage, floods, avalanches, and biological changes, the risks of these consequences appear to be low if the technology is used correctly (Weisbecker, 1974). The results of ecological research to date indicate that a catastrophic ecological impact should not be expected, although some relatively small damage could occur to some plant and animal species. However, there is some indication that the impact of weather modification downwind may take years to detect (National Water Commission, 1973:349). This and other uncertainties have led the National Park Service to oppose weather modification until it can be demonstrated that the national and historic environments of the National Park System will not be damaged (U.S. DOI, National Park Service, 1978).

Vegetation Management: Vegetation management includes managing upland terrain to increase the amount of runoff. This can be achieved by reducing the density of ground cover (trees and shrubs, for example), by converting to a type of vegetation which uses less water, and by trapping or transporting snow to reduce evaporation. Upland vegetation management has widespread potential throughout our eight-state study area, although it is thought to be most effective in the forests of Montana and Wyoming (National Water Commission, 1973:357).

The potential for upland vegetation management to increase runoff is based on the quantities of water lost to evaporation and transpiration by plants—approximately 90 percent of the annual precipitation in the CRB. If evapotranspiration in the CRB were reduced by 1 percent, the surface water supply could be increased by an average of 1.75 million acre-ft/yr (Hibbert, 1977). However, upland vegetation management is only possible on about 26 million acres (about 16 percent) of the basin since the rest of the area is covered by sparse vegetation which allows little opportunity for control.

The management of vegetation also includes removing economically nonproductive vegetation (phreatophytes) from streams, stream beds, and along canals, irrigation ditches, and other man-made water courses to decrease losses to evapotranspiration. Phreatophytes, such as salt cedar and mesquite, draw significant quantities of water from groundwater and reduce the stream flow and the volume of springs. Since phreatophytes occupy large areas of land in the West, removing them could be used as an augmentation alternative either locally or regionally.

Although it has been estimated that phreatophyte removal could save several million acre-ft/yr on a nationwide scale (National Water Commission, 1973), it is unclear how extensively or effectively it could be used in the West. However, experiments in the Southwest indicate that about 14,000 acre-ft/yr could be added by removing 6,000 acres of salt cedar and mesquite along the Rio Grande and about 40,000 acre-ft/yr could be added by removing 11,000 acres of phreatophytes along the Gila River in southern Arizona (National Water Commission, 1973:353–54).

Upland vegetation management has been estimated at about $1 per acre-foot for commercial forests in Montana, Wyoming, Colorado, Utah, New Mexico, and Arizona (National Water Commission, 1973:357) and up to $50 per acre-foot in Chapparal areas. Phreatophyte removal costs have been estimated at $14 per acre-foot for certain areas in Colorado, Utah, Arizona, and New Mexico (National Water Commission, 1973:357).

However, the largest costs—and risks—

for vegetation management are environmental. Phreatophytes provide habitats for wildlife, stabilize the shorelines, shade streams to control temperature fluctuations, and provide a food base for fish populations (National Water Commission, 1973:354–58). Upland vegetation management in forest areas, particularly clearcutting forests, could also threaten wildlife habitats and degrade scenic and aesthetic values (Hibbert, 1977). Thus, vegetation management of either sort is likely to generate considerable opposition from environmental groups.

Summary of Water Augmentation Alternatives: Table 3-4 summarizes the augmentation alternatives for the five criteria identified in Chapter 2. This information indicates that if increased water availability were the sole criterion, augmentation alternatives can be effective throughout the eight-state study area. Significant quantities of water can be added by each specific alternative and particularly by interbasin transfers. Most alternatives appear more effective in particular locations—interbasin transfers are particularly suited to the northern states (Montana, Wyoming, and Utah), weather modification to the central Rocky Mountain states (Utah, Wyoming, and Colorado), and phreatophyte removal to the Southwest (southern Utah, Arizona, and New Mexico). Uplands vegetation management and groundwater mining are generally applicable throughout the region. However, it is not clear whether these alternatives would be effective over the long term. Groundwater mining appears to be the most limited in this regard.

When economic and environmental costs and risks, and social and political values are considered, many of these choices become less attractive. Interbasin and intrabasin transfers and, to a somewhat lesser extent, groundwater mining and storage will require high and generally irreversible commitments of capital and would provide water at costs higher than current water prices throughout most of the West. Although the energy industries could probably afford these increased costs, agricultural interests almost certainly could not afford to irrigate. In the end, the taxpayer or energy consumer would pay at least part of these economic costs. In addition, interbasin and intrabasin transfers and storage projects raise serious environmental questions because of threats to scenic and wilderness areas and because of the reduction of stream flows. Transfers also generate considerable political opposition and are constrained by current laws and interstate compacts. Hence,

Table 3-4: *Summary Evaluation of Augmentation Alternatives*

Criteria	Measures	Findings
Effectiveness: How much water can be added?	Acre-feet per year	Interbasin transfers can add millions of acre-ft/yr over the long term. Groundwater mining may add millions of acre-feet for the short term. Weather modification can add large supplies, particularly in the Rocky Mountains. Intrabasin transfers, groundwater storage, upland vegetation management, and phreatophyte removal apparently can add hundreds of thousands acre-feet, although none of these options appear applicable throughout the regions. Considerable uncertainty exists regarding the long-term effectiveness of all options.
Efficiency: What are the economic and environmental costs, risks, and benefits?	Dollars per acre-foot	Interbasin and intrabasin transfers are the most costly options, with costs ranging from $50 to $300 per acre-foot for interbasin and $30 to $100 per acre-foot for intrabasin. Costs of groundwater mining range from $30 to $50 per acre-foot; groundwater storage can cost as much as $100 per acre-foot if aquifers are deep. Weather modification ($2 to $3 per acre-foot) and upland vegetation management in commercial forests ($1 per acre-foot) appear to be the least expensive. Phreatophyte removal has been estimated at $14 per acre-foot.
	Environmental costs and risks	Damage may occur to scenic and wilderness areas from transfers, impoundments, vegetation management, and phreatophyte removal. The most widespread environmental and ecological risks are probably associated with vegetation management and phreatophyte removal. Groundwater mining and surface impoundments will reduce stream flows in some areas. Environmental and ecological risks also are associated with weather modification, although they appear minimal, given current knowledge.
Equity: How will costs, risks, and benefits be distributed?	Distribution of costs and risks	Taxpayers are likely to pay for transfers and groundwater storage and mining since capital costs will require government subsidies. Recipient states will share some of these capital costs if a new water policy is adopted by Congress. Future generations will pay for groundwater mining and face the risks of weather modification. Environmental interests generally face the greatest risks associated with these choices.
	Distribution of benefits	Depending on the mix of alternatives chosen, energy development interests and UCRB states are likely to be prime beneficiaries. If interbasin transfers are used, the entire region is more likely to benefit.
	Ability to pay	If users are asked to pay the cost, energy industries could probably afford the increased prices. Agricultural interests could probably afford only the costs of weather modification and vegetation management of commercial forests.

Table 3-4: (continued)

Criteria	Measures	Findings
Flexibility: Can augmentation alternatives be applied under different conditions and over time?	Reversibility	Capital-intensive alternatives, including transfers, impoundments, and possibly groundwater storage and mining, require long-term commitments and relatively inflexible resource-allocation procedures. Weather modification, upland vegetation management, and phreatophyte removal offer the greatest flexibility in capital expenditures. Upland vegetation management and phreatophyte removal can have a relatively irreversible or long-term impact on streams and forests.
	Adaptability	Upland vegetation management and weather modification have widespread applicability across the West. Phreatophyte removal appears most effective in the arid southwest. Groundwater mining and storage is limited to suitable and accessible hydrologic areas. New intrabasin transfers are probably not possible in most areas of the Colorado River Basin. Interbasin transfers are generally limited to the northern states.
Implementability: How difficult will it be to enact and implement?	Degree of innovation or change in existing institutions, laws or regulations	DOI is prohibited from studying projects to transfer water from the Columbia River to the Colorado, and transfers from the Yellowstone require the consent of both Wyoming and Montana. The Wild and Scenic Rivers Act limits some transfers, particularly in the UCRB. Underground mining and storage is constrained by existing groundwater law in some states. Legal uncertainties may constrain weather modification.
	Acceptability	Weather modification, upland vegetation management, and phreatophyte removal are expected to generate substantial opposition from environmental groups. Transfers and impoundments will generate national and regional political opposition.

although transfers, impoundments, and diversions appear to be the most effective choices, they are, in fact, problematic.

Upland vegetation management and phreatophyte removal are constrained primarily by environmental considerations and will probably be unacceptable choices for many. Phreatophyte removal threatens the habitats of native species, some of which may be endangered; upland vegetation management is most environmentally damaging if forest and chapparal lands are clearcut.

Weather modification also will necessitate trade-offs, but apparently to a lesser degree than most of the other options. Weather modification is economically efficient, can provide environmental benefits by increasing stream flows, and appears to be more flexible and easier to implement than most augmentation alternatives. However, weather modification does involve a high degree of uncertainty about its long-term impact and damage liability and has generated opposition from those who oppose modifying natural conditions and processes.

Conservation Alternatives

Choice of Energy Technology: Perhaps the most comprehensive conservation choice is to build the least water-intensive energy conversion facilities. As discussed in the first section of this chapter, synthetic fuel processes generally require much less water per unit of energy produced than does electric power gen-

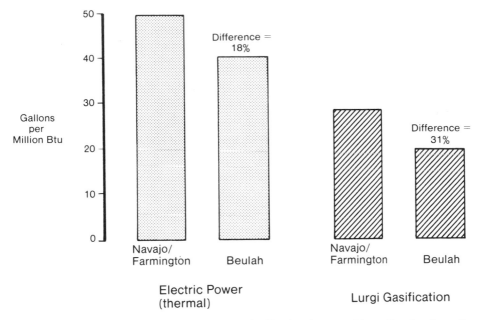

Figure 3-6: *Water Requirement Differences by Technology and Location (wet cooling)*

eration. If a high degree of wet cooling is used, electric power generation can require from about 30 percent to 100 percent more water than gasification and liquefaction (see Table 3-5). Among synthetic fuel technologies using coal, Lurgi gasification requires from 21 to 53 percent more water than Synthoil liquefaction to produce an equivalent amount of energy at the same site. Thus, some water savings can be attained by choosing less water-consumptive technologies.

Some water problems may also be avoided by the careful selection of locations for energy facilities. As discussed above, water requirements for the same facility can vary considerably with location. An indication of this is given in Figure 3-6, showing estimates of water requirements for electric power genera-

Table 3-5: *Water Consumption and Potential Savings for Coal Conversion Technologies*

Technology	Water Consumed For High Wet Cooling[a] (gallons per million Btu)	Water Saved[b]			
		Intermediate Wet Cooling		Minimum Wet Cooling	
		Percent	Gallons per Million Btu	Percent	Gallons per Million Btu
Power Generation					
Btu (electric)	125–55	67–77	96–106	. . . .	. . . .
Btu (thermal — 34 percent efficiency)	42–53	67–77	33–36	. . . .	. . . .
Lurgi Gasification	20–29	21–32	6.1–7.0	27–42	7.9–9.3
Synthane Gasification	32–36	23–24	7.4–8.1	27–29	8.9–9.8
Synthoil Liquefaction	15–19	16–19	2.7–3.0	21–25	3.6–4.0

[a]Range in amount of water used is due to site-specific variations.

[b]These data are from WPA, 1978.

tion and Lurgi gasification at two of our hypothetical scenarios. This information suggests that water requirements at Beulah, North Dakota, will be about 18 percent less for power plants and about 31 percent less for Lurgi gasification than at Farmington, New Mexico. This finding is attributable to several factors, such as the cooler climate of the Northern Great Plains which reduces cooling water requirements. The water content of the coal found in these areas is also important to the Lurgi process which can use the moisture content of the coal in the conversion process. Since coal typically found in the Beulah area contains more than twice the water (about 36 percent) of coal found in the vicinity of Farmington (about 16 percent water), a Lurgi plant in Beulah will require less water from other sources.

Obviously, site selection for energy facilities is done on the basis of many factors in addition to the availability of water. The advantages and disadvantages of alternatives for site selection are discussed in more detail in chapters 11 and 12.

Cooling Technologies for Energy Facilities: Water requirements for developing energy resources may also be reduced by using a combination of wet and dry cooling and by instituting designs which minimize the use of water in conversion facilities. For synthetic fuel technologies that use coal, cooling is required for steam-turbine condensers and for gas compressors; for electric power plants (which do not have gas compressors), cooling is required only for the steam-turbine condensers. In the high wet cooling cases, wet cooling is used for all the cooling requirements in both the synfuel and electric power plants (see Table 3-5). In the intermediate case, wet cooling is used for 10 percent of the load on the steam-turbine condensers and all of the load on the interstage coolers. In minimum wet cooling (which in this analysis applies only to synthetic fuel plants), wet cooling handles 10 percent of the load on the

condensers and 50 percent on the gas compressor coolers.

Table 3-5 summarizes the amount of water saved as a function of the degree of wet cooling for five coal conversion technologies. Intermediate wet cooling can reduce the total plant requirements for water by as much as 32 percent for Lurgi gasification and 77 percent for electric power generation. This means that between 96 and 106 gallons per million Btu's could be saved depending on where the power plant is located. Minimum wet cooling in synfuel processes can save from 25 percent (Synthoil) to 42 percent (Lurgi). Based on our Low Demand scenario, minimum wet cooling could save about 500 thousand acre-feet of water annually by the year 2000.

The economic costs, risks, and benefits of water conservation for energy production will largely depend on the price of water and how strictly conservation is defined. Water conservation by intermediate wet cooling for conversion facilities can often be economical, but opting for minimum wet cooling will generally increase the costs of energy products. As shown in Figure 3-7, high wet cooling appears to be the most economical choice for all coal conversion facilities if water costs less than about 25¢ per 1,000 gallons ($81 per acre-foot). (Water rights typically cost $10 to $100 per acre-foot or 3¢ to 30¢ per 1,000 gallons.) For synthetic fuel facilities, intermediate wet cooling becomes economical at between 25¢ and $1.50 per 1,000 gallons ($81–$490 per acre-foot); and minimum wet cooling is economical when water costs more than $1.50 per 1,000 gallons ($490 or more per acre-foot). For power plants, however, intermediate wet cooling does not become economical until water costs from $3.65 to $5.87 per 1,000 gallons ($1,190–$1,910 per acre-foot). The range of "break-even" water costs shown in Figure 3-7 is due to variations between various sites in the West (for more details, see Gold and Goldstein, 1979).

Although the costs of water rights are now relatively small, the cost of transporting water to a given site may not be. Except for facili-

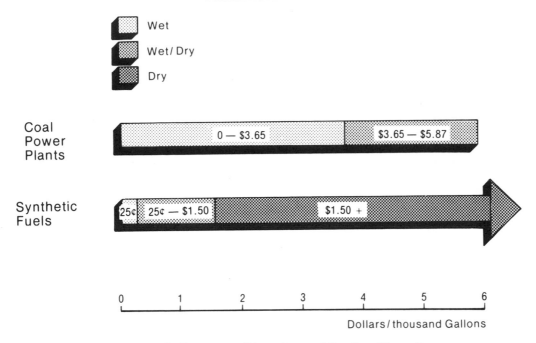

Figure 3-7: *Breakeven Water Costs of Cooling Alternatives*

ties located near the main stem of major rivers or near large reservoirs, the cost of transporting water is usually in excess of 25¢ per 1,000 gallons. Thus, intermediate or minimum wet cooling is economically desirable for synthetic fuel plants at many of the energy development sites in the West. However, for power plants, high wet cooling is usually the most economical since the cost of transporting water is much less than $3.65 per 1,000 gallons, the break-even point for intermediate wet cooling.

The information summarized in Table 3-6 suggests the economic costs to energy developers if they are forced to use water-saving cooling technologies. The economic penalties for synfuel facilities will be very small. For example, in the case of Lurgi, 1,920 to 2,140 acre-ft/yr can be saved by changing from high to minimal wet cooling at a cost not exceeding 1.5¢ per million Btu's of gas. Assuming synthetic gas costs $3.00 per million Btu's, intermediate wet cooling increases the costs by 0.3 percent, and minimum wet cooling increases the costs by 0.5 percent. In contrast,

the costs of conserving cooling water would be much higher for power plants. Assuming electricity costs 2.5¢ per kilowatt-hour (kWh) in 1978 dollars, the costs would increase by about 4 to 7 percent.

Environmentally, conserving water in energy conversion facilities will have a positive effect. While water savings for a single facility are not large, they could have a significant effect on streams with low flows or on a basin-wide scale if many energy conversion facilities were constructed.

A critical question, of course, is how water conservation would be implemented. In general, variations in water problems and in the potential influence of new technologies suggest that any implementation procedure would need to be flexible enough to allow states and localities to meet particular demands and characteristics of specific areas. For example, if a policy requiring the "best available conservation technology" were applied uniformly, it would overlook local differences in water availability. Indeed, conservation is likely to be needed in different proportions in various

Table 3-6: *Economic Costs of Conserving Cooling Water*[a] *(1978 dollars)*

Technology	Changing From High Wet to Intermediate Wet Cooling		Changing From High Wet to Minimum Wet Cooling	
	Water Saved (acre-ft/yr)	Cost	Water Saved (acre-ft/yr)	Cost
Power Plant (cents/kWh)	6,710–20,340[b]	0.11–0.18	. . . .	. . . .
Lurgi Gasification (cents/million Btu)	1,490–1,670	0.15–1.03	1,920–2,140	1.18–1.32
Synthane Gasification (cents/million Btu)	1,790–1,970	0.18–1.22	2,150–2,380	1.33–1.47
Synthoil Liquefaction (cents/million Btu)	1,690–2,100	0.14–1.00	2,200–2,640	1.06–1.27

[a]These calculations assume water rights cost 20¢ per 1,000 gallons; estimates are derived from Gold and Goldstein (1979).

[b]Ranges reflect site-specific differences.

areas of the West, depending on the quantity of water locally available and on demands.

Flexible implementation also includes the need for state and local responsibility in making policy decisions, something which has been highlighted by recent conflicts between the western states and the Carter administration regarding federal water policy. Thus, federal attempts to mandate uniform conservation policies which do not include a strong role for the states in both formulation and implementation are likely to be very difficult to enact.

Agricultural Irrigation Efficiency: Agricultural conservation is also a logical conservation choice since so much surface water in the West is diverted to irrigation. Suggestions for conserving agricultural water include removing marginal lands from production; improving cultivation; initiating planting practices such as avoiding cultivation that interferes with root systems, controlling weeds, and mulching with crop residues; and improving irrigation technologies such as changing from flood-and-furrow methods to mechanical applications, such as sprinkler or trickle irrigation. Agricultural water diversions can also be reduced by lining canals to prevent loss to groundwater and to transpiration by weeds. Each of these specific alternatives may improve

the efficiency of irrigation. Irrigation efficiency is defined as the ratio of water applied to the amount actually used for plant growth and soil leaching—the higher the efficiency, the less runoff. Improved efficiency, however, is not synonymous with conservation, which means reducing the amount of water consumptively used (that is, removed from further beneficial use). Thus, when water is "lost" by soaking into the ground, it may or may not be consumed depending on the availability of that groundwater for further use.

There is a considerable amount of uncertainty about water savings in agriculture. Despite the fact that much research has been done on irrigation efficiency, it is not clear how much consumptive use could be reduced by switching to more efficient irrigation technologies. It does appear that irrigation efficiency can be improved. For example, it has been estimated that conversion from flood-and-furrow methods to sprinkler methods can raise efficiency from below 50 percent to as high as 80 percent. In the Upper Colorado River subbasins, such a move would reduce water diversions for each acre by an average of 2.2 acre-feet (Utah State Univ., Utah Water Research Lab., 1975:60). Another estimate suggests that for alfalfa, a very water-intensive crop, average per acre water diversions can be reduced by 3.6 acre-feet in Arizona (Pritchett and Boesch, 1977). For the three sub-basins above Farmington, New Mexico, it has been estimated that water diversions could be reduced about 250,000 acre-feet by improved irrigation, based on the uses and the quantities consumed in 1975 (Utah State Univ., Utah Water Research Lab., 1975).

These studies, however, do not quantify the amount of water actually conserved or saved. Some studies have shown that little if any water is actually saved by improving irrigation efficiency because the runoff is ultimately used downstream (Wendt et al, 1977). Other studies suggest that improving irrigation efficiency saves water by reducing the amount lost both in evaporation and when return flows move directly to a saline lake or inaccessible groundwater systems (USDA, Soil Conservation Service, 1976). For six of the eight states in our study area (excluding the Dakotas), it has been estimated that 9 percent of annual water diversions to agriculture are irrecoverable (U.S. DOI, BuRec and BIA, 1978:27).

The economic costs of improving irrigation efficiency depend on the type of irrigation system, crop, acreage, and local and seasonal conditions. One estimate for the CRB is that the annualized cost of installing and operating a sprinkler irrigation system would be about $50 per acre (Utah State Univ., Utah Water Research Lab., 1975:249). However, estimates for specific areas can be much higher—for example, a total annualized cost of $64 to $148 per acre, depending on the size of the farm, in southern Arizona (Wade et al, 1977). Using the estimate of $50 per acre and assuming an average reduction in per acre diversions of 2.2 acre-feet for the UCRB, the average cost would be about $22 per acre-foot. In the example mentioned above for the three subbasins above Farmington, the costs could be less—$13.89 to $26.32 per acre-foot to reduce withdrawals by 250,000 acre-feet (Utah State Univ., Utah Water Research Lab., 1975). However, if water savings represent only ten percent of diversions, then the dollar costs per acre-foot of water saved

would range from $139 to $262 per acre-foot for the examples cited above.

Other factors influence the efficiency of agricultural conservation as an alternative. One of the long-term benefits of sprinkler irrigation systems appears to be economic—that is, the savings from reduced labor requirements. Although little data is available for evaluation, the Soil Conservation Service estimates that labor requirements can be reduced by about 50 percent with the use of mechanical irrigation (Pritchett and Boesch, 1977). Also, in some cases the cost of salvaged water may be less than buying new water. However, this can be very difficult to determine since irrigation water charges are often based on the number of acres irrigated rather than the number of acre-feet of water delivered. In addition, costs vary considerably across the West: they range from between $3 and $5 per acre for low value crops to between $25 and $30 per acre for high value crops, such as orchards (U.S. DOI, BuRec Staff, 1977).

Nevertheless, the basic economic problem for farmers appears to be the capital investment rather than the annual cost. Since irrigated agriculture in many areas of the West is only a marginally profitable enterprise, capital costs are likely to be prohibitive for many farmers unless additional subsidies are provided. Furthermore, under the current system there is a considerable risk in spending large amounts to improve irrigation efficiency: water appropriation systems may not allow a user to retain the right to salvaged water, since a downstream user may depend on the return flow.

Improving irrigation efficiency will have a positive environmental effect by reducing saline runoff and by reducing the salt-concentrating effect of consumptive use. Indeed, improved water quality, rather than water savings, may be the primary benefit of more efficient irrigation systems. In the UCRB, agricultural irrigation accounts for about 60 percent of withdrawals from surface waters, and the return flow from this diversion contributes about 1.5 million tons of salt per year. It has been estimated that irrigation runoff accounts for about 40 percent of the salt concentration at Lees Ferry. Based on a study by Utah State University, by the year 2000 sprinkler irrigation and canal lining could reduce salt loading in the CRB by 0.5 MMtpy and 1.75 MMtpy, respectively. In addition, to the extent that these measures reduce consumptive use of water, salinity is improved by allowing for more dilution of existing saline flows. Some estimates for the economic costs (1975 dollars) on a per mg/l reduction of salinity at Imperial Dam are as follows: sprinkler irrigation—$1.9 million to $4.1 million; removal of marginal agricultural lands in Grand Valley—$138,000 to $372,000 (see Utah State Univ., Utah Water Research Lab., 1975:256–62; and Leathers and Young, 1975). Therefore, as indicated in terms of salinity control, sprinkler irrigation can cost many times more than other measures. While such estimates are broad approximations, it appears as though these costs will often exceed the combined economic benefits for municipal, industrial, and agricultural water users which is estimated at $200,000 per mg/l reduction at Imperial Dam (see Chapter 4 for a more detailed discussion).

If ability to pay is considered, conservation costs appear to outweigh the benefits, from the perspective of irrigated agriculture. For example, if water prices are increased as an incentive to improve irrigation efficiency, many farmers will then be unable to afford an adequate water supply. One estimate suggests that if water costs $200 per acre-foot (60¢ per 1,000 gallons), agriculture could experience a total cost increase of about 400 percent, whereas the production costs of energy would increase only one to eight percent (Andersen and Keith, 1977:161). Although the profitability of irrigated agriculture varies considerably in the West, it will almost certainly be less able to afford price increases than will energy resource development. In the CRB, water for agriculture is estimated to have a maximum value of about $25 per acre-foot (Anderson et al, 1973).

Some estimates suggest that the relatively high average annual costs of mechanical irrigation systems might be profitable if the water saved were used to irrigate idle lands, thus increasing total farm acreage (Wade et al, 1977). However, as discussed above, not only could very small increases in production costs be prohibitive, but in many states farmers would be unable to retain the rights to "salvaged" water since this would reduce return flows. Thus, unless policymakers want explicitly to decrease or eliminate the agricultural use of water, conservation efforts will probably require direct subsidies for new irrigation systems.

Switching Crops: Water can also be conserved by changing the crops that are planted. Since different crops are known to have different water requirements, water can be saved by switching from water-intensive crops like alfalfa to crops such as vegetables or wheats which use either less water per acre or less water per dollar of crop value.

Exactly how much water can actually be saved by switching crops is unclear. In Arizona, shallow and deep vegetables require less than 2.0 acre-ft/yr per acre, and wheats, barley, and oats require about 1.9 acre-ft/yr per acre. In contrast, cotton (3.0), crop pasture (3.1), and alfalfa (5.0) all require substantially more water per acre (USDA, Soil Conservation Service, 1976). Because water intensive crops, such as alfalfa and other forage, are commonly grown in western states, switching crops could be an effective option, although several risks are involved; these are discussed below.

The economics of switching crops are also difficult to estimate. In general, the economic return for vegetables is greater than for forage crops on a per acre basis. However, a comparison of the economic return must include the risks associated with switching crops. First, it is unclear whether sufficient markets exist to support a large-scale increase in vegetable production. So, although some farmers may be able to obtain an economic advantage from switching crops, market-related risks discourage this option as a broad-based alternative. Second, vegetable crops are more susceptible to weather damage (such as late freezes, floods, etc.) than are forage crops. Hence, in some years, vegetables might provide a good economic return, but in years of adverse weather, vegetables may not return anything. Third, economic return is not the only factor to be considered. For example, many farmers have grown forage crops for decades and are not likely to be enthusiastic about radical shifts in their crop patterns. Fourth, the end use of forage crops helps explain their abundance: alfalfa, for example, is often grown to support dairy and cattle industries within the state or the area.

Thus, the market-related risks of switching crops appear to outweigh the potential economic benefits to the individual farmer. In addition, these risks would increase as more farmers switched to vegetable crops. Since the effectiveness of this conservation option would depend on its broad utilization—for example, within an entire subbasin—the overall economic efficiency appears to be negative. Furthermore, the same institutional constraints which may not allow a farmer any legal right to water saved by improving irrigation efficiency would apply also to switching crops.

Municipal Conservation: Municipal conservation techniques generally are not well established. Because there are so few large municipalities in our eight-state study area, this option is not likely to be one of the region's primary conservation methods. However, the Carter Administration has suggested that municipalities can conserve 15 percent of their water, and many western areas have had experience in water conservation, having dealt with recent drought conditions.

A 15 percent savings could mean an important difference in the UCRB, where approximately 20 percent of the annual withdrawal of water goes to Denver via interbasin transfer. If all of the 15 percent savings in Denver's

water use were subtracted from this interbasin transfer, between 96,000 and 111,000 acre-ft/yr would be saved. This estimate assumes a total yearly withdrawal in the Upper Basin of 3.2 to 3.7 million acre-feet, approximately 20 percent of which (640,000 to 740,000) goes to Denver (U.S. DOI, Water for Energy Management Team, 1974).

Although municipal conservation is a limited option in the West, whatever slight impact does occur is likely to be positive because of reduced stream withdrawals. There is little data available for estimating the economic costs associated with a 15 percent reduction in water use.

Summary of Conservation Alternatives: Table 3-7 summarizes and compares the various water conservation alternatives. Among the five options evaluated, conservation in the development of energy resources appears to be the most promising. Water requirements for energy facilities can be reduced by about 75 to 85 percent by opting for synthetic coal facilities rather than coal-fired steam-electric power plants—keeping in mind, of course, that the product outputs are different. Among the synthetic fuel technologies, Synthoil liquefaction requires less water than do either the Lurgi or Synthane gasification processes to produce an equivalent amount of energy.

Water-saving cooling technologies can also reduce the large percentages of water required for coal conversion. For synthetic fuel plants, intermediate wet cooling can reduce water requirements by about 15 to 30 percent, and minimum wet cooling can reduce the total requirements to about 20 to 40 percent of those needed for a high degree of wet cooling. For power generation, the amount of water saved can be as much as 80 percent. Water-saving cooling technologies are particularly attractive because for some technologies, economic costs will not be increased substantially. For synthetic coal facilities, increases in the cost of the energy product will be less than 1 percent, even using minimum wet cooling. The percentage increase in electric power

Table 3-7: *Summary Evaluation of Conservation Alternatives*

Criteria	Measures	Findings
Effectiveness: How much water can be saved?	Percentage reduction in water requirements for energy facilities	Water requirements based on an equivalent amount of raw coal can be reduced by as much as 75 percent by constructing synthetic coal facilities rather than electric power plants. Among synthetic fuel facilities, Lurgi requires about two-thirds and Synthoil requires about half as much as Synthane to produce an equivalent amount of energy. Large percentages of water can be saved if coal conversion facilities use water-saving cooling technologies rather than wet cooling. Water requirements for electric power plants can be reduced by about 75 percent by intermediate cooling; water requirements for synthetic fuel plants can be reduced by 16 to 32 percent (intermediate wet cooling) and 21 to 42 percent (minimum wet cooling).
	Gallons or acre-feet per year	400,000 acre-ft/yr can be saved by the year 2000 in the eight-state study area if water-saving cooling technologies are used. Although irrigation efficiency can be improved substantially by mechanical methods, actual savings are uncertain. Water savings can be attained by switching crops and by mechanical conservation, but amounts are uncertain.
	Problems avoided or mitigated	Conservation by itself probably will not eliminate either water shortages or conflicts over water rights. However, many conflicts can be mitigated, particularly if conservation measures are instituted in energy development.
Efficiency: What are the economic and environmental costs, risks, and benefits?	Dollar costs per unit of energy produced	For synthetic fuel facilities, costs will increase .15¢ to 1.2¢ per million Btu's produced if intermediate wet cooling is used and 1.1¢ to 1.5¢ per million Btu's if minimum wet cooling is used. For power plants, intermediate wet cooling will increase costs .1¢ to .2¢ per kWh.
	Percentage increase in cost of energy product	Water-saving technologies will increase the cost of synthetic gas by a maximum of .5 percent and the cost of electricity by 4 to 8 percent.
	Economic costs of agricultural production	Mechanical irrigation will increase production costs by about $60 to $150 per acre, or about $22 per acre-foot of water saved. These costs appear prohibitive for many farmers. Switching crops from forage to vegetable production would be economically beneficial, assuming adequate markets exist.
	Acceptability of risks	Agricultural conservation appears to require unacceptable risks: farmers may lose rights to water saved; the availability of markets for vegetables is uncertain. Risks of conservation in the development of energy resources seem to be acceptable — either because the percentage of cost increases are small or because the costs would be passed on to consumers.
	Environmental quality	Conservation generally will have positive overall environmental effects. The largest environmental benefit will be from improving irrigation efficiency which will significantly reduce saline contamination and could increase instream flows locally. Conservation for energy also will increase instream flows by reducing withdrawals.

Table 3-7: *(continued)*

Criteria	Measures	Findings
Equity: How will these costs, risks, and benefits be distributed?	Ability to pay	In general, irrigated agriculture will be unable to afford the costs and risks of conservation. If water prices are increased, the costs of agricultural production would increase at a much higher rate than would the costs of energy production. Conservation will almost certainly be paid for by the consumer — either directly by increased energy costs or indirectly by increased government subsidies.
	State and regional impacts	Increased interregional conflicts can be anticipated as the West develops resources to meet the needs of other regions and as other regions pay higher prices either for energy or to ensure adequate water supplies. Increased conflict among Upper and Lower Basin states of the Colorado River can be anticipated as the Upper Basin begins to use its full allotment.
Flexibility: Can conservation alternatives be applied under different conditions and over time?	Adaptability to local differences	Uniform and equitable implementation strategies will probably sacrifice the flexibility required to address local situations.
	Degree of reversibility	Water-saving cooling technologies and mechanical irrigation systems are largely irreversible. Crop switching and municipal conservation seem to be reversible.
Implementability: How difficult will it be to enact and implement the alternative?	Degree of innovation required	Comprehensive policies and innovative institutions may be required to adequately deal with the complex mix of water problems that are likely to arise.
	Acceptability	Users will almost certainly respond negatively to comprehensive changes. This is particularly true regarding agricultural conservation which could threaten state appropriation systems and BuRec pricing policies. Conservation alternatives for the development of energy resources may not generate as much resistance.
	Openness of decision-making processes	Current institutional arrangements do not allow broad public participation, thus contributing to the complexity and uncertainty about water rights and resource management.

costs will be larger (about 8 percent); thus, intermediate wet cooling for generating electricity will not be economic until the price of water increases to more than $1,000 per acre-foot.

Water conservation in agriculture will provide some benefits, particularly regarding environmental quality. However, these alternatives will be very difficult to implement. Although many institutional arrangements contribute to the current complexity of water policies, the state appropriation systems and the federal pricing policies may be the most constraining to agricultural conservation. This is because even if farmers could conserve large quantities of water, the "use it or lose it" aspects of these mechanisms mean that water "saved" by the individual farmer cannot always be retained by him. Although several states have incrementally changed their appropriation systems and Utah has proposed broad changes, the long and complex development of these systems suggest that any regional change will be very difficult.

Water conservation in the development of energy resources may be the easiest to enact and implement because these alternatives may not involve many of the difficult implementation questions. Thus, conservation in this area can save large amounts of water; it has less uncertainty associated with its effectiveness and efficiency than do conservation efforts in other areas; energy industries are probably better able than other water users to afford the associated cost increases; and it appears to be easier to implement new regulations in connection with the development of energy resources than with agriculture.

If conservation is broadly applied as a means to ensure adequate water supplies for our future energy needs, the citizen ultimately pays the bill. This may occur either directly in the form of increased costs for energy products (for example, if minimum wet cooling is used) or indirectly in the form of subsidies to irrigated agriculture. While these costs may or may not be large enough to be burdensome by themselves, to the individual citizen they are likely to be only part of the increased costs associated with developing the West's resources. Moreover, these costs raise other political issues, such as the political feasibility of further subsidies for irrigated agriculture. While agricultural subsidies have been a long-term federal policy, they have recently been the subject of increasing political debate, and choosing conservation alternatives may force policymakers to decide just how valuable irrigated agriculture is. Furthermore, interregional conflicts may be increased. Since the development of energy resources in the West would be used primarily to meet the needs of other regions, an equitable alternative may well be one which requires other regions to pay the largest share of the cost of providing an adequate supply of water and ensuring that enough water is still available for current nonenergy uses.

Summary and Comparison of Water Availability Alternatives

Table 3-8 summarizes and compares alternatives for augmenting and conserving water

Table 3-8: Summary and Comparison of Water Availability Alternatives

Criteria	Measures	Augmentation	Conservation
Effectiveness: How much water can be saved or added?	Acre-feet per year	Millions of acre-feet per year can be added regionally and to specific locations by augmentation alternatives.	Less water can be saved than can be added by augmentation; more uncertainty is associated with the effectiveness of most conservation options than with augmentations.
	Long term or short term	Interbasin transfers, weather modification, and vegetation management can provide long-term supplies. Groundwater mining is an effective short-term option.	Conservation in the development of energy resources can alleviate large percentages of the new water demands that have been projected for the next 20 to 25 years.

Criterion		
Efficiency: What are the costs, risks, and benefits?	Dollar cost per acre-foot	Capital-intensive projects cost up to $200 per acre-foot of water (interbasin transfers). Weather modification ($2 to $3 per acre-foot) and upland vegetation management ($20 per acre-foot or less in most areas are relatively inexpensive.	Costs of agricultural conservation will generally be much higher than the value of the water saved ($25 per acre-foot). Water-saving cooling technologies can be economical or, at worst, represent only small percentage increases in the cost of energy, except for electric power generation.
	Environmental costs and risks	Nearly every option has environmental costs and risks, but especially for water transfers, phreatophyte removal, and upland vegetation management. Risks of ecological damage from weather modification are uncertain.	Environmental effects are generally beneficial since surface flows will be increased. However, some wetlands will be threatened by agricultural conservation.
Equity: How will costs, risks and benefits be distributed?	Who will pay?	Taxpayers and energy consumers will ultimately pay most of costs for capital-intensive projects.	Agriculture, in most cases, will be unable to afford the costs and risks of conservation.
	Distribution among states and regions	There is a potential for inequities in donor basins and subbasins if transfers are used. Perceptions of inequitable distribution of water among the basins of Colorado River will increase.	Increased conflict between the West and other regions can be anticipated as other regions pay higher energy prices and taxes and as the West's water is used for the energy needs of other regions.
Flexibility: Can the alternatives be applied under different conditions and at different times?	Reversibility	Several of the options are inflexible because of long-term financial commitments. Commitments to weather modification can easily be reversed, but its impact may be long term.	Most conservation options require long-term commitments either because of financial investments or because of the need for new technologies or equipment.
	Applicability	Except for upland vegetation management and weather modification, augmentation alternatives are applicable only in specific locations.	If alternatives are implemented by uniform regulations, the flexibility to deal with local circumstances will be sacrificed. Most alternatives can be applied regionally.
Implementability: How difficult will it be to enact and implement?	Institutional constraints	Transfers face severe legal and institutional constraints. Groundwater laws in some states will constrain groundwater storage and mining. Legal uncertainties may constrain weather modification.	Comprehensive changes in institutional and regulatory structures will be required, particularly in the state appropriation systems.
	Acceptability	Except for groundwater storage, most alternatives have already generated strong opposition from environmental groups.	Agricultural interests and most western states will oppose changes in current water management practices.

resources in order to reduce both the likelihood of shortages and conflicts over water use. This evaluation indicates that substantial quantities of water can be added to existing supplies, perhaps enough to avoid water shortages in many areas of the West, if both augmentation and conservation are employed. The most effective augmentation choices to meet this goal, which are potentially capable of adding millions of acre-ft/yr, seem to be interbasin transfers, weather modification, vegetation management, and phreatophyte removal. Although conservation alternatives seem to be generally less effective than augmentation, and although more uncertainty may be associated with estimating the effects of each conservation alternative, water-saving cooling technologies for energy facilities can save as much as 400,000 acre-ft/yr by the year 2000.

Although augmentation and conservation alternatives potentially can supply large quantities of water, serious risks and real costs are associated with some of the alternatives. Several alternatives will produce substantial environmental costs and risks; this is more likely to be the case for augmentation alternatives, particularly transfers (interbasin and intrabasin), impoundments, phreatophyte removal, and upland vegetation management. In contrast, substantial environmental benefits, largely associated with preserving instream flows, are derived from most of the conservation alternatives.

Augmentation and conservation also result in increased economic costs. In fact, because of their economic costs, many of these alternatives would seem to require federal subsidies. This is especially true for transfers, impoundments, underground storage, and conservation in irrigated agriculture—all largely because of construction costs. The most effective conservation option, intermediate wet cooling for electric power plants, appears to be economically infeasible in most cases at the moment. However, this alternative could still be used to save large quantities of water if costs were passed on to the consumer.

Conclusion

The central question addressed in this chapter is whether enough water exists in the West to support traditional users as well as the rapidly expanding demands of energy developers, Indians, environmental interests, and others. While demands from any one of these groups by itself will not necessarily create a water shortage, the essential fact of water resources in the West seems to be that not all of the legitimate demands for water can be met. While this is particularly evident in the Colorado, which has largely reached the limit to which physical controls can be used to augment and transfer supplies, it is also becoming a reality in the Yellowstone and other rivers of the UMRB.

Three deficiencies exist in the current management of water problems. First, state appropriation systems tend to settle disputes over water use only after damage (or impairment) has occurred. The present system is designed primarily to identify the legality of water rights and not to manage the resource. One result is that it is usually very difficult, if not impossible, to determine the actual patterns of use over a broad area.

Second, the system for devising and implementing water policies is fragmented. The most obvious example of this is that the system for allocating water is not tied to salinity control. Even though two of the major causes of salinity are water withdrawals, which tend to concentrate salinity levels, and irrigated agriculture runoff, which picks up salt as the water returns to the stream, there are little or no regulatory or administrative links that allow policymakers to consider the two issues together. Thus, the consequences of allocation policies on water quality are often ignored.

Another example of fragmentation is the lack of coordination between surface water policy and groundwater policy. Although clear hydrological relationships exist between the two, western water law traditionally has treated them separately. In several western states, groundwater is being mined, and this directly affects the flow of surface water. However, it is very difficult in the current system to address this kind of situation.

Third, the current system of setting policies tends to discourage change. This can be attributed largely to the long and complex history of water policies and to the reliance on court rulings both to determine water policy and to define and clarify how water is used. As a result, water policies are often made up of a series of court cases which seldom address issues beyond the specific questions of a given case. This legal and regulatory system is very difficult to understand or to change, and it often creates incentives for water use which increase rather than decrease existing problems. This is exemplified in doctrines such as "use it or lose it" and non-impairment which often discourage individual users from minimizing use or protecting water quality.

Our analysis does not lead automatically to a single alternative—a mix of approaches will be required to deal with future water problems. Policies designed to address deficiencies in the current system will be difficult to enact and to implement because they will affect so much of western society. However, attempts to change the current system should be based on three basic principles:

1. Water availability and water quality are inherently related; *future water policies should consider them together.*
2. Although states are primarily responsible for water resource policy, water problems and issues are regional; *future water policies should encourage basin-wide approaches to water management.*
3. The causes of water problems and the implications of policy alternatives vary considerably across the West, within river basins, and within states. Thus, although regional mechanisms are important for the overall management of water, they are only part of the solution. *Future water policies, even if established regionally, should be implemented in ways that allow for local differences.*

REFERENCES

Andersen, Jay C., and John Keith. 1977. "Energy and the Colorado River." *Natural Resources Journal* 17 (April):157–68.

Anderson, Mark H., et al. 1973. *The Demand for Agricultural Water in Utah,* PRWG100-4. Logan: Utah State University, Utah Water Research Laboratory.

Arizona v. *California,* 373 U.S. 546 (1963), Decree 376 U.S. 340 (1964).

Belle Fourche River Compact of 1943, 58 Stat. 94 (1944).

Boulder Canyon Project Act, Pub. L. 70–642, 45 Stat. 1057 (1928).

Colorado River Basin Project Act, Pub. L. 90-537, 82 Stat. 885 (1968).

Colorado River Basin Salinity Control Forum. 1978. *Second Annual Progress Report: Water Quality Standards for Salinity in Colorado River System.* Salt Lake City, Utah: Colorado River Basin Control Forum.

Colorado River Compact of 1922, 42 Stat. 171, 45 Stat. 1064, declared effective by Presidential Proclamation, 46 Stat. 3000 (1928).

CQ Weekly. 1977. "Water Projects." July 2, p. 1377.

Davidson, Craig. 1977. "Mesita Battleground for Water Dispute." *Denver Post,* December 19.

Denver Post. 1977a. "EDF to Sue for Water Salinity Control." April 15.

Denver Post. 1977b. "Limits on Irrigation Opposed in Wyoming." November 16.

Denver Post. 1978. "Colorado Water Project Endangered by Fish." April 20.

Fischer, Ward H. 1977. "Coloradans Are Terrorized by Federal Water Policy Report." *Denver Post,* September 4.

Gill, Douglas. 1977. "Man, Nature Share Blame for Colorado River's Salinity." *Denver Post,* April 24.

Glenn, Bruce, and Kenneth O. Kaufman. 1977. "Institutional Constraints on Water Allocation." Paper presented at the Energy, Environment, and Wild Rivers in Water Resource Management Conference, Moscow, Idaho, July 6–8.

Gold, Harris, and D. J. Goldstein. 1979. *Wet/Dry Cooling and Cooling Tower Blowdown Disposal in Synthetic Fuel Steam-Electric Power Plants.* Washington, D.C.: Environmental Protection Agency.

Hibbert, Alden R. 1977. "Vegetation Management for Water Yield Improvement in the Colorado River Basin." Paper presented at the Annual Meeting of the Colorado River Water Users Association, Las Vegas, Nevada, December.

Ingram, Helen, and J.R. McCain. 1977. "Federal Water Resources Management: The Administrative Setting." *Public Administration Review* 37 (September/October): 448–55.

James v. *Barker,* 99 Colo. 551, 64P. 2d 598 (1973).

Kirschten, J. Dicken. 1977a. "Draining the Water Projects Out of the Pork Barrel." *National Journal* 9 (April 9):540–48.

Kirschten, J. Dicken. 1977b. "Turning Back the Tides of Long-Time Federal Water Policy." *National Journal* 9 (June 11): 900–903.

Kirschten, J. Dicken. 1978. "The Quiet before the Shootout Over 'The Water Law of the West.'" *National Journal* 10 (January 28):149–53.

Laws of Utah, Chapter 23 (1976).

Leathers, K. L., and R. A. Young. 1975. "Economic Evaluation of Non-Structural Measures to Control Saline Irrigation Return Flows." Paper presented at the Western Agricultural Economics Association Meeting, Las Vegas, Nevada, July.

MacDonald, J. F., et al. N.d. *Weather and Climate Modification: Problems and Prospects,* 2 vols. Washington, D.C.: National Academy of Sciences/National Research Council.

MacMeekin, Daniel H. 1971. "The Navajo Tribe's Water Rights in the Colorado River Basin." Unpublished report.

Missouri Basin Inter-Agency Committee. 1971. *The Missouri River Basin Comprehensive Framework Study,* 7 vols. Denver: U.S. Department of the Interior, Bureau of Land Management.

Montana Department of Natural Resources and Conservation (DNRC), Water Resources Division. 1975. *Final Environmental Impact Statement for Water Reservation Applications in the Yellowstone River Basin,* 2 vols. Helena: Montana DNRC, Water Resources Division.

Montana Department of Natural Resources and Conservation (DNRC), Water Resources Division. 1976. *Which Way? The Future of Yellowstone Water,* Draft Report. Helena: Montana DNRC, Water Resources Division.

National Academy of Sciences (NAS)/National Research Council (NRC). 1973. *Weather and Climate Modification: Problems and Progress.* Washington, D.C.: NAS.

National Petroleum Council (NPC), Committee on U.S. Energy Outlook. 1972. *U.S. Energy Outlook.* Washington, D.C.: NPC.

National Water Commission. 1973. *Water Policies for the Future,* Final Report. Washington, D.C.: Government Printing Office.

The Plains Truth. 1979. "Meanwhile, Back at the Ranch. . . ." 8 (February): 4–5.

Price, Don, and Ted Arnow. 1974. *Summary Appraisals of the Nation's Ground-Water Resources—Upper Colorado Region,* Geological Survey Professional Paper 813-C. Washington, D.C.: Government Printing Office.

Pritchett, Harold R., and Bruce E. Boesch. 1977. *Irrigation Improvement Program for the Wellton-Mohawk District.* Wellton Ariz.: U.S. Department of Agriculture, Soil Conservation Service.

Radosevich, George E. 1978. *Western Water Laws and Irrigation Return Flow.* Ada, Okla.: U.S. Environmental Protection Agency, Office of Research and Development, Robert S. Kerr Environmental Research Laboratory.

Reclamation Act of 1902, Pub. L. 57-161, 32 Stat. 388.

Reclamation Project Act of 1939, Pub. L. 76-260, 53 Stat. 1187.

Revised Codes of Montana (R.C.M.) §89-892(3), H.B. 83, Ch. 338, L. 1975.

Saile, Bob. 1977. "Minimum Stream Flows Sought." *Denver Post,* January 20.

Salt River Valley Water Users Association v. *Kovacovich,* 411 F. 2d 201 (1966), 3 Arizona app. 28.

Stockton, Charles W., and Gordon C. Jacoby, Jr. 1976. *Long-Term Surface Water Supply and Streamflow Trends in the Upper Colorado River Basin,* Lake Powell Research Project Bulletin Number 18. Los Angeles: University of California, Institute of Geophysics and Planetary Physics.

Strain, Peggy. 1977. "Water, Land, Life—It's All One in Valley Pipeline Debate." *Denver Post,* November 13.

Strain, Peggy, and Gary Cook. 1977. "Panelists Will Stress States' Water Rights." *Denver Post,* October 19.

Strain, Peggy, and Leonard Larsen. 1978. "Water Policy Victory for West." *Denver Post,* June 7.

Swenson, Frank A. 1974. *Possible Development of Water from Madison Group and Associated Rock in Powder River Basin, Montana-Wyoming.* Denver: Northern Great Plains Resources Program.

Tipton and Kalmbach, Inc. N.d. *Water Supplies of the Colorado River.* In U.S. Congress, House of Representatives, Committee on Interior and Insular Affairs. *Lower Colorado River Basin Project. Hearings* before the Subcommittee on Irrigation and Reclamation, 89th Cong., 1st sess., 1965.

Treaty between the United States of America and Mexico Respecting Utilization of Waters of the Colorado and Tijuana Rivers and of the Rio Grande, February 3, 1944, 59 Stat. 1219 (1945), Treaty Series No. 994.

U.S. Army, Corps of Engineers, Phoenix, Arizona. June 1978. Personal communication.

U.S. Congress, Senate Committee on Energy and Natural Resources. 1978. *Water Availability for Energy Development in the West. Hearings* before the Subcommittee on Energy Production and Supply, 95th Cong., 2nd sess., March 14.

U.S. Department of Agriculture (USDA), Soil Conservation Service. 1976. *Crop Consumptive Irrigation Requirements and Irrigation Efficiency Coefficients for the United States.* Washington, D.C.: USDA.

U.S. Department of the Interior (DOI), Bureau of Reclamation (BuRec). 1972. *Appraisal Report on Montana-Wyoming Aqueduct.*

Billings, Mont.: BuRec.

U.S. Department of the Interior (DOI), Bureau of Reclamation (BuRec). 1975. *Westwide Study Report on Critical Water Problems Facing the Eleven Western States.* Washington, D.C.: Government Printing Office.

U.S. Department of the Interior (DOI), Bureau of Reclamation (BuRec). 1978. *Operation of the Colorado River Basin, 1977, and Projected Operations, 1978.* Washington, D.C.: Government Printing Office.

U.S. Department of the Interior (DOI), Bureau of Reclamation (BuRec), Engineering and Research Center Staff, Denver, Colorado. June 28, 1977. Personal communication.

U.S. Department of the Interior (DOI), Bureau of Reclamation (BuRec) and Bureau of Indian Affairs (BIA). 1978. *Report on Water Conservation Opportunities Study.* Washington, D.C.: Government Printing Office.

U.S. Department of the Interior (DOI), Bureau of Reclamation (BuRec), and Department of Agriculture (USDA), Soil Conservation Service. 1977. *Final Environmental Impact Statement: Colorado River Water Quality Improvement Program,* 2 vols. Denver: BuRec, Engineering and Research Center.

U.S. Department of the Interior, Geological Survey (USGS). 1975. *Plan of Study of the Hydrology of the Madison Limestone and Associated Rocks in Parts of Montana, Nebraska, North Dakota, South Dakota, and Wyoming,* Open-File Report 75-631. Denver: USGS.

U.S. Department of the Interior (DOI), National Park Service, Glen Canyon National Recreation Area, Page, Arizona. April 1978. Personal communication.

U.S. Department of the Interior (DOI), Water for Energy Management Team. 1974. *Report on Water for Energy in the Upper Colorado River Basin.* Denver: DOI.

U.S. v. *District Court in and for County of Eagle,* 401 U.S. 502 (1971).

U.S. v. *District Court in and for Water Division No. 5,* 401 U.S. 527 (1971).

Upper Colorado River Basin Compact of 1948, Pub. L. 81-37, 63 Stat. 31 (1949).

Utah State University, Utah Water Research Laboratory. 1975. *Colorado River Regional*

Assessment Study, part 2: *Detailed Analyses: Narrative Description, Data, Methodology, and Documentation.* Logan: Utah Water Research Laboratory.

Wade, James C., et al. 1977. "Sprinkle Irrigation Technologies and Energy Costs: A Comparative Analysis of Southern Arizona Irrigated Agriculture." Paper presented at the American Water Resources Association Conference, Tucson, Arizona, October 31–November 3.

Water Purification Associates (WPA). 1978. *Aspects of Water Impact Analysis in Coal Conversion.* Cambridge, Mass.: WPA.

Weisbecker, Leo W. 1974. *The Impacts of Snow Enhancement.* Norman: University of Oklahoma Press.

Wendt, C.W., et al. 1977. "Effect of Irrigation Systems on Water Use Efficiency and Soil-Water Solute Considerations." In *Proceedings of National Conference on Irrigation Return Flow Quality Management,* edited by James P. Law and G. V. Skogerboe. Fort Collins: Colorado State University.

White, Michael D. 1975. "Problems under State Water Laws: Changes in Existing Water Rights." *Natural Resources Lawyer* 8 (No. 2): 359-76.

Wild and Scenic Rivers Act of 1968, Pub. L. 90-542, 82 Stat. 906.

Winters v. *U.S.,* 207 U.S. 564 (1908).

Wyoming Statutes Annotated (W.S.A.) §41-41.

Yellowstone River Compact of 1950, 65 Stat. 663 (1951).

CHAPTER 4

Water Quality

Water quality has long been an issue in the West. Traditionally, attention has been centered on pollution from agricultural runoff and from natural sources. The concern has intensified recently, however, in part because the quality of water is closely related to its availability: as water consumption increases, the quality can deteriorate because existing pollutants become more concentrated. In addition, conflicts over water quality can be traced to increased demands for water for municipal use, increased awareness of the environmental impact of pollution, and the potential threat to water quality from the development of energy resources. These concerns over water quality have created intergovernmental conflicts as well as conflicts among water users. For example, over the past several years serious disputes have arisen among the federal government, the states of the CRB, and environmental interest groups over appropriate policies and regulations for dealing with salinity problems in the Colorado River.

Developing energy resources will increase conflicts over water quality for several reasons:

- Energy extraction and conversion processes can cause pollution of both surface water and groundwater;
- Consumptive use of water by energy facilities will generally increase salinity concentrations downstream; and
- Rapid population increases associated with energy development can overload the sew-

age-treatment capacity of small communities.

The 1977 Clean Water Act (CWA) limits the direct discharge of industrial pollutants into surface waters.[1] Nevertheless, energy extraction and conversion pose a potential threat to water quality both from the disruption of land and from the possible seepage or leaching of contaminants from discarded liquid and solid wastes. The second problem is related to the effects of water consumption and not to the discharge of pollutants. In most river basins salinity levels generally increase downstream. Therefore, the consumption of water upstream will reduce the water available for dilution, thus increasing salinity concentration. While salinity is a concern throughout the western region, the issue is especially important at the present time in the CRB. Finally, as the population of an area rapidly increases, inadequately treated or untreated municipal wastes can degrade surface streams. And the increased costs for sewage-treatment facilities are already creating conflicts among levels of government about who should pay.

---

[1] The 1977 CWA amended the Federal Water Pollution Control Act (FWPCA) of 1972. The CWA includes the previous FWPCA legislation. In this chapter, provisions enacted prior to 1977 are referred to as part of the FWPCA and modifications enacted in 1977 are referred to as part of the CWA.

The Consequences of Energy Development

As indicated in Chapter 3, both surface water and groundwater are likely to be affected by the development of energy resources. In the UMRB, the surface waters of the Yellowstone, Belle Fourche, and Little Missouri tributaries will be especially affected. The Upper Colorado tributaries will be affected by developments in Colorado, Utah, and New Mexico. In addition, groundwaters in both the Upper Colorado and Upper Missouri basins may be contaminated by the effluents and solid wastes from energy facilities, and aquifers may be disrupted by mining. For example, the water quality in the Madison aquifer and several shallow aquifers in the Fort Union Coal Formation (Wyoming, Montana, and North Dakota) has already been affected by energy development in the UMRB, although only on a highly localized basis so far.

In analyzing the impact of energy development, several technological and locational factors were identified that can contribute to surface water and groundwater pollution. Technological factors include:

- Quantities and composition of solid and liquid wastes produced by conversion facilities;
- Waste-disposal techniques;
- Labor intensity (of both construction and operation); and
- Type of mining employed.

Locational factors include:

- Present water quality;
- Soil permeability;
- Aquifer depth;
- Size of the nearby communities; and
- Capacity of existing municipal wastewater-treatment facilities.

Because of both technological characteristics and the potential magnitude of future energy operations in the West, the greatest concerns

Waste Holding Ponds at a

about water quality are associated with the development of coal, oil shale, and uranium.

Waste Disposal

One of the most significant factors influencing the quality of water near energy development sites is the quanity of effluents produced. The effluents removed as either dissolved, wet, or dry solids are listed in Table 4-1 for three energy resources—coal, oil shale, and uranium. Quantitative estimates of the wastes produced by the other three resources considered in this study (oil, natural gas, and geothermal energy) are not known precisely, but they are generally much smaller in magnitude. As indicated in Table 4-1, the effluents from surface retorting of oil shale are the largest, over 16 MMtpy; most of this is dry solids in the form of spent shale. Of the coal conversion facilities, Synthoil liquefaction produces the highest volume of total effluents (about twice as much as other synthetic fuel processes), based on a standard-size facility. If judged on a Btu-electric basis, electric power generation produces more effluents per million Btu's than any other facility.

In coal, oil shale, and uranium facilities, dissolved and wet solids will be diverted to evaporative holding ponds and later depostied in a landfill. Dry solids may be treated with water to prevent dusting and then deposited in a landfill. As a result, problems from effluent disposal do not arise from routine, direct discharge into surface water but rather are the result of accidental releases from holding ponds (e.g., flooding or breaking of a holding-pond berm), seepage from holding ponds into groundwater, and runoff or percolation of rainfall from landfills.

Coal Conversion Wastes: The content of effluents from coal-fired power plants and synthetic fuel facilities varies according to

Coal-Fired Power Plant

Table 4-1: *Effluents from Energy Conversion Technologies*

Technology	Size	Total Effluents[a]	
		(million short tons/year)	(Pounds per 10⁶ Btu)
Coal			
Power Generation	3,000 MWe	0.526-2.554	
Per Btu electric			15.63-79.04
Per Btu thermal			5.48-26.60
Lurgi Gasification	250 MMcfd	0.448-1.680	10.91-40.91
Synthane Gasification	250 MMcfd	0.447-1.964	10.88-47.83
Synthoil Liquefaction	100,000 bbl/day	0.827-3.651	8.99-39.69
Oil Shale			
TOSCO II Oil Shale	50,000 bbl/day	16.187	365
Modified *In Situ* Oil Shale Processing	57,000 bbl/day	U	U
Modified *In Situ* Oil Shale Processing with Surface Retort	57,000 bbl/day	8.8	176
Uranium			
Underground Mine	1,100 mtpd (ore)	N	N
Surface Mine	1,100 mtpd (ore)	N-0.002	N-0.01
Mill	1,000 mtpy (yellowcake)	0.367	2.45
Solutional Mine-Mill	250 tpy	0.003-0.004	0.08

U = unknown. Mtpd = metric tons per day. N = negligible. tpy = tons per year.

[a]Effluents include dissolved, wet, and dry solids; the range of values is that found at the six sites analyzed.

the elemental content of the source coal and
the process. Accumulations of wet solids containing heavy metals, trace elements, and
aromatic hydrocarbons (HC) in holding ponds
could produce acute effects in local surface
waters if the wastes are released accidentally
(see box). The design of holding-pond berms
must be site-specific, and failures are common
in areas where previous design experience is
not available (for example, see Smith, 1973:
358). The quantities of wastes involved can
be quite large; at a typical site such as Gillette,
Wyoming, the operation of a power plant and
of Lurgi, Synthane, and Synthoil synthetic
fuel facilities would produce more than 68
million tons of effluents over a 25-year period.

In addition to possible berm failures which
would allow pollution of surface waters, seepage from holding ponds can also contaminate
groundwater. The degree of contamination
depends on a number of factors: the composition of materials in the ponds, the pond design, the liner design, pond-management techniques, and the characteristics of nearby aquifers and of the soil overlaying the aquifers.
Once contaiminated, aquifers may introduce
pollutants into local springs, seeps, and
streams. The problem may be a long-term one.
Although the quality of water in a polluted
surface stream will usually improve dramatically within one to two years after the sources
of pollution are eliminated, polluted aquifers

require much longer periods to cleanse themselves, depending on local geologic and soil
conditions.

Spent-Shale Disposal: Probably the most
potentially serious water-quality problem associated with the surface retorting of oil shale
is the enormous volume of spent shale which
must be disposed of. Over a 30-year period,
the production of 400,000 bbl/day would
result in enough spent shale to cover roughly
24 square miles to a depth of 100 feet. Because the total quantities are so large, the
potential for water contamination is great.
Salt mobilization is roughly four times greater in spent shale than in unprocessed shale
(depending on the retorting method used),
resulting in a potential long-term source of
saline surface water and groundwater. The
spent shale from the TOSCO II process also
contains between 1 and 2 percent organic
matter with identified but unknown proportions of carcinogenic hydrocarbons.

Because of the volumes involved, energy
producers using surface retorts have only two
options: to store all the spent shale above
ground or to return part of the waste to the
mine and store the remainder above ground.
Currently, all surface retorting projects propose to use aboveground storage (Crawford,
et al, 1977:24–55). According to the plans,
the spent shale will be accumulated in can-
yons behind strategically placed dams to force
runoff water from the pile to flow through
special drainage channels into catchment basins; other channels will allow drainage from
surrounding land to bypass the storage areas.
However, these runoff controls will cause a
portion of the natural watershed to be effectively removed from the basin, thus causing
a decrease in the flow of local streams.

The spent shale can potentially affect water
quality in two ways: leaching into surface
runoff and percolation of surface and ground
water through the shale pile. If the controls
on surface runoff (described above) are maintained properly, then the quality of surface
water should be adequately protected while

Piceance Creek Oil Shale Area in Western Colorado

the plant is in operation. However, this still leaves the long-range problem. After an oil shale facility has shut down, the surface-water quality will depend on the stability of the retention dams and on how successfully the spent-shale pile can be revegetated. Although there has been a considerable amount of research on the subject, and despite several researchers' claims of successful revegetation, considerable uncertainty still remains about the long-term ability to revegetate spent shale piles. Likewise, the potential for groundwater contamination is unclear. On the one hand, recent research has shown that compacting the spent shale with steam rollers used in road construction results in a mass of low permeability (DRI, 1979:sect. 7, p. 27). The permeability would be equivalent to that for "impervious clay linings" required by the Re-

source Conservation and Recovery Act of 1976 (RCRA) for ponds handling hazardous wastes. However, it is not known whether such techniques will be used (no regulations currently require it) and what the long-term effectiveness will be. In short, continuing research is needed to better assess the long-range risks to water quality and to provide the information needed to dispose of processed shale in an environmentally safe manner.

Total Quantities of Wastes: Table 4-2 shows the estimated amounts of solids and wastewater that will be created both by energy facilities and by related increases in population in the year 2000. The range of values represents differences between the Low- and Nominal-Demand cases.

In general, the population-related produc-

Table 4-2: *Projected Solid Residuals and Wastewater: Year 2000*[a]

Source	Upper Colorado River Basin		Upper Missouri River Basin	
	Solids (MMtpy)	Wastewater (thousand acre-ft/yr)	Solids (MMtpy)	Wastewater (thousand acre-ft/yr)
Energy facilities	131-553	30- 64	39-57	61- 88
Population[b]	0.01-0.04	16- 56	0.04-0.06	58- 84
Total	131-553	47-120	39-57	119-172

[a]The range in values represents the Low-Demand and Nominal-Demand cases.

[b]Wastewater at 100 gallons/person/day and 500 milligrams per liter solids.

tion of solids (almost entirely sewage sludge) is negligible compared to those produced by energy facilities. The quantity of wastewater generated by the population, however, is not negligible; it is about 50 to 90 percent as large as that generated by energy facilities.

In the UCRB, development of oil shale (both the TOSCO II and modified *in situ* processes) contributes between 85 percent (in the Low-Demand case) and 95 percent (in the Nominal case) of the total solids produced by energy development in the basin. As previously noted, this is due primarily to spent shale. Overall, solid effluents in the Nominal case are more than four times greater than those in the Low-Demand case, and wastewater is more than two times greater. In the UCRB, between 37 and 58 percent of the total wastewater generated by energy facilities comes from the uranium mills.

In the UMRB, solid effluents generated by energy facilities are estimated to range from 39 to 57 MMtpy. The quantity of wastewater produced ranges from 119,000 to 172,000 acre-ft/yr, 80 percent of which is attributed to power plants and gasification facilities.

Land Disturbance

The other major threat to water quality from energy development is caused by disturbances to the land, including both: (a) surface disturbances (e.g., surface mining, roads, and processing facilities), which can increase erosion, and the dissolved solids content (salinity) and other pollutants in runoff; and (b) disruptions of aquifers (e.g., by surface or underground mining, oil and gas production, and *in situ* oil shale recovery). Table 4-3 summarizes some of these risks and possible controls for each of the six energy resources we have considered.

Salinity Impacts: Salinity control from disturbed lands is one of the most difficult problems to deal with. Although salinity effects are difficult to quantify and can vary significantly from site to site, a recent study provides several valuable insights (see Rowe and McWhorter, 1978). In examining surface coal mining in northwestern Colorado, annual salt loading from the disturbed land was estimated at between 2.13 and 2.37 tons per acre, a 500 percent increase above the premining rate.[2] The analysis also suggests that groundwater seepage from the disturbed areas accounts for more than 99 percent of the salt load from those lands. This is explained by the fact that the salinity concentration and the volume of groundwater seepage are both large relative to the figures for overland runoff. Therefore, spoil-management practices

[2]The authors also note, however, that this is still relatively small compared to the annual salt loading from some irrigated agricultural lands which has been reported as high as 12 tons per acre in Colorado (Skogerboe and Walker, 1972).

that promote overland runoff at the expense of percolation and groundwater seepage will decrease the adverse effects of salinity caused by mining. Other considerations, however, such as maximizing soil moisture for revegetation, are not necessarily compatible with a practice of enhancing overland runoff. As discussed below, there are some regulatory programs whose purpose is to control these problems, especially in the case of coal. However, "Notwithstanding that on-site management practices contribute to the minimization of salt loading from disturbed areas, it is unlikely that increased salt loading can be eliminated" (Rowe and McWhorter, 1978: 337).

Erosion: Sediment loading of streams as a result of the erosion of a mine's exposed spoils or overburden has also long been recognized as a major surface-water-quality problem. Sedimentation can have an adverse impact on aquatic ecosystems, and in some situations it may change the morphology and stability of stream channels and flood plains so as to increase the frequency and severity of flooding. Sedimentation is a potential problem in all regions of the U.S., especially in southern Appalachia where the slopes are steepest and precipitation is abundant and frequently intense. However, the northern Great Plains and Rocky Mountain region can experience severe sedimentation problems re-

Table 4-3: *Summary of Major Risks to Water Quality from Land Disturbances*

Extraction Technique	Risks to Surface and Groundwater	Possible Controls
Coal Mining	Disturbance of land surface	Control reclamation and sedimentation ponds
	Dewatering of mines Excavation of aquifer Introduction of overburden and spoils into aquifer	Adjust groundwater movement and percolation Dispose of minewater properly
Oil Shale Mining	Disturbance of land surface Excavation of aquifer Dewatering of mines	Control reclamation and drainage Dispose of minewater properly
Modified *In Situ*	Excavation of aquifer Mobilization of hazardous materials within aquifer	Seal spent retorts
Uranium Mining	Exploration drill holes Disturbance of land surface Excavation of aquifer Dewatering of mines	Control drainage Dispose of minewater properly
Solvent Extraction	Release of hazardous substances Introduction of recovery agents	Control solvent front Inject neutralizing agents Flush formation
Oil and Gas Drilling	Boreholes between aquifers Introduction of enhanced recovery agents (brines, acids, solvents)	Case and cement properly
Geothermal Drilling	Boreholes between aquifers Reinjection of brines	Case and cement properly

Sources: General Electric, 1973; White et al., 1979a; and Grimshaw et al., 1978.

lated to intense rainfall, even though annual precipitation there is low. Sedimentation can be controlled or minimized during and after mining by a number of practices. These include diversion of surface runoff from active mining areas, the use of sedimentation traps or ponds, terracing to reduce long slopes, and the use of fast-growing grasses and other types of vegetation. Properly designed and maintained sediment ponds can help control the movement of sediment into streams during the time spoils are most susceptible to erosion. After mining is completed, revegetation can stabilize the spoils and prevent further erosion if the ground cover is adequate. (Land reclamation is discussed in Chapter 6.)

Groundwater Pollution: The pollution of groundwater due to mining is primarily the result of changes in percolation or movement of groundwater through newly exposed soluble or toxic materials. For *in situ* recovery of uranium or oil shale, pollution is due to the hazardous new materials that may be introduced into the groundwater. At each of the six hypothetical site-specific scenarios we considered (see White et al., 1979b), some disruption of the aquifers was projected to occur. There have already been problems of aquifer disruption, for example, in the Grants Mineral Belt area of New Mexico (New Mexico Environmental Improvement Agency, 1977), in eastern Utah and western Colorado in areas of coal, oil shale, and petroleum development (Price and Arnow, 1974; Bishop et al., 1975), and in coal- and uranium-rich areas of Wyoming and other states in the northern Great Plains (Montana Department of Mines, 1979).

A survey of eight surface coal mines located in New Mexico, Colorado, Wyoming, and Montana has determined that strip mining can increase levels of carbonates, sulfates, clays, and sulfides even in low rainfall areas by the increased movement of water through the mine's distrubed overburden (Hounslow et al., 1978:2). However, if the coal is lo-

cated above the aquifer, mining operations typically do not cause changes in the quality of groundwater unless significant precipitation filters through the soil (Hounslow et al., 1978:179).

The specific increase in dissolved solids depends on local conditions, principally groundwater flow, and on the mineral composition of the aquifer and the disturbed overburden. In several mines in the northern Great Plains, for example, the groundwater has experienced an increase in total dissolved solids (TDS) content of 8,000 to 10,000 parts per million (ppm), making the water unsuitable for all domestic and most industrial uses. These levels, generally caused by increased concentrations of magnesium, calcium, and sulfate, can in some cases be great enough to render the water unusable for any purpose. Thus far, such changes have not been detected in groundwater outside the mined areas. Nevertheless, as increased coal development occurs in the western United States, some counties will experience an area-wide disturbance to aquifers with increased salinity that can preclude some uses. For example, in Campbell County, Wyoming, according to our Nominal-Demand scenario, up to an estimated 18 percent of lands would be disturbed by the year 2000.

As described in Chapter 2, oil shale extraction can be accomplished by mining the shale and then retorting it in facilities on the surface or by *in situ* (or modified *in situ*) methods. Both types of extraction processes can have an effect on the quality of groundwater. The extraction of oil shale in Colorado, for example, occurs in the Mahogany Zone, which separates two aquifers.

With oil shale mining, water quality could be affected in three ways. First, to get an idea of the magnitude of the mining operations, it is necessary to recall that the production of 100,000 bbl/day from surface retorting will require mining approximately 60 million tons of oil shale annually. After 30 years of operation this will mean that roughly 12,000 acres (or 19 square miles) will be affected by the underground mining. This level of mining could result in mine dewatering of 900 to 7,500 acre-ft/yr, depending on the location of the mines. Depending on how it is treated and disposed of, the quality of surface water could be adversely affected since the mine water may have a TDS content of up to 40,000 milligrams per liter (mg/l) and high levels of some trace elements. For example, fluoride levels of 10 mg/l to 20 mg/l have been measured; this exceeds federal drinking water standards by a factor of ten (Crawford et al., 1977:121). Options for disposing of mine water include reinjecting the water or spray irrigating the excess water on the spent-shale pile. A second potential problem is that groundwater from lower aquifers with salinities up to 40,000 mg/l could intrude into the fresher, dewatered aquifers. This may reduce the quality and use of the fresher groundwater and surface water of streams such as Piceance Creek and Parachute Creek in Colorado. A third but highly uncertain problem is the release of trace elements in mined formations due to oxidation and leaching.

With *in situ* extraction, the disposal of minewater and the intrusion of saline groundwater into fresher aquifers are also potential problems. In addition, during the operational phase, shale oil and related organic and inorganic compounds can all be considered contaminants that could enter the groundwater system. After operations cease, the *in situ* retorts may become saturated by the water table, releasing contaminants into the water and lowering water quality. The higher permeability of the retort area can also be expected to affect the flow relationship between upper and lower bedrock aquifers. The combustion of the oil shale in these retorts may mobilize trace elements, organic substances, or other potentially hazardous materials that could subsequently be leached by groundwater after operations cease and the water table reestablishes itself.

Problems and Issues

Problems and issues related to water quality are closely tied to those of water availability, and therefore, many of the interests at stake in water-quality control are the same as, or are related to, those discussed in Chapter 3. In the private sector, goals promoted by environmental interest groups often conflict with those championed by energy industries and agricultural interests. For example, there are already conflicts over controlling the discharge of pollutants and the impact of various water uses on water quality.

Water quality is already an important problem in the eight-state study area, and energy development will intensify existing conflicts. Specific water-quality problems and issues include:

- Control of changes in the quality of surface water and groundwater due to energy activities;
- Adequacy of municipal wastewater-treatment facilities; and
- Control of river basin salinity.

In the public sector, many anticipated effects of energy development on water quality will be regulated by both the states and the federal government. Although the federal government, especially the Environmental Protection Agency (EPA), has taken the lead in this area, intergovernmental relations have already been strained by the federal-state partnership requirements of the CWA (1977), by conflicts over salinity control, and by disputes related to channeling federal funds for wastewater treatment to the states.

The current status of water-quality control is mixed. On the one hand, since passage of the FWPCA (1972), water-quality control has greatly reduced the discharge of industrial pollutants into surface water. On the other hand, the remaining sources of pollution appear to be the most difficult and expensive to eliminate; and the control of municipal pollution has been much less successful than the control of industrial pollution.[3] Furthermore, as our knowledge grows, it becomes increasingly clear that water-quality control is intricately tied to water supply and its use.

Pollution from Energy Facilities

Current federal and state laws limit the direct discharge of pollutants from energy facilities into surface streams. Nevertheless, energy production and conversion processes potentially can pollute both surface water and groundwater through the disposal of waste products, the disturbance of land surfaces, or the disruption and contamination of aquifers during mining and in situ recovery operations. Given current federal and state regulations and the level of scientific uncertainty, it appears as though existing environmental regulations and control techniques are inadequate to ensure that long-term or irreversible damage to the quality of surface water and groundwater in the West does not occur.

As described in the previous section, the energy production and conversion technologies considered in this study pose a threat to water quality for two reasons. First, these activities—such as coal, uranium, or oil shale mining; oil or gas production; and *in situ* oil shale recovery—can disturb land surfaces and aquifers and, thus, degrade surface water and groundwater. Second, producing and converting energy often generates large quantities of wastes which represent a threat to water quality because of possible seepage, leaching, and runoff.

There are a variety of federal and state laws aimed at protecting the nation's surface water and groundwater resources. In terms of energy development, these legal controls cover

[3]For example, in 1976 EPA estimated that 85 percent of all major industries, but only about 30 percent of the nation's cities and towns, would meet 1977 goals for initial water-quality improvements (see CEQ, 1977).

two categories: they control both the discharge of pollutants into surface water or groundwater and the disposal of wastes; and they control pollution from land disturbances.

Pollution Discharges and Waste Disposal: At the federal level, effluent discharges from industrial sources (including energy facilities) into interstate or navigable waters are primarily regulated through the CWA (1977). The program to regulate the sources of pollution under the CWA is known as the National Pollutant Discharge Elimination System (NPDES). Under this program, no effluent

Table 4-4: *Federal Effluent Regulations*[a]

Type of Facility or Pollutant	Treatment Required
Industrial facility	"Best practicable control technology currently available" as defined by the EPA administrator by July 1, 1978.[b] "Best available technology economically achievable" as determined by the EPA administrator by July 1, 1984.[b] Limits are based on categories or classes of industries. National performance standards — including zero-discharge, if practicable — for each new category of source.
Toxic pollutants (seriously harmful to human or other life)	Effluent limitations including prohibition of discharge, if needed, to provide "an ample margin of safety" set by the EPA administrator.
Thermal discharge	Effluent limitations set to ensure a balanced population of fish, shellfish, and wildlife.
Oil or hazardous substances	No discharge into U.S. waters, adjoining shorelines, or contiguous zone waters.

[a]FWPCA, 1972. Adapted from Congressional Quarterly, 1973:797.

[b]Data here reflect the change in the law made by the CWA, 1977.

can be discharged by a source (such as an energy facility) without a permit which sets the conditions under which the discharge may be made. Table 4-4 summarizes the regulations affecting such discharges. Permits are issued by EPA or the state, if the state program has been approved by EPA. Four of the eight states in our study area—Colorado, Montana, North Dakota, and Wyoming—have EPA-approved programs. In those states that do not have EPA-approved permit programs, the regional EPA office issues the permits. However, even when it has approved a state program, EPA still retains control: it can veto any individual permit proposal and can withdraw its approval of a state's entire permit program.

Before 1972, when the FWPCA was passed, the major federal water-quality program (Water Quality Act, 1965) required the states to adopt quality standards for stream water that met the approval of the secretary of the interior (the administrator of EPA after EPA was created). These standards are now enforced under the FWPCA requirement that all dischargers meet the technology-based effluent standards discussed above or more stringent limitations (CWA, 1977). Streams or segments of streams can be designated as either "effluent limited" or "water quality limited" (EPA, 1974:8); discharges into a "water quality limited" stream are subject to more stringent limitations.

However, in some cases these controls are inadequate—for example, due to legal challenges by the uranium industry. In the past, six uranium companies have challenged EPA's authority under the CWA primarily on the basis, first, that the streams receiving the discharges are not waters protected by the NPDES permit system and, second, that the effluent limitations are arbitrary. In addition, regulations seeking to monitor and establish permits for discharges into groundwater have also been described as inadequate (see box, "Monitoring Inadequate").

All these requirements affect the development of energy resources in the western

---

Monitoring Inadequate

In New Mexico, for example, more than 30 million gallons per day of water are discharged from uranium mines, affecting both mine water quality and groundwater and surface waters below the discharge. The New Mexico Environmental Improvement Division has documented elevated levels of radiation and trace elements in groundwater from these activities, and has indicated that "adequate monitoring of the movement and the seepage and of the injected wastes is not underway. . . ." This has led staff members of the agency to indicate the "imminent necessity for adequate monitoring and surveillance and the implementation of adequate regulatory controls to control unchecked degradation of the environment resulting from uranium mining and milling actixities."—EPA, n.d.

U.S. The limitations on effluents have the effect of requiring that discharges be cleaned up or controlled in holding ponds. However, the costs of supplying water and of cleaning up discharges to meet the standards may make it more economical for the developer to continue to treat and recycle the water as long as possible and to discharge the pollutants into evaporative ponds rather than to discharge treated effluents into surface waters. However, as described previously, this approach does not necessarily solve the water-quality problem since these wastes accumulate in the ponds and can create potentially significant surface water and groundwater problems. Thus, the CWA requirements aimed at protecting the quality of surface water contribute to the decision to use alternative means of effluent disposal that can, in turn, lead to other, potentially serious water-quality problems. Specifically, the accumulated pollutants can be accidentally released into surface waters or can leach into groundwater; and these pollutants will constitute a problem of waste disposal and land use long after the energy facility has been shut down.

Until recently, the disposal of such wastes was largely unregulated. However, the RCRA

(1976) represents the closing of the last major portion of the biosphere to unregulated disposal of pollutants; the nation's land, like our air and water resources, are now subject to federal environmental control. In addition to attaining the goal of conserving virgin resources, RCRA addresses the need to control pollution of the land by developing a nationwide system for managing solid and hazardous wastes and for recovering resources from such wastes.

At the present time the act's regulations dealing with hazardous wastes and with state and regional solid-waste plans are still being developed, so that it is not yet possible to assess what the specific effects on energy production and conversion will be. By far the most important unknown factor is whether some of the solid wastes generated by the energy resources considered in this study will be classified as hazardous. The definition of "hazardous waste" contained in Section 1004(5) reads:

> The term "hazardous waste" means a solid waste, or combination of solid wastes, which because of its quantity, concentration, or physical, chemical, or infectious characteristics may (A) cause, or significantly contribute to an increase in mortality or an increase in serious irreversible, or incapacitating reversible, illness; or (B) pose a substantial present or potential hazard to human health or the environment when improperly treated, stored, transported, or disposed of, or otherwise managed.

Regulations recently proposed by EPA include classifying fly ash, bottom ash, and scrubber sludge found hazardous to human health or the environment as "special wastes." For special wastes, EPA has proposed postponing the announcement of standards until June 1982 and their effective date to June 1983 to give the agency more time to study the composition and characteristics of special wastes. If they are eventually subject to strict controls as hazardous wastes, the cost of waste

disposal by energy industries could increase from the present rate of about $2 a ton to as much as $90 a ton (*EPRI Journal,* 1978:38). If both ash and sludge are determined to be hazardous, another estimate puts the cost of disposal up by as much as 84 percent. In addition, disposal methods may have to be changed, and processes in addition to, or other than, holding ponds—such as incineration or drying and landfilling—may become necessary.

EPA is also authorized through the Toxic Substances Control Act (1976) to regulate the disposal of toxic materials by their manufacturers, users, transporters, and commercial disposal firms. This act is primarily directed toward control of the chemical manufacturing industry, but it does provide controls related to energy processes, including synthetic fuels plants.

Land Disturbances: Regulations established to control the environmental effects of surface coal mining exist in order to protect groundwater resources. However, there is still considerable uncertainty as to how effective the controls will be and how rigidly the regulations will be enforced in the future. In addition, there are no similar, comprehensive regulations for ensuring water-quality protection for other resources, such as uranium and oil shale mining and geothermal energy production.

The Surface Mining Control and Reclamation Act (1977) has several sections that specifically protect water quality from coal mining. For example, exploratory drill holes must be cased to prevent surface water and groundwater contamination (30 C.F.R. 816.13–15); and if needed, surface water and shallow groundwater may be diverted around mined areas to minimize the need for water-treatment technology within the mine (30 C.F.R. 816.43). In general, certain steps must be taken to protect the hydrologic balance and quality of groundwater (30 C.F.R. 816.50–51). These provisions include: first, the proper placement of backfilled materials to minimize contamination of groundwater; second, the appropriate location and design of cuts and excavations to preserve groundwater quality for postmining use; and third, the restoration of the aquifer to approximate premining recharge capacity. Special protection for alluvial valley floors is also required. This provision is primarily applicable to the arid and semiarid areas of the country (30 C.F.R. 822) and, to obtain a permit for mining, places the burden of proof directly on the mining company, so that "surface coal mining and reclamation operations shall not cause material damage to the quality or quantity of water in surface or underground water systems that supply alluvial valley floors."

North Dakota serves as one example of state implementation of the federal surface-mining reclamation act. In April 1979 the governor signed into law the Surface Mine Reclamation Act (N.Dak. Century Code, Ch. 38–14.1). It provides procedures, for example, to declare lands unsuitable for mining if operations would result in a substantial loss or reduction in productivity of the long-range water supply, including aquifer and aquifer-recharge areas (N.Dak. Century Code, Ch. 38–14.1–05). When a mining permit is obtained, the applicant must indicate the probable hydrologic consequences of the mining and reclamation operations both on and off the mine site, including changes in the quality of groundwater under seasonal flow conditions (N.Dak. Century Code, Ch. 38–14.1–14.1.o). A description of protective measures is also required, including processes to assure protection of the quality of groundwater (N.Dak. Century Code, Ch. 38-14.1–14.2.i). A bond is required to obtain a mining permit, and this bond can be released only after the mining commission determines the extent of groundwater pollution and, if needed, the cost of abating the pollution (N.Dak. Century Code, Ch. 38-14.1–17). Alluvial valley floors are protected, and the Mining Corporation can delete certain areas from a strip mining permit if it finds that water pollution may result. Specific reclamation procedures are required if groundwater is threatened (N.Dak. Century Code, Ch. 38-14.1).

Federal and state regulations do exist to protect groundwater quality from the land disturbances that are caused while developing energy resources, although compared to those affecting coal mining most of the regulations are quite limited. Section 208 of the CWA establishes procedures under which states or regional agencies are required to establish "nonpoint source" regulatory programs. Examples of nonpoint sources include irrigated and nonirrigated agriculture, mining, urban runoff, and rural sanitation. However, the overall effectiveness of these programs has been criticized, and their application to energy projects in the West has been especially limited. The Safe Drinking Water Act (1974) is another statutory tool that could be used to help control the results of energy development, at least as they might affect the quality of drinking water. However, the regulations under this act have not been widely applied to energy production operations. Furthermore, the act explicitly states that regulations will not be established that would interfere with the production of oil or natural gas unless such regulations are deemed essential to prevent endangering underground sources of drinking water (Shaw, 1976:532).

States have jurisdiction for protecting groundwater in areas of oil and gas operations and of uranium development. However, uranium mining operations occur largely on patented or private lands, and control by state agencies is limited. The statutes in most states typically do not provde the authority to control mine dewatering or subsurface waste injection. For example, the Water Quality Control Division of the Colorado Department of Health (1979) has recommended that new monitoring programs be implemented for hardrock mining, including uranium mining. For oil and gas development, control is typically vested in oil and gas commissions or in a corporation commission that is primarily responsible for "efficient production."

In summary, not only is there scientific uncertainty as to the long-term impact of energy development on water quality in the West, but there are also gaps and uncertainties in existing regulatory programs. As one state water-quality agency put it, one of the greatest needs in protecting the quality of water near active mining operations is to develop regulatory coordination and program consistency between the several state and federal agencies (Colo. Dept. of Health, WQCD, 1979:24).

Municipal Wastewater Pollution

Rapid, relatively large fluctuations of population in small western communities can cause water pollution due to the inadequate treatment of municipal sewage. Without increased assistance, only a few of the small communities in the West that are affected by energy development will be able to afford the cost either of upgrading their capacities to meet new demands or of installing the secondary and tertiary treatments required by the CWA.

The labor requirements of energy development can contribute to problems of water quality in proportion to the size of the facility's construction and operational work forces. These rapid and large population increases will impose heavy demands on the wastewater-treatment plants of small, western communities. Without proper advance planning and financial support, inadequate sewage treatment or even the bypassing of sewage treatment can occur. Our analysis and past experiences suggest that in many instances existing sewage-treatment plants are likely to be quickly overloaded by energy-related growth.

To get an idea of the magnitude of the problem, we have estimated that an oil shale industry in western Colorado and eastern Utah with a 400,000 bbl/day capacity by 1990 would result in increased municipal wastewater amounting to some 3 million gallons per day (MMgpd) during the peak construction period and 2 MMgpd during operation. Using a rough estimate of capital costs for water supply and sewage-treatment facilities of $1.76 million per 1,000 additional people, the capital costs for western communities, as outlined above, would be near $50 million in 1975 dollars.

Although, as we have discussed earlier in this chapter, the quantities of population-related effluents are much less than those associated with energy facilities, several potential problems can occur. First, poorly treated effluents can cause degradation of surface waters. Second, many communities will be unable to afford the cost either of upgrading capacities to meet new demands or of installing secondary treatment, as required by the CWA. One complicating factor is that the need for sewage treatment is usually greater during the construction phase of a facility than during its operation. Thus, it may be impractical to build sewage-treatment plants to serve short-term peak demands because they will be underutilized later. Third, insufficient sewer systems may affect other local problems; for example, new housing may be delayed pending adequate sewers, and community health standards may be violated.

The 1977 CWA requires municipal wastewater-treatment facilities to apply secondary treatment by July 1, 1978, and the best practicable technology by July 1, 1984. Secondary treatment is currently defined by EPA under the NPDES permit system as a biochemical oxygen demand not exceeding an average of 30 milligrams of oxygen per liter, with suspended solids not exceeding an average of 30 mg/l and a pH between 6.0 and 9.0. The best practicable technology is now considered to be a tertiary treatment that removes such chemical forms as nitrates and phosphates.

Many communities in the West that are affected by energy development currently do not meet the secondary standards established by the CWA. Over half of these towns either lack a water or a sewer system or have reached capacity in their existing systems. Upgrading facilities to meet the new standards will add to the financial burdens of local governments.

Efforts to upgrade treatment facilities are aided by the Wastewater Construction Grants Program, which is administered by EPA through the states. However, various bottlenecks and complicated administrative procedures have been blamed for delays in meeting the 1977 deadline (CEQ, 1975:71–72; Kirschten, 1977). The Construction Grants Program, the nation's largest recent public works program, is intended to help communities meet the costs of new treatment works. But through fiscal year 1976, the eight western states had received only 1.55 percent of the total national funds for the program; these same states account for about 4 percent of the nation's population. Unfortunately, the priority system established by the states for distributing EPA sewage-treatment funds to nonmetropolitan communities places towns that have recently received funds far down on the list for future funds (depending on the specific state program). Water and sewage systems in towns where growth has continued, such as Gillette, Wyoming, remain overburdened (Enzi, 1977; Pernula, 1977). Also, long-term planning by the states for wastewater-treatment construction is hampered by yearly variations in congressional funding. However, communities affected by energy development could be given some special consideration for funding, depending on the system used to establish priorities in that state.

When EPA funds are not available to meet continuing needs, towns may sometimes obtain assistance from other agencies, such as the Economic Development Administration, the Title V Regional Commissions, and state governments.

Salinity Control

Salinity has already been singled out for regulatory control by the federal government and by each of the states in the CRB; *it is of increasing concern in the Yellowstone River Basin. The major sources of salinity currently are natural salt flows and runoff from irrigated agriculture. Increases in salinity concentration as a result of energy development are expected to be small, relative to existing levels (which are primarily the result of the concentrating effects of consumptive water use), although the salt-loading effects of mining are highly uncertain. Energy development is likely to intensify conflicts over salinity control, and thus energy production could be held up by salinity standards if adequate controls are not established.*

Background: The level of salinity is of concern in both of the major river basins in our study area. However, salinity control is already a major issue in the CRB, whereas it has only been in the last few years that salinity has been recognized as a problem by some states (Such as Montana) in the UMRB. Therefore, in the following discussion the emphasis will be on the CRB. The increasing levels of salinity are important for both environmental and economic reasons. It was estimated in 1972 that each mg/l of increased salinity at Imperial Dam had a cost—both in environmental terms and in terms of lost agricultural production—of about $230,000 (U.S., DOI, BuRec, and USDA, Soil Conservation Service, 1974). This figure is currently being updated; given inflation and other changes, it is expected to increase substantially (GAO, Comptroller General, 1979:29).[4]

Along the Colorado River, salt concentrations and loadings generally increase downstream, mainly because of agricultural runoff and highly saline springs. Figure 4-1 shows the major rivers of the CRB and gives 1976 salinity concentrations at a few selected points (U.S., DOI, 1979). In the UMRB, the subbasin of most concern with regard to salinity

---

[4]BuRec has recently been using a cost of $343,000 per mg/l salinity increase at Imperial Dam. This estimate is under review and has not yet been approved.

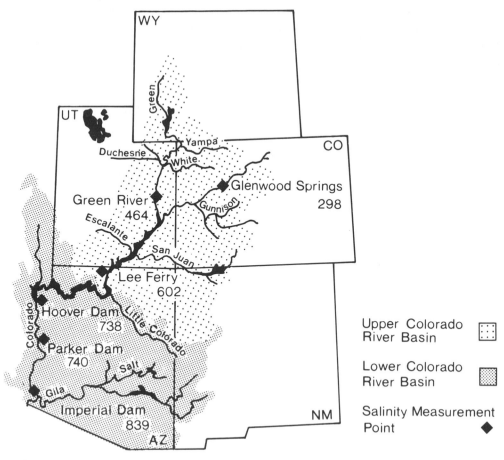

Figure 4-1: *Salinity in the Colorado River Basin (mg/l)*

is the Yellowstone River Basin. The UMRB is shown in Figure 4-2, including recent average salinity concentrations for various sampling locations (Klarich and Thomas, 1977). Salinity problems are most severe in the Powder River and in the late summer and early fall during periods of low flow and high irrigation return flow.

Man's development of the CRB has been the main cause of the increased salinity that has occurred in recent decades. This has come mainly from the 2.5 million acres of irrigated agriculture in the basin as well as from industrial and municipal uses. It has been estimated that in 1970 natural sources contri-

buted about 68.6 percent of the total salt load in the basin; agriculture contributed about 30.2 percent; and municipal and industrial sources about 1.3 percent (Utah State Univ., Utah Water Research Lab., 1975a:66).

Effects of Energy Development: Energy development can contribute to increased salinity in two ways:

- Salt loading due to runoff from surface disturbances (such as shale disposal, mining, and roads) and disturbances to aquifers that feed surface streams; and
- The concentrating effects of water consumption by energy conversion facilities.

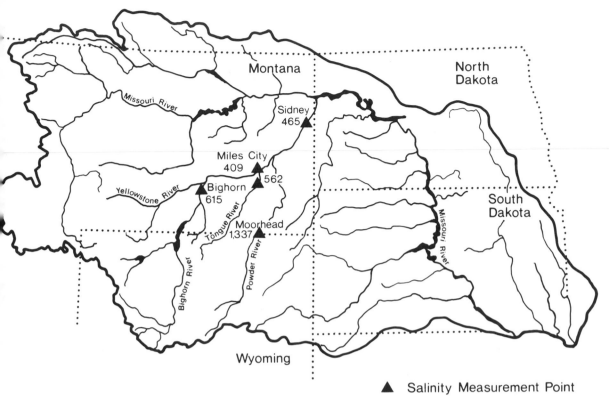

Figure 4-2: *Salinity in the Missouri River Basin (mg/l)*

There is great uncertainty concerning the salt-loading effects of energy development for several reasons (discussed above). Salt loading will depend on such factors as type of soil, levels of precipitation, degree of reclamation success, and degree of runoff diversions and controls from mined areas. In surface coal mining, as discussed above, the runoff from reclaimed areas is diverted into catchment basins and then either treated or placed in holding ponds to minimize salt additions to surface water. With oil shale, the initial level of salinity in runoff from spent-shale piles is estimated to be about 2.5 times that for normal surface runoff. However, the level of salinity decreases significantly with time, and in addition, most shale development plans call for controlling runoff from disposal areas. Thus, it is expected that proper controls can keep salt additions to a low level. However,

this is far from certain. There is concern, for example, that the dewatering of fresh aquifers during oil shale mining might allow the saline groundwater in deeper aquifers to intrude into the dewatered aquifer. This saline water could then possibly flow into surface streams, depending on the aquifer's characteristics. Despite such uncertainties, it generally is believed that increased salinity due to salt loading in surface waters will be small compared to the concentrating effects of consumptive water use.

Several salinity models have been developed to estimate the increases in salinity for the CRB due to consumptive water use by energy facilities. However, the results of these salinity models must be used with caution, primarily because they are based on certain assumptions which may not accurately portray the river system. Current salinity models may

Table 4-5: *Predicted Effects on Salinity at Imperial Dam for Alternative Energy Development Futures*[a]

Case	Water Use[b]		Salt Loading (concentration)[c]		
	Agriculture	Energy	1977	1983-85	1990-2000
A	M	M	920 (770)	918 (895)	780 (1045)
B	M	H	920 (780)	915 (940)	760 (1140)
C	L	H	920 (905)	910 (905)	750 (1055)
D	L	M	920 (875)	913 (875)	775 (1000)

[a]Adapted from Bishop, 1977:667.

[b]Resource utilization level: H = high, M = medium (most likely), L = low. The development levels projected in this study and the amounts of water used by agriculture, energy, and other sectors were not given. Instead, flow at Lees Ferry (million acre-ft/yr) under two development levels were presented as follows:

Case	1977	1983	1990
A	10.471	9.924	9.177
B	10.468	9.821	8.740

[c]Salt loading is expressed in thousands of tons per year; concentration in mg/l.

overestimate or underestimate the effects of salinity by as much as 100 percent. Nevertheless, such models are instructive in indicating the magnitude of effects from either a single energy facility or from energy development on a regional basis.

For example, one study estimated the effect of two coal-fired power-plant units totaling 760 megawatts (MW) planned near Craig, Colorado. These two units would consume 12,000 acre-ft/yr and were estimated to result in an average increase in salinity of 0.7 mg/l below Hoover Dam (U.S., DOI, BLM, 1976: III-3–III-9). Another study estimated that a 410 MMcfd coal gasification development in the Four Corners region of New Mexico, consuming 15,000 acre-ft/yr from the San Juan River, would increase salinity by about 2.4 mg/l at Imperial Dam (U.S., DOI, BuRec, 1978).

Still another study has comprehensively analyzed the effects of salinity for the CRB due to large-scale energy development throughout the region (Utah State Univ., Utah Water Research Lab., 1975a and 1975b). Three levels of energy development (low, medium, and high) were projected to determine the amount of water that would be used by energy,

agriculture, and for export. Based on 1972 conditions, projections were made for 1977, for 1983 through 1985, and for 1990 through 2000 at three assumed levels of virgin flow. Table 4-5 summarizes the results of this study assuming a virgin flow of about 14 million acre-ft/yr at Lees Ferry and a moderate amount of water exported from the basin. These results show the overall effect on salinity of intensive energy development and of shifting water from agriculture to energy.

Case A in Table 4-5 shows the salinity effects of a medium amount of water usage for both agriculture and energy development. If case A is the base condition, then case B— with medium agricultural water use and high energy water use—shows the impact of energy development on salinity. The salt load in case B decreases with time, relative to the base case, because of the withdrawals of water (and, therefore, of salt). In case B, the flow is projected to decline from the base case by about 1.73 million acre-ft/yr in the 1990 to 2000 period, due to increased withdrawals for energy development. Therefore, even though salt loadings are reduced, the salinity concentrations are projected to increase because of the reduced flow. Thus, in all cases, even though there is a reduction in the amount of salt that accompanies the withdrawals of water, it is more than offset by the increased concentration of salts downstream.

Comparing case A with cases C and D shows what might happen to the level of salinity if water is shifted from agriculture to energy or if agricultural water use is reduced. For cases A and C similar trends in increased salinity concentrations are projected but salt loading is less for case C. In case C the agricultural use of water is less and the consumption of water for energy development is greater that in case A. However, compared to case A in case D the salt load on the river basin is decreased even more than in case C; in D increased energy consumption of water is moderate and agricultural usage of water is low.

A similar projection of future levels of salinity has also been made for the UMRB (Klarich and Thomas, 1977). The general conclusions of this analysis are that in the eastern portion of the Yellowstone River Basin, energy development and the resulting reduction in stream flow can have a major adverse effect on surface water. Increased salinity—both in amount and in concentration—is predicted to be most severe in the Tongue and Powder River subbasins under any level of development; in the Lower Yellowstone subbasin in Montana (from Miles City to Sidney), it will be severe if a high level of development occurs. In the western portion of the Yellowstone Basin, the effects of energy development on salinity are not projected to be nearly as severe. In the Upper and Mid-Yellowstone subbasins and the Bighorn subbasin the only projected problem with salinity will occur during late summer and early fall, if there is a high level of water use, and during dry years.

The Salinity-Control Policy System: Most of the salinity problems with surface waters in our eight-state area are in the CRB. Consequently, the regulatory structure in this basin is the most developed, and the basin's political/social system is the most sensitive to the issue of salinity control (salinity levels in surface water have not been much of a public issue in the UMRB until recently).

Section 103 of the 1972 FWPCA required that the states of the Colorado River (and others) establish a mechanism for interstate cooperation in setting numerical criteria for salinity control. In response the CRB Salinity Control Forum was formed in November 1973 by the seven basin states. Shortly thereafter these states agreed not to allow salinity in the river at Imperial Dam to increase above the 1972 level. This standard was subsequently approved by EPA.

Also in 1973, the U.S. entered into an agreement with Mexico, the effect of which was to limit the salinity of water from the Colorado River flowing into that country. The agreement requires that the salinity level of water delivered to Mexico shall be no greater than 115 ppm (plus or minus 30 ppm) over the annual average salinity of water entering Imperial Dam (International Boundary and Water Commission, 1973).

In 1974, Congress passed the Colorado River Basin Salinity Control Act, which authorized the development of a major salinity-control project near Yuma, Arizona. This project includes a desalination plant and other activities such as lining irrigation canals and reducing irrigated acreages. This project is being handled by the BuRec and is designed mainly to help the U.S. comply with the agreement with Mexico. In addition, the act authorized four other salinity-control projects (Paradox Valley and Grand Valley, Colorado; Crystal Geysers, Utah; and Las Vegas Wash, Nevada) and formed the Colorado River Water Quality Improvement Program to investigate the feasibility of 12 additional salinity-control projects. However, the projects have been delayed and encountered large cost increases. For example, the estimated cost of the Yuma desalination plant alone has risen by 187 percent (from $62,080,000 to $178,400,000) between 1974 and 1977; this increase comes despite a decrease in the capacity planned for the plant to the minimum size allowed by the act. The delays are of considerable concern to the Salinity Control Forum and may retard attainment of salinity goals for the river.

Several other federal programs have been established for controlling pollution (including salinity) from agricultural lands and other "nonpoint" sources. The responsibility for controlling salinity on federal lands belongs to the bureau of land Management (BLM) of the DOI. Within the Department of Agriculture (USDA) there are several programs that deal with salinity. For example, the Soil Conservation Service (SCS) is responsible for nonstructural activities to control salinity on agricultural lands.[5] The USDA also administers the Rural Clean Water Program to correct salinity problems associated with "nonpoint" sources of pollution from rural (but not necessarily agricultural) lands (CRB Salinity Control Forum, 1978:78–79).

In addition to these independent federal actions, each of the five CRB states in our study area cooperates with federal and regional salinity-control efforts and has also implemented its own water-quality program, which includes salinity standards as set forth by the CRB Salinity Control Forum (1975).

More generally, the Salinity Control Forum is currently preparing draft "baseline values" for salinity at 12 monitoring points throughout the Colorado River system, identifying and evaluating changes in the river system that may occur upstream of the monitoring sites, overseeing the progress of salinity-control projects, and providing the member states with an overview of progress and problems in salinity control (CRB Salinity Control Forum, 1978:27–29). Recently, the Salinity Control Forum adopted a policy for industrial sources of "no-salt return" whenever practicable and a limit to the incremental increase in salinity

from municipal discharges in any portion of the river system of 400 mg/l or less (CRB Salinity Control Forum, 1978:A-1–A-10).

Although salinity standards at three points (Hoover, Parker, and Imperial dams) and for industrial and municipal discharges have been agreed upon by the states in the CRB and approved by EPA, it is unclear just how these standards will be enforced and how the states will respond to increased salinity. For example, even if the salinity standards are exceeded at any of the three points, there is no monitoring and enforcement mechanism to determine the cause (or causes) of the increased salinity and to control it.

The baseline values being developed by the states are intended to provide guidelines for monitoring increases in salinity throughout the basin; but at present these values are not intended as standards, and no enforcement mechanism is planned. Specific state and local activities for controlling salinity have not yet been clearly defined, although some general activities have been identified.

The slow progress towards implementing an effective salinity-control plan in the CRB has been criticized by environmentalists, and the EDF has filed suit against EPA for its failure to comply with FWPCA requirements for stream standards and development of an adequate control plan, including a compliance schedule. The EDF has suggested that salinity standards be set for five additional stations, located at and upstream of Lees Ferry, Arizona, to ensure control, monitoring, and guidance for future development. The Salinity Control Forum, however, argues that the present three stations and the anticipated baseline stations are adequate and that setting standards at additional stations is unnecessary. The forum does not consider state or subbasin standards to be the most cost-effective or consistent regional approach to development of the river (CRB Salinity Control Forum, 1978: 1–14).

---

[5] Irrigation return flows were removed from point sources definition by the CWA in 1977.

Summary of Problems and Issues

Water quality can be adversely affected by energy development in several ways. First, energy extraction and conversion activities can disrupt and contaminate aquifers and generate large amounts of wastes which must be disposed of to minimize leakage or leaching into groundwaters. However, these potential water-quality problems are difficult to quantify, and current scientific knowledge about the risks and possible controls is inadequate. The problems arise not because energy development activities will discharge large amounts of pollution directly into surface waters but rather because the magnitude of the operations is huge and there is a need to isolate a wide variety of chemical contaminants from the environment over the long term. A second water-quality problem is salinity. Although the major sources of salinity are natural salt flows and agricultural runoff, energy development can contribute to the problem somewhat by increasing salt loading and by consuming water, which in turn can increase the concentration of salinity downstream. Finally, the rapid population growth in small western communities near the sites of energy development poses a very real threat to water quality due to the possibility of inadequate sewage treatment.

Water-quality problems and issues have already led to several political conflicts that are likely to increase as the development of energy resources makes increasing demands on water resources. The most critical of these conflicts may occur among the states of the Upper and Lower CRB over the appropriate uses of Colorado River water, in general, and of salinity control, in particular. Intergovernmental conflicts are also likely to increase between the western states and EPA over control of salinity and of effluents from municipalities and energy conversion facilities. Furthermore, each of the water-quality problems and issues identified is likely to exacerbate other conflicts among environmental, industrial, municipal, agricultural interests.

Policy Alternatives for Water Quality

The discussion in the previous two sections indicates that policymakers are likely to face several water-quality problems and issues as the development of western energy resources progresses. The basic question they must address is how to make present goals for water quality and large-scale energy development compatible. As has been noted in both this chapter and Chapter 3, water use and water quality are intimately related. Thus, policies aimed primarily at dealing with the availability of water can also have important implications for its quality. For example, water conservation will reduce the problem of salinity caused by the concentrating effects of an energy plant's consumption of water. However, the purpose of this chapter is to address policy options whose primary aim is to resolve water-quality issues. We will examine three such options:

- Improve water-quality control plans for energy development, including predevelopment monitoring, technological controls, and research;
- Allow temporary sewage-treatment measures for communities affected by energy development; and
- Construct salinity-control projects.

Since the long-term risks to water quality from energy development are so uncertain, the first policy option we will consider is primarily aimed at improving our scientific understanding. Increased monitoring and research activities will allow the design of better technological controls for both resource extraction and waste disposal. The second policy option addresses problems of municipal sewage treatment in small communities that are affected by energy development projects. The goal of the policy alternative we considered is to allow temporary treatment measures (namely, sewage lagoons) until more conventional secondary and tertiary treatment facilities can be constructed. The final policy option deals with problems of salinity. Although it does not address energy development directly, we have included it to help in understanding the possible measures that can be taken to offset any increased salinity resulting from development (either salt loading or the effects of concentration). Constructing salinity-control projects—which can include desalination plants, diverting flows around areas of high salt pickup, or containing and evaporating highly saline flows—is a primary strategy that is currently being followed in the CRB. It should be noted that some of the policy options discussed in Chapter 3, dealing primarily with issues of water availability, also can have important implications for water quality. For example, improvements in the agricultural use of water will have real benefits in terms of salinity control.

Following the same format as other chapters in this part of the study, these three policy alternatives will be described and analyzed below. We will conclude this chapter with a brief summary and comparison of the alternatives.

Improved Water-Quality Controls for Energy Development

The goals of improving water-quality controls are two-fold: first, to improve the regulatory controls on current energy extraction and conversion activities so as to minimize the threat of surface water and groundwater pollution;[6] and second, to increase our scientific knowledge so that over the longer term there will be a better understanding of the environmental risks and the adequacy of technological controls. This option includes three elements:

---

[6]It should be noted that many past mining activities are now causing serious water-quality problems. However, since this study is concerned with future energy development, we have not specifically addressed policy options for controlling or correcting past bad practices.

- Predevelopment monitoring of surface water and groundwater quality;
- Water-quality control plans for each energy facility; and
- Monitoring and research.

Because of the high degree of uncertainty concerning the effects of energy development on the quality of surface water and groundwater, monitoring and research are especially important. The public, government regulators, and the engineering and scientific community are unsure about the necessary degree of protection, the effectiveness of various pollution controls, and the long-term assurance of a continued supply of high-quality groundwater. Monitoring is essential to understand geohydrologic behavior and to verify the effectiveness of control technologies. The data available only through a comprehensive monitoring program are indispensable to national environmental monitoring and research laboratories and to other research establishments. Through this research and development process, more effective technologies to protect groundwater can be developed.

In describing and evaluating the three elements of this policy option, each element might be considered as a separate policy choice; however, taken together they constitute a comprehensive water-quality control policy. Some of the elements have already been included in one way or another in existing federal and state legislation (though jurisdiction and the degree of implementation varies considerably between surface water and groundwater, among the six energy resources, and among the eight states).

Predevelopment Monitoring: One of the biggest uncertainties concerns the quality of groundwater before any development of energy resources. For example, in some locations in the Grants Mineral Belt in New Mexico, the levels of selenium and radioactive isotopes in the groundwater are significantly higher than in other areas of the Southwest. It is not clear, however, whether these elevated levels

Synthetic Liner on Waste Holding Pond

are due to mining activities. The vast variations in the local quality of groundwater before any disturbance presents a real problem in establishing regulatory guidelines. Predevelopment monitoring of groundwater can provide essential baseline data to determine appropriate technologies for control, standards for permits and monitoring, and liability for subsequent changes in groundwater quality.

Requirements for predevelopment monitoring of water quality in both surface waters and aquifers associated with energy resources could be added to state or federal statutes. General water-quality information is now part of the environmental impact statement (EIS) required for most major energy development projects. However, specific predevelopment monitoring is not a part of the permit process for mining, drilling, or siting energy conversion facilities; but monitoring could be added as a permit, lease, or statutory requirement.

The cost of such data collection efforts will depend on several site-specific factors: for example, with groundwater it will depend on such variables as the number, character, depth, and extent of aquifers. However, the cost would represent only a very small fraction of the total cost of a mine or an energy conversion facility. Since the quality of groundwater does not fluctuate rapidly and since a lengthy period of measurements prior to development would not be practical, it would not be necessary to take frequent measurements over an extended period of time (as is the case, for example, with air-quality baseline data). On the other hand, with surface flows it would be desirable to collect several years of data.

Water-Quality Control Plan: A water-quality control plan (covering both surface water and groundwater) would be part of a regulatory program that indicated the existing hydrologic resources, potential disturbances, control measures, monitoring practices, and restoration options for all energy development projects. This plan would be submitted by energy developers to the relevant local, state, or regional resource-management agency.

A water-quality control plan would be directed primarily to protect hydrologic resources within the geographic boundaries of a developer's holdings, but would also include the effects on hydrologic resources beyond the immediate area. This would provide information to planners and government officials, enabling them to assess the potential effects of aggregated development on the regional hydrologic system.

As discussed previously in this chapter, such a plan is already required for surface mining of coal under federal law (Surface Mining Control and Reclamation Act, 1977) and various state laws. In order to obtain a permit, the federal law basically places the burden of proof on the mining company to show that groundwater quality will not be significantly harmed. However, at the present time no similar comprehensive plans are required to protect water in connection with either the extraction of the other five resources we have considered or the disposal of wastes from energy processing and conversion facilities, although the legal basis for such regulations may exist. (The federal law does call for studies of the reclamation needs of other minerals.) A plan, such as the one described here, would require comprehensive planning to minimize the degradation of water quality, covering all energy resources and encompassing both extraction and conversion.

For mining, the data and performance requirements for this water-quality control plan

could be modeled after the permit requirements for surface coal mining. The plan could also contain sections similar to the effluent discharge plan now required by the state of New Mexico. The New Mexico plan differs from the NPDES permit in attempting to provide information on the fate of pollutants that are within or adjacent to energy development but that may not necessarily come under the NPDES permit system (i.e., discharges that may not enter navigable or interstate waters). In general, the water-quality control plan would include identification of potential water pollution from both "point" and "non-point" sources and would project the effects on surface water and groundwater quality.

For both mining and waste disposal, the technological measures available now are in the early stages of development and testing. Because these technologies generally are both labor- and materials-intensive with significant capital and operating costs, strategies for implementation need to be carefully designed. For example, in one demonstration project on the handling of overburden, it was estimated that the selective burial operations cost about 1.1 to 1.5 times as much as the normal operation of a surface mine (Dollhoph et al., 1977).

Implementation of control measures by specifying design standards may have considerable disadvantages such as restricting available options or requiring technical specifications that are inappropriate to some locations. Implementation based on technical review and performance standards would avoid delay, excessive costs, and the possible deployment of inappropriate controls to protect groundwater quality. Efficient implementation could be achieved by requiring a review of groundwater-protection techniques and restoration options and could also call for the development of appropriate site-specific plans, based on performance standards. Examples of these general performance standards would include:

(1) Use of the best practicable surface and

subsurface mining practices to minimize changes in groundwater quality;

(2) Specification of the maximum permis-
sable seepage from waste-disposal areas
(which would allow the industry to
choose, for example, the amount of
pretreatment or the type of pond liner
in order to meet the standard);

(3) Removal of contaminated groundwater
where feasible to enhance the immediate beneficial use of aquifers; and

(4) Use of the best practicable aquifer-
restoration technology to preserve the
long-term beneficial use of groundwater.

In addition to increasing the cost of energy
production, comprehensive controls could also slow the pace of development somewhat
both because of the planning requirements
and because some controls may require procedures which in themselves slow operations.
For example, the selective handling of overburden layers, depending on its toxicity (as
is being done in some surface coal mines),
can mean that trucks and shovels must be
used instead of draglines. Because of these
factors, industry resistance can probably be
expected. However, the comprehensive planning described here represents an extension
of the controls that now exist for surface coal
mining, and its application to all energy development activities would ensure a higher
degree of protection for the West's water
resources.

Monitoring and Research: An expanded
monitoring and research program would lead,
over the longer term, to a better understanding
of the risks to water quality (especially groundwater quality) from energy development in
the West. At the present time, publicly available data on the quality of groundwater are
limited. Monitoring is being conducted by
the energy industry, but the data are largely
proprietary. The state of New Mexico has
established periodic monitoring requirements
for all surface and subsurface discharges, but

other states in our eight-state study area do
not have this comprehensive monitoring requirement.

The cost of monitoring is largely dependent
on the number of wells and the frequency of
sampling. Monitoring can be conducted by
the energy companies in compliance with
standardized analytical techniques. The cost
is highly dependent on local conditions, but
may be small (less than $10 per sample, for
example) if existing wells are used and if
sampling our analysis does not require significant additional manpower. However, a major
sampling program could range in cost from
$10,000 to $100,000 per year for a single
mine or waste-disposal site, depending on
well investment, personnel, and analytical
equipment needed.

In terms of research, there are already
energy-related research programs on groundwater quality within various federal agencies
(primarily the Department of Energy [DOE],
EPA, and the Geological Survey [USGS]),
state governments, and universities. However,
the current level of research does not appear
adequate to provide the needed information
on a timely basis. To be most effective, a wide
range of studies is needed to examine problems connected with the several important
energy resources (but mostly coal, oil shale,
and uranium) in a variety of western environments. These studies should examine potential risks and alternative controls for mining,
in situ recovery, and waste disposal. For example, listed below are some of the types of
studies needed on the subject of holding ponds.

• *Failure rates of holding ponds:* Effort is
needed to gather historical data on the
kinds of failures, their significance, and
the variables that are important, including rainfall, floods, and structural failure.

• *Movement of pollutants:* Field and laboratory research is needed to better understand the migration of pollutants through
various types of pond liners during normal operations and over the long term.

Studies by EPA and the USGS have begun to assess this problem. Coordinated studies in a variety of western locations need to be initiated.

- *Water Consumption by Holding Ponds:* Evaporation losses from existing holding-pond technology are not adequately documented. Some state agencies suggest that holding ponds may increase the consumption of water by energy facilities between 10 and 20 percent due to evaporation of water that could be cleaned, treated, and conserved with more costly technology. Information on the design requirements and economic cost of these alternative technologies could be made more available to decisionmakers.

Generally, these studies of holding ponds reflect the need to apply technological criteria to the unique water resources and hydrological and environmental conditions found in the Rocky Mountains and Northern Great Plains states. Currently there is inadequate historical documentation of the magnitude and kinds of problems experienced, inadequate attention to technical alternatives for achieving goals for zero-discharge of pollutants and their associated environmental and economic trade-offs, and inadequate dissemination of existing information to local, state, and federal decisionmakers.

Of course, the major obstacle to such a comprehensive research program is the cost. While it would depend very much on the scale of effort, a moderate-sized program of the type being described here is judged to cost between $20 million and $50 million over the next ten years. Much of this research would probably require support from the federal government; but to be most efficient and effective, it should be undertaken cooperatively with the states and energy developers.

Allowing Temporary Sewage-Treatment Measures

As discussed earlier in this chapter (and in more detail in chapters 7 and 8), western communities near large energy development projects will experience rapid increases in population. After construction is completed, however, the population may decrease (since less manpower is needed for operation than for construction) or the population may continue to increase as additional energy projects and associated activities are located in the area. Since wastewater-treatment plants in many western communities are already at or near capacity, a rapid and relatively large population increase is likely to result in inadequately treated sewage and, thus, in increased pollution of receiving waters. Also, few small communities in the West affected by energy development will be able to afford their share of the cost of upgrading and/or expanding sewage-treatment facilities to meet the requirements of the CWA. The need is for an inexpensive, rapid method to expand the capacity to treat wastewater. The policy alternative considered here would allow the use of sewage lagoons (or, waste-stabilization ponds) as a temporary measure until more conventional secondary treatment facilities can be constructed.

Waste-stabilization ponds are currently used by many communities in the West, especially the smaller ones. The main advantages of these ponds include low capital and maintenance costs and simplicity of operations. Although both labor and capital costs for constructing and operating waste-stabilization ponds are very small compared to conventional secondary treatment facilities, both require wastewater-conveyance systems which may cost anywhere from $176 to $1,195 per capita (1976 dollars) in small communities (Dames and Moore, 1978:6–18). The capital costs for construction of waste-stabilization ponds include purchase of the land, purchase of wastewater-conveyance systems, and construction of berms. Generally, local manpower and skill is sufficient both for construction and for operation. Land is readily available in most of the areas of the West where energy-related boomtown conditions are likely to occur. However, often the most desirable location for a pond is also desirable farm land. This may also be the case for conventional treatment facilities, although in some areas ponds may require more land.

In addition to lower costs, ponds can be designed and constructed much more rapidly than can conventional secondary treatment facilities and can therefore be available in a more timely fashion. Also, since capital costs for ponds are less, if energy development does not occur as expected, the community will not have lost as big an investment. And waste-stabilization ponds are more readily reversible than conventional systems; in most cases, the land for sewage lagoons can be easily drained, filled in, and converted back to its original use. It is more difficult and costly to restore the land used for conventional systems.

The major disadvantage of waste-stabilization ponds compared to conventional waste-water-treatment facilities is their apparent inability to consistently meet EPA's secondary sewage treatment standards. In the West difficulties may occur with excess suspended solids during the late summer and with anaerobic conditions during deep ice cover in the winter. In fact, during early 1979 EPA and several western states relaxed the suspended solids limitation for sewage lagoons to allow their continued use. However, in many areas of the West where net evaporation rates are high and the necessary land is readily available, waste-stabilization ponds can often be constructed to completely retain all of a municipality's wastewater without having any discharge (as is encouraged in Utah). In northern and higher elevation areas of the West, ice cover limits the effectiveness of lagoons during the winter. In such areas, ponds may have to be quite large to retain the wastewater during these periods (as is currently required in North Dakota). Of course, a major cost of the complete retention of wastewaters

is the increased consumptive use of water. For example, as discussed earlier, a 400,000 bbl/day oil shale industry was estimated to result in an increased wastewater flow during construction of 3 MMgpd (or 3,000 acre-ft/yr). Thus, use of sewage lagoons with complete retention would increase water consumption related to energy development by this amount.

Another consideration regarding the use of sewage lagoons is that if such ponds are not constructed while temporary and rapid growth occurs, the existing sewage facilities are likely to be overloaded or bypassed altogether, resulting in even greater pollution of receiving waters. Requiring capital-intensive facilities before, or at the time of, the population increase puts a very large financial strain on a community already struggling to provide sufficient services. However, the construction of ponds, at least on a temporary basis, would allow small western communities to provide economical wastewater treatment while collecting the necessary initial financing from the expanded tax base to build more conventional facilities.

Salinity-Control Projects

Much of the salt from man-made or natural sources may not originate at specific (point) sources or may be difficult to manage by point-source controls. Therefore, salinity-control alternatives can include such measures as constructing desalination plants, diverting flows around areas of high salt pick-up, eliminating saline flows by evaporation (including agricultural return flows), and improving irrigation practices.[7] These approaches are currently being used to control salinity in the Colorado River. The CRB Salinity Control Act (1974) established a comprehensive program, including the au-

[7] Improvements in water use for irrigation were discussed in Chapter 3; they are included here in terms of their effect on salinity.

Table 4-6: *Projects Authorized for Construction under P.L. 93-320*

Project	Description	Current Status
Yuma Unit, Arizona	Includes desalting plant and construction of a bypass drain and lining a canal. Other activities include increased irrigation efficiency, acreage reduction, and protective groundwater wells along Mexican border.	Size of desalting plant reduced to 96 MMgpd; construction of various components delayed.
Paradox Valley Unit, Colorado	Groundwater pumping to lower the freshwater-brine interface below the river channel; pumping to an evaporation pond is being considered to remove about 180,000 tons of the 205,000 tons of salt that annually seep into the Dolores River.	Test wells begun; data for design are being collected.
Grand Valley Unit, Colorado	Improved irrigation distribution systems (canal and lateral lining) and management to increase efficiency are planned to reduce the salt entering the Colorado River by 410,000 tpy.	Preliminary investigations completed; initial phase monitoring to be established.
Crystal Geyser Unit, Utah	This abandoned oil test-well contributes about 3,000 tpy of salt; the flow was to be collected and evaporated.	Construction deferred indefinitely due to low cost effectiveness.
Las Vegas Wash Unit, Nevada	Originally, interception of groundwater and evaporation was planned for Stage I and a desalting plant was planned for Stage II. Lining of ponds in the area has resulted in a changed hydrology.	Construction delayed pending further study.

Source: U.S. DOI, 1979:71-85; Comptroller General, GAO 1979:35-45.

thorization for construction of 5 salinity-control projects and for the study and planning of 12 other projects. The projects authorized for construction by the CRB Salinity Control Act are described in Table 4-6. Table 4-7 shows the projects being studied or planned by BuRec under the CRB Salinity Control Act; these projects have not yet been approved for construction.

Desalination techniques include chemical pretreatment and several methods for separating dissolved solids from water, such as freezing, distillation, and filtering through membranes to remove brine (reverse osmosis). For example, the salinity-control complex being constructed near Yuma, Arizona (described briefly above), which was authorized by the 1974 CRB Salinity Control Act, includes a desalination plant that will use a combination of chemical pretreatment and membrane systems. Now scheduled for completion in 1981, this plant would treat about 107,500 acre-feet of saline agricultural runoff each year and remove about 515,000 tons of dissolved solids annually (GAO, 1979:

35–45; Kaufman, 1980). Brine discharged from this plant would be diverted from Yuma to a drainage channel, terminating in the Gulf of California. In other locations distant from marine water, brines could be disposed of by evaporation or by removal to isolated groundwater formations. Because it is near the Mexican border, the Yuma desalting project will have no effect on salinity above Imperial Dam. It is included here for comparative purposes since similar desalting plants have been discussed for the Colorado River above Imperial Dam.

The practices of diverting streams around areas of high salt pick-up or intercepting highly saline groundwaters can also control salt loading; and as indicated in Table 4-6 and Table 4-7, several diversions and interceptions are included in plans for the Colorado River. Impoundments are used to prevent highly saline waters from flowing into a stream in which salinity is being controlled; and canals are used to circumvent saline seeps or formations. This option is being considered for the LaVerkin Springs unit,

Table 4-7: *Projects for Salinity Control Being Studied under Title II of P.L. 93-320*

Project	Description
Irrigation-Source Control Projects	
Lower Gunnison Basin Unit, Colorado	Irrigation management and improved water systems are planned for 160,000 acres contributing about 1.1 million tons of salt per year. Initial program has begun.
Uintah Basin Unit, Utah	Irrigation management and improved water systems are planned for 170,000 acres contributing about 450,000 tons of salt per year. Initial program has begun.
Colorado River Indian Reservation Unit, Arizona	Irrigation management and improved water systems are planned for about 80,000 acres. A study found no salt is contributed to the Colorado River by this unit, and development of the project has been discontinued.
Palo Verde Irrigation District Unit, California	Irrigation management and improved water systems are planned for about 91,400 acres contributing about 152,000 tons of salt per year.
Point-Source Control Projects	
LaVerkin Springs Unit, Utah	River flow will be diverted around the springs; the saline water will be pumped to a desalting plant. About 103,000 tons of salt per year would be removed from the stream.
Lower Virgin River Unit, Nevada (replaces Littlefield Springs Unit)	Wellfield or barriers dam and solar evaporation to collect irrigation return flow and saline groundwater are planned.
Blue Springs Unit, Arizona	Feasibility studies to control 160,000 acre-ft/yr that contribute about 550,000 tpy of salt were planned, but due to high costs and environmental problems these plans have been canceled.
Glenwood-Dotsero Springs Unit, Colorado	Treatment to remove most of the 250,000 tpy of salts contributed to the river is planned.
Diffuse-Source Control Projects	
Big Sandy River Unit, Wyoming	Interception of saline groundwater seeps is planned to remove about 80,000 tons of salt per year in 6,000 acre-feet. Evaporation in a pond is being considered.
Price River Unit, Utah	Selective withdrawal, farm management, and other improvements in the water systems are planned to control 100,000 tpy of salt.
San Rafael River Unit, Utah	Selective withdrawal, farm management, and other improvements in the water systems are planned to control 80,000 tpy of salt.
Dirty Devil River Unit, Utah	Selective withdrawal, farm management, and other improvements in the water systems are planned to control 80,000 tpy of salt.
McElmo Creek Unit, Colorado	Planning has not progressed far enough to identify potential alternatives.
Meeker Dome Unit, Colorado	Collection of salt wells and seeps with treatment or utilization is being considered.

Source: CRB Salinity Control Forum, 1978:42-49; U.S., DOI, 1979:85-99.

the Price River unit, and several others of the Colorado River Salinity Improvement Program (CRB Salinity Control Act, 1974: Title II).

The technique of eliminating saline flows by constructing storage ponds and allowing the saline water to evaporate can be applied to naturally occurring saline streams and to saline agricultural runoff. However, given the volume involved with agricultural return flows, many large ponds would be required. Nevertheless, this is an option in some cases, and this approach is planned for saline streams in the San Rafael River and Dirty Devil River units of the Colorado River Salinity Control Program.

As shown in Table 4-6 and Table 4-7 there are several projects currently underway to reduce salinity contributions from irrigated agriculture. These include the Grand Valley, Lower Gunnison Basin, Uintah Basin, Colorado River Indian Reservation, and Palo Verde Irrigation District units. Activities planned to reduce salinity contributions from these sources include improving irrigation management and making improvements in the water systems. Irrigation management is a nonstructural approach and includes such things as scheduling of delivery and application of water. Structural actions that are part of improving the water systems include lining canals, using pipe systems, upgrading diversion and measurement structures, and using sprinkler systems. The effect of such actions on water availability is discussed in Chapter 3.

How Much Can Salinity Be Reduced? It is meaningless at this time to project the total reduction in salinity that is likely to occur if all the projects authorized for construction were in operation because several of them have been delayed indefinitely or are being substantially revised. However, as Table 4-8 shows the effects of some selected salinity-control projects (as presently planned) can be projected. The combined effect of these seven projects alone would result in an estimated

70.2 mg/l reduction of salinity concentration at Imperial Dam. Another study has shown that the Grand Valley unit will reduce salinity at Imperial Dam by about 43.0 mg/l (GAO, Comptroller General, 1979:29–30). The implementation of the other salinity-control projects would reduce salinity even further. However, the projects currently planned have been chosen largely because they are thought to be the most cost-efficient and to control the major sources of salinity in the basin. Additional projects may be more expensive in proportion to their effect on salinity at Imperial Dam.

What Are the Costs and How Are They Distributed? Along the southern part of the Colorado River, constructing desalination plants and evaporating highly saline waters can remove a ton of salt for approximately $30 to $40 (U.S., DOI, BuRec, 1978:195–211). The costs of flow containment or diversion can vary greatly depending on the specific project. For example, the Paradox Valley program which includes evaporation of highly saline water is estimated to cost about $10 per ton of salt removed.

Table 4-9 shows estimates of the costs per mg/l reduction at Imperial Dam for four of the major projects originally authorized by the CRB Salinity Control Act (1974). Although these mg/l reductions and their associated costs are difficult to compute, and therefore there is a great degree of uncertainty, this measure is more relevant for com-

Table 4-8: *Estimated Salt Load and Salinity Concentration Reduction by Selected Salinity-Control Projects*

Project	Estimated Salt Removal (tons)	Estimated Salinity Reduction at Imperial Dam (mg/l)
Paradox Valley Unit, Colorado	180,000	18.2
Uintah Basin Unit, Utah	100,000	10.0
LaVerkin Springs Unit, Utah	103,000	9.0
Big Sandy River Unit, Wyoming	80,000	7.0
Price River Unit, Utah	100,000	10.0
San Rafael River Unit, Utah	80,000	8.0
Dirty Devil River Unit, Utah	80,000	8.0

Source: U.S., DOI, 1979:85-99.

Table 4-9: *Estimated Costs of the Four Projects Authorized by Title II of P.L. 93-320*[a]

Project	Salinity Reduction at Imperial Dam (mg/l)	Annual Equivalent Cost	Annual Equivalent Cost per mg/l Salinity Reduction at Imperial Dam
Paradox Valley	18.2	$ 3,507,000	$192,700
Grand Valley	43.0	10,824,000	251,700
Las Vegas[b] Wash	9.0	8,727,000	969,700
Crystal[b] Geyser	0.3	234,000	780,000

[a]GAO, Comptroller General, 1979:29-30; the date of these cost estimates was not provided.

[b]These units have been delayed indefinitely but are included for comparative purposes to estimate the range of costs likely to be found in salinity-control projects.

parison than simply estimating the total tons of salt removed or the cost per ton. As indicated, these costs range from about $200,000 to $1 million per mg/l salinity reduction. These costs are for some of the area's largest sources of salinity and may only be generally indicative of the costs to be expected in controlling other sources of salinity.

An additional cost of some of these projects will be the loss of water in the Colorado River. For example, wells for the Paradox Valley unit will remove about 3,260 acre-ft/yr from the river which will be pumped to an evaporation pond. The LaVerkin Springs unit in Utah is planned to include pumping about 2,300 acre-ft/yr of brine to a solar evaporation pond. The Big Sandy River unit might evaporate about 6,000 acre-ft/yr as brine in a pond (U.S., DOI, 1979:71–99).

Since these technological controls are all capital-intensive, they will almost certainly require public (largely federal) financing. Thus, taxpayers living in regions other than the West will pay a large proportion of the costs. Since agriculture is a major cause of salinity and since loss of agricultural production is one of the major costs of high levels of salinity, then a significant share of such expenditures can reasonably be viewed as a subsidy for irrigated agriculture. While schemes might be developed to assess water users for the cost of the project, such strategies would be difficult to implement and,

unless carefully designed, could cause downstream users to pay for problems largely originating upstream.

The benefits of improving the quality of surface water will go primarily to westerners. For example, agricultural benefits of each mg/l decrease in salinity have been estimated to range from about $46,000 to $108,000. Municipalities are estimated to benefit by about $120,000 for each mg/l reduction in salinity by, for example, not having to replace appliances and water-treatment and distribution facilities as frequently. Benefits to industry are estimated to be about $1,500 for each mg/l decrease in salinity (Utah State Univ., Utah Water Research Lab., 1975a: 166–68). Environmental benefits including the protection of fish and other aquatic life as well as recreational uses of these waters will also result, but dollar estimates are not possible.

Are There Implementation Obstacles? The treatment of surface water is generally constrained by the high capital cost, particularly for desalination plants. Nevertheless, the legal and regulatory basis has already been established for each of these alternatives, so implementation may be less difficult than for many other ways to control the sources of pollution. For example, as discussed previously, five salinity-control projects were originally authorized by the CRB Salinity Control Act in

1974, and commitments were made to 12 others if the planning reports proved favorable. Hence, this alternative could be implemented in the CRB with existing programs and agreements. However, as also described previously, these projects have been the subject of considerable controversy, and they have all encountered delays. The major opposition has arisen from some interests who feel that the environment will be harmed by some of the specific projects and from other interests who feel that the benefits do not justify the large public expenditures and/or that more cost-effective methods are available (GAO, Comptroller General, 1979).

Summary and Comparison of Water-Quality Alternatives

Energy development can contribute to water-quality problems in three primary ways:

- Contamination of surface water and groundwater due to land disturbance and waste disposal;
- Increased salinity concentration downstream due to consumptive use of water; and
- Inadequate sewage treatment in energy boomtowns.

In this chapter we have described these problems and evaluated three policy options for dealing with them. Table 4-10 lists the alternatives and summarizes some of the key findings for each. In addition, because problems of water quality and water availability are often inextricably linked, several of the policy alternatives considered in other chapters can also have important implications here. Especially significant in this respect is the option for improving the efficiency of water use in agriculture (see Chapter 3).

The three policy options considered in this chapter do not compete; each could be used to contribute in some way to the protection of water quality in the West. Therefore,

these options are not directly comparable. The first alternative, improved water-quality controls for energy development activities, consists of three primary elements: predevelopment monitoring of water quality, water-quality control plans, and monitoring and research. These items do not represent a radically new program, but rather represent an expanded and more comprehensive approach to ongoing activities. The goals of this option are to ensure, first, that the best available practices are currently being used and, second, that over the longer term we can improve our understanding of geohydrology and of the behavior of potential contaminants so that we can identify the risks of energy development and the effectiveness of various control measures. Without such a program, the chances will increase that serious and, in some instances, irreversible damage will occur to the West's water resources, especially its groundwater. The disadvantages of this option are that the cost of energy could increase due to stricter control measures (although the increases should not be large) and that greater public (mostly federal) expenditures will be required for research. However, given the high level of scientific uncertainty concerning the potential effects and the likelihood of greatly expanded production of western energy resources, the costs appear to be worth paying.

The alternative we have considered for dealing with municipal wastewater problems would allow the use of sewage lagoons (waste-stabilization ponds) in boomtowns as a temporary measure until more conventional secondary sewage-treatment facilities can be constructed. The main advantages of this approach are that capital and operating costs would be substantially lower and that the lead time for construction would be shorter. Thus, sewage lagoons provide a flexible, economical means of providing municipal wastewater-treatment capacity until the population stabilizes and the necessary financing from the expanded tax base is available. The major disadvantage is the inability to consistently

Table 4-10: *Summary and Comparison of Water-Quality Alternatives*

Criteria	Improve Water-Quality Controls for Energy Development	Install Temporary Sewage-Treatment Measures	Salinity-Control Project
Effectiveness: Achievement of policy objective	Would provide needed baseline data Would minimize long-term, sometimes irreversible, effects on groundwater quality	May not meet current standards On the other hand, may reduce pollution if inadequate treatment or bypassing would otherwise occur	Can remove large amounts of salts (e.g., two proposed projects can reduce salinity by 18 and 43 mg/l, respectively, at Imperial Dam)
Efficiency: Costs, risks, and benefits	Predevelopment monitoring costs small Water-quality plans would increase cost of energy somewhat and could slow development Research costs estimated at $20 to $50 million over 10 years	Substantially lower capital and operating costs than conventional secondary treatment Short lead time for construction Complete retention would increase water consumption	Large capital and operation costs (e.g., operating costs between $0.2 million and $1.0 million per mg/l reduction in salinity) Economic benefits estimated at between $0.17 and $0.23 million for each mg/l salinity reduction Will reduce stream flows
Equity: Distribution of costs, risks, and benefits	Research program would probably require federal support Energy users would pay increased production costs Western water users would receive benefits	Would lower requirements for federal expenditures for small western communities Downstream water users would suffer any potential degradations in water quality	Taxpayers will pay a large portion of costs Will represent a subsidy to irrigated agriculture since it is a major contributor to salinity and a major beneficiary of salinity reductions
Flexibility: Adaptability to locational needs and changes over time	Water-quality control plans should be kept flexible to meet specific needs	Major advantage in flexibility in face of uncertain population fluctuations — short lead time and easily reversible	High capital cost of some projects makes alternatives essentially irreversible
Implementability: Institutional constraints and acceptability	Basically represents an expansion and coordination of existing program Increased energy costs and federal expenditures may create opposition from the public and industry	Lower costs would create some public support Would require revisions of official EPA policy	Institutional and regulatory basis already exists in CRB Some opposition due to large federal cost and the environmental impact of some projects

meet EPA's secondary treatment standards. While this problem could be dealt with by complete retention of wastes, it would do so at the cost of increasing consumptive water use.

Salinity problems are already an important issue in the CRB and some streams in the Yellowstone River Basin. Although the primary sources of salinity are natural salt flows and agriculture, energy development will exacerbate the problem through salt loading and the downstream concentrating effects of consumptive water use. In addition to the first policy option discussed, which would reduce the risks of salt loading from energy production facilities, this chapter considers a range of salinity-control projects currently being planned in the CRB. Controlling salinity has both environmental and economic benefits. The economic benefits for each mg/l decrease are estimated at between $46,000 and $108,000 for agriculture, $1,500 for industry, and $120,000 for municipalities. These comprehensive projects can have a great effect on salinity levels. For example, two of the projects authorized by the CRB Salinity Control Act (Grand Valley and Paradox Valley) are estimated to decrease salinity concentrations at Imperial Dam by about 43 and 18 mg/l, respectively. The disadvantages are that large capital outlays by the federal government are required and that water availability is decreased. Cost estimates for these two projects range between $200,000 and $1 million per mg/l salinity reduction at Imperial

Dam (1975 dollars). Although there is some opposition to these projects—and although the authorized projects have been subject to considerable delays—the legal and institutional basis for such an approach already exists in the CRB through the CRB Salinity Control Act and the CRB Salinity Control Forum.

Conclusion

In summary, our evaluation reinforces the concept that water quality cannot be separated from water availability. None of the alternatives considered in this chapter would, by itself, be a successful response to water-quality concerns. For example, if the scarce water resources of the West are continually depleted, none of the alternatives considered here will help preserve the region's water quality.

Policymakers may choose to trade one set of water-related problems for another; for example, choosing to limit agricultural production in many areas of the West in favor of developing energy resources could reduce salinity problems but increase the risk of more toxic pollution from energy production. However, regardless of which policy alternatives are chosen, many similar problems can be anticipated. These include more public expenditures for water-quality control and more conflicts among potential water users, among states of the various water basins, and between the West and other regions.

REFERENCES

Bishop, A. Bruce. 1977. "Impact of Energy Development on Colorado River Water Quality." *Natural Resources Journal* 17 (October).

Bishop, A. Bruce, et al. 1975. *Water as a Factor in Energy Resources Development.* Springfield, Va.: National Technical Information Service.

Clean Water Act (CWA) of 1977, Pub. L. 95-217, 91 Stat. 1566.

Code of Federal Regulations (C.F.R.), Title 30, §816, as promulgated in *Federal Register* 44 (March 13, 1979):15395.

Code of Federal Regulations (C.F.R.), Title 30, §822, as promulgated in *Federal Register* 44 (March 13, 1979):15450.

Colorado Department of Health, Water Quality Control Division (WQCD). 1979. *Water Quality and Mining: A Process to Identify and Control Water Pollution from Past, Present and Future Mining Activities in the State of Colorado.* Denver: Colorado Dept. of Health, WQCD.

Colorado River Basin (CRB) Salinity Control Act of 1974, Pub. L. 93-320, 88 Stat. 266.

Colorado River Basin (CRB) Salinity Control Forum. 1975. *Proposed Water Quality Standards for Salinity: Colorado River System.* Salt Lake City, Utah: CRB Salinity Control Forum.

Colorado River Basin (CRB) Salinity Control Forum. 1978. *Proposed 1978 Revision, Water Quality Standards for Salinity, Including Numeric Criteria and Plan of Implementation for Salinity Control: Colorado River System.* Salt Lake City, Utah: CRB Salinity Control Forum.

Congressional Quarterly, Inc. 1973. *Congress and the Nation, Vol. 3:1969-1972.* Washington, D.C.: Congressional Quarterly.

Council on Environmental Quality (CEQ). 1975. *Environmental Quality,* Sixth Annual Report. Washington, D.C.: Government Printing Office.

Council on Environmental Quality (CEQ). 1977. *Environmental Quality,* Eighth Annual Report. Washington, D.C.: Government Printing Office.

Crawford, K. W., et al. 1977. *A Preliminary Assessment of the Environmental Impacts from Oil Shale Developments.* Washington, D.C.: Environmental Protection Agency.

Dames & Moore, Water Pollution Control Engineering Services. 1978. *Construction Costs for Municipal Wastewater Conveyance Systems: 1973-1977.* Washington, D.C.: Environmental Protection Agency, Office of Water Program Operations.

Denver Research Institute (DRI). 1979. *Predicted Costs of Environmental Controls for a Commercial Oil Shale Industry,* Working Draft Report. Denver: University of Denver, Research Institute.

Dollhopf, D. J., et al. 1977. *Selective Placement of Coal Stripmine Overburden in Montana, II. Initial Field Demonstration,* Research Report 125. Bozeman: Montana Agricultural Experiment Station, Reclamation Research Program, Montana State University.

Environmental Protection Agency (EPA). N.d. *Areawide Wastewater Management Program.* Washington, D.C.: EPA.

Environmental Protection Agency (EPA). 1971. *The Mineral Quality Program in the Colorado River Basin,* Summary Report. Washington, D.C.: Government Printing Office.

Environmental Protection Agency (EPA). 1974. *Clean Water: Report to Congress — 1974.* Washington, D.C.: EPA.

Enzi, Michael B. 1977. "Energy Boom: Wyoming's Coal Veins Just Bring Troubles." *Los Angeles Times,* May 15, p. V-5.

EPRI Journal. 1978. "Disposal and Beyond." 3 (October):36–41.

Federal Water Pollution Control Act (FWPCA) Amendments of 1972, Pub. L. 92-500, 86 Stat. 916.

General Accounting Office (GAO), Comptroller General. 1979. *Report to Congress on the Colorado River Basin Water Problems: How to Reduce Their Impact.* Washington, D.C.: GAO.

General Electric Corporation, TEMPO. 1973. *Polluted Groundwater: Some Causes, Effects, Controls, and Monitoring.* Springfield, Va.: National Technical Information Service.

Grimshaw, Thomas W., et al. 1978. *Surface-Water and Ground-water Impacts of Selected Energy Development Operations in Eight Western States.* Austin, Tex.: Radian Corporation.

Hounslow, Arthur, et al. 1978. *Overburden Mineralogy as Related to Ground-Water Chemical Changes in Coal Strip Mining.* Springfield, Va.: National Technical Information Service.

International Boundary and Water Commission. 1973. "Permanent and Definitive Solution to the International Problem of the Salinity of the Colorado River," Minute 242. *Department of State Bulletin* 69 (September 24):395–96.

Kaufman, Ken. January 1980. Personal communication.

Kirschten, J. Dicken. 1977. "Plunging the Problems from the Sewage Treatment Grant System." *National Journal* 9 (February 5): 149–53.

Klarich, Duane A., and Jim Thomas. 1977. *The Effect of Altered Streamflow on the Water Quality of the Yellowstone River*

Basin, Montana, Technical Report No. 3, Yellowstone Impact Study. Helena: Montana Department of Natural Resources and Conservation, Water Resources Division.

Kurtz, Howie. 1979. "N.M., Ariz. Waters Probed for Effect of Radioactive Spill." *Denver Post,* September 2, p. 61.

Montana Department of Mines, Billings. April 1979. Personal communication.

New Mexico Environmental Improvement Agency, Staff. November 1977. Personal communication.

North Dakota Surface Mine Reclamation Act, North Dakota Century Code, Chapter 38-14.1.

Pernula, Dale. 1977. *City of Gillette/Campbell County: 1977 Citizen Policy Survey.* Gillette, Wyo.: Gillette/Campbell County Department of Planning and Development.

Price, Don, and Ted Arnow. 1974. *Summary Appraisals of the Nation's Ground-Water Resources—Upper Colorado Region,* U.S. Geological Survey Professional Paper 813-C. Washington, D.C.: Government Printing Office.

Resource Conservation and Recovery Act (RCRA) of 1976, Pub. L. 94-580, 90 Stat. 2795.

Rowe, Jerry W., and David B. McWhorter. 1978. "Salt Loading in Disturbed Watershed— Field Study." *Journal of the Environmental Engineering Division, Proceedings of the American Society of Civil Engineers* 104 (No. EE2, April):323–38.

Safe Drinking Water Act of 1974, Pub. L. 93-523, 88 Stat. 1660.

Shaw, Bill. 1976. *Environmental Law.* St. Paul, Minn.: West.

Skogerboe, G. V., and W. R. Walker. 1972. *Evaluation of Canal Lining for Salinity Control in Grand Valley.* Washington, D.C.: Environmental Protection Agency.

Smith, E. S. 1973. "Tailings Disposal— Failures and Lessons." In *Tailing Disposal Today,* edited by C. L. Aplie and G. O. Argall. San Francisco: Miller Freeman.

Surface Mining Control and Reclamation Act of 1977, Pub. L. 95-87, 91 Stat. 445.

Toxic Substances Control Act of 1976, Pub. L. 94-469, 90 Stat. 2003.

U.S. Department of the Interior (DOI). 1979. *Quality of Water, Colorado River Basin,* Progress Report No. 9. Washington, D.C.: DOI.

U.S. Department of the Interior (DOI), Bureau of Land Management (BLM). 1976. *Northwest Colorado Coal: Final Environmental Statement,* 4 vols. Washington, D.C.: Government Printing Office.

U.S. Department of the Interior (DOI), Bureau of Reclamation (BuRec). 1978. *Status Report: Yuma Desalination Test Facility.* Boulder, Colo.: BuRec.

U.S. Department of the Interior (DOI), Bureau of Reclamation (BuRec); and Department of Agriculture (USDA), Soil Conservation Service. 1974. *Colorado River Water Quality Improvement Program: Status Report.* Washington, D.C.: BuRec.

Utah State University, Utah Water Research Laboratory. 1975a. *Colorado River Regional Assessment Study,* part 1: *Executive Summary, Basin Profile, and Report Digest,* for National Commission on Water Quality. Logan: Utah Water Research Lab.

Utah State University, Utah Water Research Laboratory. 1975b. *Colorado River Regional Assessment Study,* part 2: *Detailed Analyses: Narrative Description, Data, Methodology, and Documentation,* for National Commission on Water Quality. Logan: Utah Water Research Lab.

Water Quality Act of 1965, Pub. L. 89-234, 79 Stat. 903.

White, Irvin L., et al. 1979a. *Energy From the West: Energy Resource Development Systems Report,* 6 vols. Washington, D.C.: Environmental Protection Agency.

White, Irvin L., et al. 1979b. *Energy From the West: Impact Analysis Report,* 2 vols. Washington, D.C.: Environmental Protection Agency.

White, Irvin L., et al. 1979c. *Energy From the West: Policy Analysis Report.* Washington, D.C.: Environmental Protection Agency.

CHAPTER 5

Air Quality

The conflict between increasing energy development and protecting the West's generally pristine air has affected and will continue to affect development of the region's energy resources, especially large coal and oil shale reserves. Because the development of uranium, geothermal energy, oil, and natural gas has much less impact on air quality, their development is not likely to be affected significantly by air-quality regulations. Accordingly, in this chapter we will focus on the development of coal and oil shale.

National concerns about the deterioration of the quality of the air led to federal air-quality regulations calling for reduced SO_2 emissions. Installing flue gas desulfurization (FGD) units to clean local, high-sulfur coal was considered by many nonwestern utilities to be a risky and an expensive way to meet federal standards. The use of low-sulfur western coal was an attractive alternative for reducing SO_2 emissions. Together, these regulations and a national energy policy emphasizing increased reliance on domestic resources led to a boom in western coal and to increased exports of both raw coal and the electricity generated by coal-fired power plants located at or near western coal mines.

More recent changes in federal air-quality regulations, however, promise to dampen this boom as well as to affect the development of oil shale. For example, regulations to prevent any significant deterioration of air quality in designated "clean air" areas can limit the production of synfuels and electricity within the region. Policies that require use of the "best available control technology" (BACT) on all new facilities and that promote the use of local coal—even when it has a higher sulfur content—can also lessen the demand for western coal.

In large part the debate over air quality, both nationally and regionally, revolves around a complex and constantly changing set of federal and state regulations. Federal and state standards will play a key role in influencing the rate, location, and type of technology used to produce energy from western coal and oil shale. These regulations are a vital concern to everyone in the region, those who favor accelerated energy development as well as those who wish to see such development limited.

In exploring the issues involved in this debate, we will first present a summary description of the impact on air quality created by the development of western energy resources. Second, we will identify and discuss the major issues which either have arisen or can be expected to arise. Finally, we will focus on alternative policies and strategies for achieving a balance between developing energy resources and protecting the quality of the air. Much of the material in this chapter requires at least a general familiarity with air-quality regulations. For readers unfamiliar with this subject, a brief summary is provided in the third section of this chapter.

Consequences of Energy Development

The extent and type of effects that the development of energy resources has on air quality can vary significantly depending upon which production and conversion technologies are used and where they are sited. Air emissions, emissions controls, and labor requirements are the technological factors that can cause the greatest variation in the impact on the quality of the air. The characteristics of a resource (such as the sulfur content of coal), the background levels of pollutants, the terrain, meteorological conditions, and the proximity of an energy facility to "clean air" areas (such as national parks) are key factors: the impact of energy development on air quality will depend very largely on how these technological and locational factors are combined. In the following section, we will describe some of the most important technological and locational factors and then summarize a few findings from the site-specific and regional-impact analyses. (For a further discussion, see White et al., 1979.)

Air Emission Rates

Coal-fired steam-electric power facilities emit more SO_2, particulates, and NO_x per Btu of energy produced than any of the other technologies considered in this study (see Table 5-1). These data are based on analyses which assume removal of 80 percent of the SO_2 and 99 percent of the particulates.

If no SO_2 controls were installed, the power plant's SO_2 emissions would be about five times greater. Hydrocarbon (HC) emissions from Synthoil coal liquefaction and TOSCO II oil shale facilities are generally higher than those from power plants. Conventional oil and gas extraction produces few residuals, although HC emissions for natural gas production are high in comparison to most of the other technologies. Enhanced oil recovery, using steam generated by burning high-sulfur oil, and geothermal electric-power generation also emit relatively high levels of sulfur, which in the case of geothermal is in the form of H_2S. The emissions from coal and uranium mining are negligible, although blowing dust can be a problem at some times.

Population Growth

The relatively large labor requirements for constructing and operating energy facilities will lead to large, rapid population increases in nearby rural western communities. (See chapters 7 and 8 for a description of some facets of the impact of these major population shifts.) In some cases, this can result in peak ground-level concentrations of particulates, SO_2, and HC (primarily from more automobiles and home heating) that exceed the emissions from the new energy facility itself. This can occur even though the total emissions from urban sources are usually only about 10 percent of what is produced by the energy facility; the urban emissions, however, are released close to the ground where little mixing or dilution of pollutants takes place.

Resource Characteristics

As shown in Table 5-2, the heat and sulfur content of western coal varies considerably with location. Although all of the coal is low in sulfur on a percent-of-weight basis, on a Btu basis, the SO_2 emissions from coal which have a low heating value are not necessarily low. For the six cases shown in Table 5-2, only three would have met previous federal NSPS of 1.2 pounds of SO_2 per million Btu's even if 20 percent of the sulfur were retained in the ash. But, for our site-specific analyses, we assumed that none of the sulfur was retained in the ash. In addition, of course, the low heating value of most western coal would require more coal to be burned and thus could increase plant costs by as much as $49–$74 (in 1975 dollars) per kilowatt (kW) of capacity (Crenshaw et al., 1976).

Table 5-1: *Air Emissions on an Equivalent Energy Basis (pounds per 10^6 Btu produced)*[a]

Conversion Facility	Particulates	SO_2[b]	NO_x	HC
Coal				
Power Plant[c]				
Btu (electric)	0.11-0.50	0.57-1.37	1.40-3.43	0.04-0.06
Btu (thermal)	0.04-0.17	0.20-0.48	0.49-1.20	0.01-0.02
Lurgi Gasification	N	0.05	0.06	0.004
Synthane Gasification	0.001	0.34	0.48	0.01
Synthoil Liquefaction	0.02-0.05	0.04-0.05	0.20-0.25	0.06-0.07
Surface Mine	0.001	0.001	N-0.01	0.001-0.008
Oil Shale[d]				
TOSCO II Oil Shale[d]	0.08	0.03	0.17	0.09
Underground Oil Shale Mine	0.006	N	0.01	0.003
Modified *In Situ* Oil Shale Processing	0.006	0.01	0.04	0.009
Modified *In Situ* Oil Shale Processing with Surface Retort[e]	0.03	0.03	0.14	0.01
Crude Oil				
Conventional Crude Oil Extraction	N	0.002	0.001	N
Enhanced Crude Oil Extraction Steam Injection	0.02-0.05	0.5	0.01	0.005
Enhanced Crude Oil Extraction CO_2 Miscible	0.005	0.15	0.04	0.001
Natural Gas				
Natural Gas Production/ Extraction	N	0.05	0.07	0.10
Uranium[f]				
Underground Mine	N	N	0.0001	N
Surface Mine	0.01	0.0003	0.004	U
Mill	0.001	N	N	N
Geothermal				
Btu (electric)	NA	0.61[g]	NA	NA
Btu (thermal)	NA	0.21[g]	NA	NA

N = negligible. U = unknown. NA = not applicable.

[a]These numbers represent the range of emissions found in the site-specific scenarios; and facilities are assumed to be operating at a full load. Equivalent energy values are based on the Btu/year for the standard size for each facility. These are, in Btu/year: power plant, 6.73×10^{13} (Btu electric), 1.92×10^{14} (Btu thermal); gasification processes and natural gas, 8.21×10^{13}; liquefaction, 1.84×10^{14}; TOSCO II oil shale, 8.87×10^{13}; *in situ* oil shale, 1.01×10^{14}; conventional crude oil, 9.198×10^{13}; underground oil shale mine, 1.67×10^{14}; uranium mine-mill, 2.99×10^{14}; coal mine, 1.92×10^{14}; geothermal, 2.24×10^{12} (Btu electric), 6.4×10^{12} (Btu thermal).

[b]Had scrubbers not been hypothesized for the power plants, these numbers would be about five times larger.

[c]99 percent of particulates removed and 80 percent of SO_2 removed.

[d]No range for oil shale is available because the processes were hypothesized at only one site.

[e]Assumes mined-out shale is processed by surface retorting facility.

[f]Emissions from solutional uranium mining are not shown, but consist of ammonium, ammonium chloride, and uranium oxide.

[g]This number is for H_2S and assumes 90 percent H_2S removal.

Coal-Fired Power Plant in Four Corners Area of New Mexico

Background Pollutant Levels

Another important factor that could affect western energy development is that the background levels of total suspended particulates (TSP), HC, and ozone, which apparently are a product of natural sources, have been measured in some areas at or above the federal ambient air standards. TSP are largely due to windblown dust composed of large particle-sized inorganic materials not considered a health hazard by the EPA. Background concentrations of HC, apparently from trees and bushes, have sometimes exceeded federal ambient air standards in northwestern Colorado (Rasmussen, 1972). The high background levels of ozone found in some parts of the West have not been explained (see, for example, Berry and Amber, 1977:529). Energy development in our eight-state study area may be restricted by these high background levels unless air-quality regulations take these existing conditions into account. Apparently, EPA and some states are moving in this direction, particularly in the case of particulates, ozone, and HC.

Characteristics of Terrain and Meteorological Conditions

In some areas of the West, characteristics of the terrain and meteorological conditions may limit the development of energy resources by contributing to high ground-level concentrations of pollutants. "Plume impactions" occur when stack plumes flow into nearby elevated terrain such as is found in southern Utah and western Colorado. This is most likely to occur when atmospheric mixing is low—that is, when wind speeds are less than five to ten miles per hour, when there is a clear sky, or in the predawn hours. The meteorological phenomenon known as a temperature inversion occurs when the air temperature increases with altitude, thereby inhibiting the vertical mixing of air and the dispersion of pollutants. When either plume impaction or a temperature inversion occurs, the ground-level concentrations of pollutants can quickly exceed ambient air-quality or PSD standards. The frequency of atmospheric inversions varies both seasonally and geographically throughout our eight-state study area; however, the oil shale region of western Colorado has one of the nation's highest frequencies.

Site-Specific Analyses

As noted earlier, we chose six site-specific scenarios as representative of the combination of technological and locational factors that can be expected when western energy resources are developed. Extensive analyses of the impact on air quality were performed for each site-specific scenario to understand the effects of (1) each individual facility, (2) various levels of emission controls on individual facilities, (3) the combination of facilities, and (4) urban emission sources due to population growth. A few selected highlights from these analyses follow.

Table 5-1 indicates that power plants emit considerably more pollutants than do other facilities and that they thus create the most serious air-quality concerns. Figure 5-1 shows the percentage of sulfur removal which would be required and identifies the limiting standard for coal-fired power plants to meet all federal and state standards at each of the six scenario sites. These results suggest that facilities in Wyoming, New Mexico, and possibly Montana are likely to have the fewest air-quality problems since applicable federal and state standards can be met in these three

Table 5-2: Heat and Sulfur Characteristics of Selected Western Coals

Site	Sulfur Content (percent by weight)		Btu Content (per pound)		SO$_2$ Emission Rate[a] (pounds of SO$_2$ per million Btu)	
	Range	Projected for Scenario	Average	Range	20 Percent Sulfur Retention	0 Percent Sulfur Retention
Kaiparowits[b]	0.2-1.4	.5	10,800	10,600-11,000	.74	.93
Farmington[c]	0.6-0.9	.7	8,580	7,930- 9,525	1.49	1.86
Rifle[d]	0.4-0.7	.6	11,220	10,830-11,410	.71	.89
Gillette[d]	0.3-0.9	.6	7,980	7,240- 8,540	1.20	1.50
Colstrip[d]	0.3-1.8	.8	8,870	6,870- 9,500	1.44	1.80
Beulah[d]	0.4-1.1	.8	7,070	6,830- 7,280	1.36	1.70

[a] The range of emission in this table reflects variations in the amount of sulfur retained in the ash. This retained sulfur does not form SO$_2$. Sulfur retention in the combustion of western coals is generally higher than for eastern coals, but the amount of retention is variable. (For additional descriptions, see White et al., 1979a.)

[b] U.S. DOI, BLM, 1976.

[c] U.S. DOI, BuRec, 1976.

[d] Ctvrtnicek, Rusek, and Sandy, 1975.

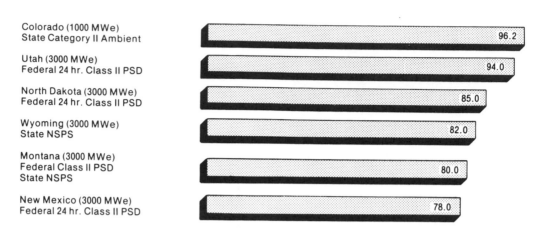

Figure 5-1: *Sulfur Dioxide Removal to Meet All Standards Except BACT*

states with the smallest percentage of SO_2 removed. But in none of the six cases can all federal and state standards be met without the use of some SO_2 emission controls with an efficiency of about 80 percent.

Locating energy development facilities in southern Utah and western Colorado will present the greatest pollution problems unless the amount of SO_2 removed is quite high. In Colorado, this is largely due to the state's SO_2 standards and, to some extent, to the nature of the terrain in the western part of the state. A combination of factors affect the air quality in southern Utah; there the poor potential for dispersion of pollutants and the complex terrain could result in a violation of Class II PSD increments as well as a violation of Class I PSD areas close to the development site.[1]

Table 5-3 shows the peak ground-level concentrations of four criteria pollutants (SO_2, NO_x, HC, and particulates) that have been calculated for power plants and a variety of synthetic fuel plants at our six scenario sites.[2] The table also shows for a point of reference, the most restrictive of the NAAQS or Class II PSD increments. These are not necessarily the most restrictive standards at any particular site, however, as either federal emission standards or a state's standards could be limiting.

As table 5-3 indicates, the Escalante power plant exceeds Class II PSD increments for 24-hour particulates, for 24-hour SO_2, and for 3-hour SO_2.[3] Therefore either more stringent emission controls or smaller plants would be required (as was indicated for SO_2 in Figure 5-1). Although it is not shown in Table 5-3, the Escalante plant would violate all Class I PSD increments except for annual particulates, and thus would have to be located suf-

[1] Class I PSD increments, designed to protect the nation's most pristine areas, allow essentially no additional amounts of SO_2 or particulates. Class II areas are designed to allow some development and growth while keeping air quality cleaner than required by ambient air quality standards. Refer to sections 5.3.1 and 5.3.2 for an elaboration of these regulations.

[2] Criteria pollutants are those for which ambient air quality standards have established peak permissible ground level concentrations. In addition to these four, criteria pollutants also include CO_2 and photochemical oxidants.

[3] The average concentration of the pollutant over these time periods must exceed the standard for a violation to occur.

ficiently far from any Class I areas to allow the dilution of emissions by atmospheric mixing, thus bringing the concentrations of pollutants down to an acceptable level before they reach such areas. Due to the complex terrain in the area, separation distances are difficult to predict, but the 3-hour SO_2 increment for Class I areas may be exceeded in Bryce Canyon National Park, located about 25 miles to the west of Escalante. The power plant at Farmington would not exceed any of the standards shown, while the power plants at Colstrip and at Beulah would violate the SO_2 Class II increments. To prevent these limits from being exceeded, either the size of the plants would have to be reduced or the levels of SO_2 control shown in Figure 5-1 would have to be instituted.

Coal gasification plants are included in Table 5-3 for Farmington (Lurgi) and for Gillette (Synthane). Neither plant violates any of the standards included in the table. However, both would violate the Class I increments for SO_2 and thus would need to be located away from Class I areas. For example, in the Gillette case a separation of less than five miles would be required.

We also considered a synthoil coal liquefaction facility in the Colstrip scenario. As the data indicate, only the 3-hour HC standard would be exceeded, but it would be by a factor of more than 100. Because the HCs are emitted at ground level, the total amount of HC need not be large to produce these high, peak concentrations (which are associated with most liquid-fuel facilities). Most of these released HCs come from leaks in valves and fittings and from storage tanks; they are termed "fugitive" emissions. These short-term violations of the HC standard are not expected to block energy development since the HC standard is interpreted by the EPA as a "guideline" for assessing potential oxidant problems rather than as a rigidly enforced maximum limit. There is concern, however, because some of the HCs emitted are known carcinogens and thus pose a risk for both workers and the public. With the present state of knowledge

Table 5-3: Peak Concentrations for Energy Facilities in the Site-Specific Scenarios[a] (micrograms per cubic meter)

Pollutant and Averaging Time	Federal[b] Standard	Escalante Power Plant	Farmington Power Plant	Farmington Lurgi	Rifle TOSCO II	Rifle In Situ w/surface retort	Gillette Synthane	Gillette Power Plant	Colstrip Synthoil	Beulah Power Plant
SO_2										
Annual	20(II)	11	3.3	0.3	1.3	0.3	2.3	2.7	2.4	1.8
24-Hour	91(II)	293	65	4.6	22	5.8	32	87	17	112
3-Hour	512(II)	1,060	454	44	122	37.7	324	657	92	692
Particulates										
Annual	19(II)	4	1.8	neg.	2.0	1.0	neg.	0.5	0.6	1.4
24-Hour	37(II)	105	36	neg.	83	32.6	0.1	17	9	26
NO_2										
Annual[c]	100(N)	29-49	6.5-11	0.3	2.8	3.6	3.3	3.6-6.0	4	12-14
HC										
3-Hour	160(N)	58	46	57	38,500	128	114	43	17,200	50

[a] The facility sizes are as follows: all power plants are 3,000 MWe; Lurgi and Synthane plants are 250 MMcfd; TOSCO II plant is 50,000 bbl/day; Synthoil plant is 100,000 bbl/day; in situ oil shale with surface retorting is 57,000 bbl/day. All power plants assume 80 percent effective SO_2 removal and 99 percent effective particulate removal.

[b] The standards listed here are the most restrictive of either federal PSD Class II increments (II) or NAAQS (N). These are provided simply for a convenient reference point.

[c] The range of NO_2 for power plants is due to assumptions about NO_x removal by scrubbers; one case assumes 40 percent removal while the other assumes no removal.

it is not possible to quantify this risk, although it is thought to be small. Although not shown in Table 5-3, the Synthoil plant would also violate Class I increments for SO_2; a separation of approximately 13 miles would be required from such areas.

Finally, Table 5-3 includes two oil-shale production facilities located at Rifle, Colorado: one is a 50,000 bbl/day TOSCO II plant and the other is a 57,000 bbl/day *in situ* production facility using surface retorting of the mined shale. Although the HC standard is greatly exceeded for the TOSCO II plant, such violations for fugitive HC emissions, as discussed above, are not necessarily indicative of a major air pollution problem. The Class II PSD increments for 24-hour particulates are also exceeded for the TOSCO II plant; and in addition, although not shown, the Class I PSD increments for 3-hour and for 24-hour SO_2 and Colorado's 3-hour SO_2 standard (100 micrograms per cubic meter [$\mu g/m^3$] all are violated. It was determined that by raising the stack height from 200 to 300 feet, the Class I increments could be met; Colorado's 3-hour SO_2 standard, however, would still be exceeded. On the other hand, the *in situ* facility would meet all federal NAAQS and PSD Class II increments. However, the Class I PSD increments for 24-hour particulates and for 24-hour and 3-hour SO_2 would be exceeded.

Visibility in the vicinity of energy development facilities will also be affected. Table 5-4 shows the worst-case, short-term effects on visibility that would result from a conversion of SO_2 to sulfates at the rate of 1 percent per hour for the various facilities and combinations of facilities considered at our six sites. As the data show, in many cases the reduction in visibility can be quite high—80 to 90 percent. These worst-case meteorological conditions can be expected to occur a few times each year. An analysis of the average reduction in visibility for the six site-specific scenarios shows about an 8 percent drop—that is, on an average day, visibility will be reduced by 8 percent.

Regional Impact Analysis

The impact on air quality for the region is dependent on the existing quality of the air and on meteorological conditions as well as the overall level of energy development and the amount of other industrial emissions. The existing ambient air quality in our study area appears to be good when considered on the basis of annual averages. However, as indicated previously, the federal 24-hour primary standard for particulates is periodically violated due to wind-blown dust. Also short-term oxidant and HC concentrations, apparently from natural sources, approach the federal standards.

Table 5-5 gives the projected emissions from the energy facilities and associated population for the Northern Great Plains subregion (Wyoming, Montana, and North Dakota) and the Rocky Mountain subregion (Colorado, Utah, and New Mexico). These emission levels for the four criteria pollutants are based on the Low Demand scenario (see Chapter 2) for the year 2000. In addition, Table 5-5 shows the increases above the 1975 levels attributable to energy development.

Overall, emission levels in the Northern Great Plains are projected to be higher than in the Rocky Mountain region. This is due primarily to our assumption that more coal-fired power plants will be sited in the Northern Great Plains than in the Rocky Mountain states. In the Northern Great Plains, the largest increase above the 1975 level will be in NO_x which is predicted to be 3.73 to 5.47 times greater in the year 2000. This range in NO_x emissions is the result of current uncertainty about how much NO_x will be removed by scrubbers; we have assumed a range of 0 to 40 percent removal. Increases in emissions in the Rocky Mountain region range from 2 percent (for HC) to between 36 and 51 percent (for NO_x) above the 1975 levels.

The previous discussion and the data in Table 5-5 deal only with emissions from energy facilities. A somewhat different picture is obtained if all major emission sources (e.g.,

Table 5-4: *Worst-Case, Short-Term Effects on Visibility*

Facility[a]	Site	Background Visibility (miles)	Worst-Case Short-Term Visibility[b] (miles)	Percent Visibility Reduction
Coal fired	Kaiparowits	70	8.6	87.7
power plant	Rifle	60	43.6	27.3
Coal-fired power	Gillette	70	9.6	86.3
plant and mine	Beulah	60	4.8	92.0
Synthane gasification	Gillette	70	59.5	15.0
and mine	Colstrip	60	48.9	18.5
Synthoil liquefaction and mine	Gillette	70	48.8	30.3
TOSCO II oil shale	Rifle	60	44.4	26.0
Lurgi and Synthane gasification, coal-fired power plant,	Farmington	60	9.3	84.5
and mines	Colstrip	60	8.1	86.5
Synthane gasification and mine (2 plants)	Beulah	60	29.4	51.0

[a]Facilities modeled are 3,000 MWe coal-fired power plants, (except for Rifle, where a 1,000 MWe plant was modeled), 250 MMcfd Lurgi gasification, 250 MMcfd Synthane gasification, 100,000 bbl/day Synthoil liquefaction, 50,000 bbl/day TOSCO II oil shale and the associated mine. The power plants were modeled with 99 percent removal of particulates and 80 percent removal of SO_2.

[b]Short-term impact on visibility was investigated using a "box-type" dispersion model. This particular model assumes all emissions occurring during a specified time interval are uniformly mixed and confined in a box capped by a lid or stable layer aloft. A lid of 500 meters has been used through the analyses. The conversion rate of SO_2 to sulfur was assumed to be 1 percent per hour.

Table 5-5: *Emissions for the Northern Great Plains and Rocky Mountain Regions: Low-Demand Scenario (thousands of tons per year)*

	Particulates	SO_2	NO_x	HC
Northern Great Plains				
In 1975	471	1,123	338	438
Projected in 2000:			1,221.5-1,809.3	
Energy facilities	124.9	788.7		43.8
Population	1.0	5.2	39.2	48.0
Total & increase above 1975[a]	596.9 (+27%)	1,916.9 (+71%)	1,598.7-2,186.5 (+370%-550%)	529.8 (+21%)
Rocky Mountain				
In 1975	414	712	472	489
Projected in 2000;				
Energy facilities	43.4	65.7	159.1-231.7	-1.7[b]
Population	0.3	1.5	11.1	13.6
Total & increase above 1975[a]	457.7 (+11%)	779.2 (+9%)	642.2-714.8 (+36%-51%)	500.9 (+.2%)

[a]The increase above 1975 (obtained by dividing the total by the 1975 value) is given in parentheses.

[b]This number is negative because oil and natural gas production will decline from the 1975 level.

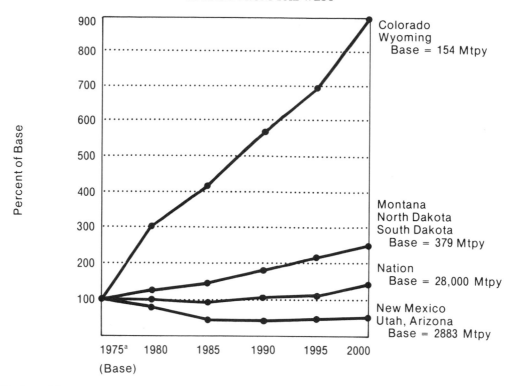

Figure 5-2: *Projected Growth of Sulfur Dioxide Emissions*

[a]1975 base SO_2 emissions were as follows: National, 28.17 million tons per year; Subregion I, 379.7 thousand tons per year (Mtpy); Subregion II, 154.1 Mtpy; and Subregion III, 2,883.0 Mtpy.
(Note: These projections are based on a Nominal-Dirty scenario defined as follows: 4.2 million bbl/day of shale oil by the year 2000 rather than the 2.5 million bbl/day assumed in the Nominal-Demand scenario described in Chapter 2. Emission control assumptions correspond to pre-1977 State Implementation Plans, with NSPS becoming effective in 1979, except in Arizona, Colorado, New Mexico, Utah, and Wyoming, where stricter state standards apply after 1979.)

smelters) are included. Figure 5-2 shows the projected levels of SO_2 emissions nationally and in the eight-state study area through the year 2000 for the highest level of coal and oil shale development considered (see the notes for Figure 5-2). These data were developed using EPA's Strategic Environment Assessment System (SEAS) model (see Ball, 1977). While emissions of SO_2 through the year 2000 are projected to increase nine-fold in the Central Mountain region and by 149 percent in the Northern Great Plains, they will grow by only 18 percent for the nation as a whole and will be cut in half in the Southwest. SO_2 emissions in Colorado will grow substan-

tially due to the development of coal and oil shale. However, these increased emissions are more than offset by the tightened controls on copper smelting (primarily in Arizona), the source of 93 percent of the 1980 SO_2 emissions. Although not shown in Figure 5-2, of the five air pollutants we studied (SO_2, NO_x, HC, H_2S, and particulates), NO_x emissions are projected to have the greatest relative increase in the eight-state study area, reaching 450 percent of their 1975 level by the year 2000.

The long-range movement of sulfates and fine particulates could extend pollution problems over considerable areas, at great distances from energy facilities, and air quality outside the region could be adversely affected. Although the extent to which such long-range movement will occur is not known, regulations to resolve air pollution problems across political jurisdictions will almost certainly add to intergovernmental conflicts over air-quality control.

In addition, changes in precipitation due to local weather modification or increased acidity may occur, particularly if several energy facilities are sited close together. However, current knowledge is inadequate to support conclusions about the probability and extent of such an impact.

Problems and Issues

Public concern about the quality of our air changed significantly during the 1960s. By 1970 a majority of Americans considered air pollution to be one of our two most important domestic problems (Jones, 1975:144–45). This public concern was largely the result of a growing awareness of the effects of air pollution on human health, property, and the natural environment. This concern increased public support for greater governmental control of air emissions and protection of ambient air quality. This has led to an evolving web of laws, regulations, and court decisions which is at once complex and controversial. And in no area of the country is the controversy greater than in the West.

It is important to note at the outset that current knowledge about the relationship between given levels of air emissions and their impact on human health, property, and the natural environment is limited. The lack of scientific agreement on acceptable levels of pollution or even on which pollutants should be regulated has resulted in a continuing debate among spokesmen for industry, environmental interest groups, and policymakers. In the absence of a complete understanding of the relationship of air pollutant emissions and the impact on air quality, the debate over an

appropriate policy tends to polarize people. On the one hand, there are those who argue that the prudent thing to do, given this uncertainty, is to set very strict standards which ensure that we do not do irreversible damage to human health and to the environment. On the other hand, there are those who argue that we should not pay the high economic and regulatory costs of strict standards until we know that they are, in fact, needed. In short, positions in the debate over air quality are based as much on values and perspectives as on hard knowledge.

The political system has responded to these conflicting pressures incrementally by modifying existing regulations and by periodically adding new ones. This has produced a complex of overlapping, sometimes contradictory regulations which makes it possible for some elected officials in the West to view the existing regulatory system as a barrier to the region's economic development while others believe that the same system provides inadequate protection. And it is clear that this complex system of air-quality regulations offers a convenient vehicle for those who wish to challenge the development of western energy resources—with their challenge zeroing in on air-quality regulations as a convenient surrogate for some other concern.

Before describing some specific issues concerning air quality that are associated with energy development, we will first briefly describe current regulations. Then we will identify some of the major problems and issues which arise because of conflicts between protecting the quality of the air and developing western energy resources.

Summary of Current Air-Quality Regulations

Although in the following discussion we will attempt to put some logic and order into the descriptions of the air-quality regulatory system, in fact it is a maze. If the reader comes away with the feeling that he understands the regulatory system, we will have failed because we will have created the illusion of order where there is, in fact, chaos.

The existing air-quality regulatory system consists of two types of standards: emission standards limit the amount of pollution that can be emitted by any source (usually expressed in pounds per unit of energy input); and air-quality standards limit the maximum concentration of certain pollutants that can exist at ground level (usually expressed in $\mu g/m^3$). Each kind of standard is identified and briefly described below, with the emphasis on those that are most important for the areas in which western energy resources are being developed.

Emission Limits:

- Federal NSPS: These standards set the maximum allowable emission rates for certain pollutants (lbs/10^6 Btu's fuel input) for each specific type of facility (e.g., coal-fired power plants and coal preparation plants). In addition, the recent CAA Amendments (1977:§109) now require use of the best available system of continuous emission reduction on new facilities regardless of the quality of fuel being used (i.e., BACT). In mid 1979 EPA promulgated the BACT regulations for SO_2 emissions from coal-fired power plants: all new plants must now remove from 70 to 90 percent of the SO_2, depending on the sulfur content of the raw coal.

- State Implementation Plan (SIP) Limits: These plans require that emission standards be set for various categories of sources. The state standards for new or modified sources must be at least as strict as the federal NSPS. (For a description of these state standards, see White et al., 1979b:chap. 2.) Five of the eight states in our study area have standards for new plants that are more strict than the previous federal NSPS (1.2 pounds of SO_2 per million Btu's).

- Offsets in Nonattainment Areas: An "offset policy" exists to avoid inhibiting economic development in areas that do not meet federal ambient air-quality standards (i.e., "nonattainment" areas). Briefly, the requirements of this policy include the use of the "lowest achievable emission rate" on new sources and reductions in emissions from other sources, so that the amount of reductions more than offset the amount of emissions from the new source (CAA Amendments, 1977:§129).

- Hazardous Air Pollutants: These regulations establish emission standards for exceptionally toxic pollutants such as asbestos, beryllium, mercury, and vinyl chloride. The 1977 CAA Amendments allow EPA to set standards for design, equipment, work practices, or operations with hazardous materials when emission standards are not feasible (CAA Amendments, 1977:§110).

- Mobile Emission Standards: These standards establish maximum emission rates for carbon monoxide (CO), HC, and NO_x from motor vehicles (lbs/vehicle-mile).

Air Quality Limits:

- National Ambient Air Quality Standards: Two categories (primary and secondary) of maximum permissible ground-level concentrations of pollutants are established by the NAAQS. At the present time, there are standards for six "criteria" pollutants: suspended particulates, SO_2, CO, photochemical oxidants, HC (nonmethane), and NO_x. The current NAAQS are listed in Table 5-6. Primary standards are defined as having the goal of protecting public health, while secondary standards are defined as having the goal of protecting public welfare—that is, property, crops, wildlife, livestock, aesthetics, etc.

- State Ambient Air Quality Standards: As

Table 5-6: *National Ambient Air Quality Standards*

Pollutant	Averaging Interval	Primary Standard ug/m³ (ppm)	Secondary Standard ug/m³ (ppm)
SO₂	Annually	80 (0.03)[a]	—
	24-hr	365 (0.14)[b]	—
	3-hr	—	1,300 (0.5)[b]
Particulate Matter	Annually	75[c]	60[c,d]
	24-hr	260[b]	150[b]
CO	8-hr	10,000 (9)[b]	10,000 (9)[b]
	1-hr	40,000 (35)[b]	40,000 (35)[b]
Photochemical Oxidants	1-hr	160 (0.08)[b]	160 (0.08)[b]
HC (nonmethane)	3-hr (6-9 A.M.)	160 (0.24)[b,c]	160 (0.24)[b,e]
NOₓ	Annually	100 (0.05)[a]	100 (0.05)[a]

Source: 40 C.F.R. §50, 1978. EPA has now promulgated primary and secondary NAAQS for air-borne lead of 1.5 ug/m³, quarterly arithmetic mean (Fed. Reg., 1978). The CAA Amendments (1977: §122) direct EPA to study radioactive pollutants, cadmium, arsenic, and polycyclic organic matter for possible inclusion as pollutants covered by the NAAQS.

[a] Arithmetic mean.

[b] Maximum concentration not be exceeded more than once a year.

[c] Geometric mean.

[d] For use as a guide to assess SIP's plans to achieve the 24-hour standard.

[e] For use as a guide in devising SIP's plans to achieve oxidant standards.

part of the SIP requirement, each state sets ambient air-quality standards that must be at least as strict as the NAAQS. Each of the eight states in our study area has complied with this requirement, and some have set standards for pollutants other than the six covered by federal regulations. Seven of the states (all except Utah) have set limits on SO₂ that are below the federal primary standards for the annual arithmetic mean and for the maximum 24-hour concentration.

• PSD Standards (CAA Amendments, 1977:§127): Three classifications of land are established by PSD standards limiting allowable increases in particulates and SO₂ in areas where the existing air quality is cleaner than is required by NAAQS. Class I areas would be allowed the smallest increases. Certain areas, such as na-

tional parks that are larger than 6,000 acres and national wilderness areas, are mandatory Class I areas. Figure 5-3 shows the mandatory Class I areas in our eight-state study region. All other areas are initially designated as Class II. They can be redesignated as either Class I or the less restrictive Class III by the states or by Indian tribes, except that some areas, such as national monuments, national preserves, and national wilderness areas, cannot be redesignated as Class III if the area in question is larger than 10,000 acres. Within two years of passage of the 1977 CAA Amendments, EPA must submit plans for PSD from NOₓ, HC, CO, and oxidants. Table 5-7 gives the specific allowable PSD increments for both particulates and SO₂ for each of the three classifications. Except for the Class I 18-day variance case, the increments may be

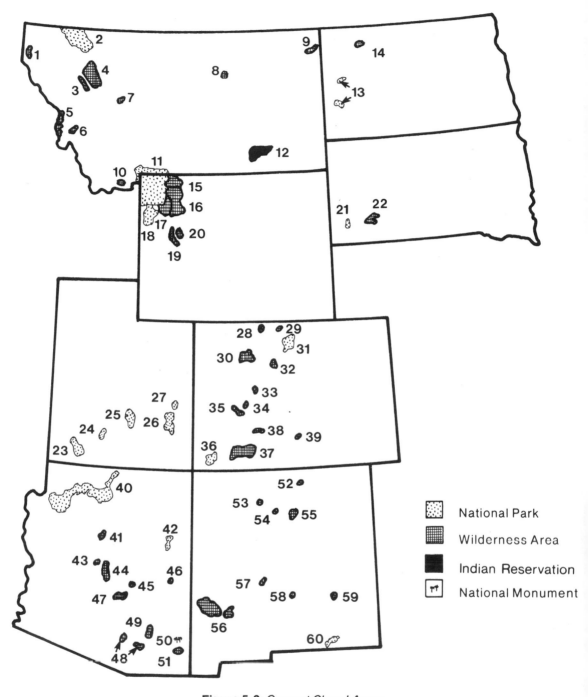

Figure 5-3: *Current Class I Areas*

Table 5-7: *Prevention of Significant Deterioration:*
Allowable Increments^a (in micrograms per cubic meter)

PSD Classification	TSP		SO$_2$		
	Annual	24-Hour	Annual	24-Hour	3-Hour
Class I	5	10	2	5	25
Class I "relief"	19	37	20	91	325
Class I 18-day variance:					
low terrain	—	—	—	36	130
high terrain	—	—	—	62	221
Class II	19	37	20	91	512
Class III	37	75	40	182	700

Source: CAA Amendments, 1977, as cited in Garvey et al., 1978.

[a] Except for the Class I 18-day variance case, the increments may be exceeded only one day per year.

exceeded only one day per year. The Class I "relief" and 18-day variance cases are allowed only after a review and approval process specified in the legislation.

- Visibility Protection for PSD Mandatory Class I Areas: Under the 1977 CAA Amendments, EPA has two years to develop regulations for remedying or preventing the impairment of visibility in mandatory Class I areas as a result of man-made air pollution.

It should be noted that these standards are not independent; the applicable standard for any facility will vary from case to case depending on the characteristics of the technology, the type of fuel, when the plant was constructed or modified, and its location. For example, emissions from a coal-fired power plant may be constrained in one site by SIP limits and in another site by PSD regulations. The regulations are even more difficult to understand and implement because they are continually being revised and reinterpreted by the states, EPA, or the courts.

In summary, the air-quality regulatory system is highly complex and controversial. Especially since the 1973 Arab oil embargo, there has been an intense public debate over the conflict between increasing the production of domestic energy and maintaining and en-

hancing air quality (and other environmental values). While many of the issues are national in scope, some are especially important to the West. In the following sections, four major areas of conflict involving air quality and the development of western energy will be described. These four are: PSD and visibility standards, emissions offset policies, BACT requirements, and the role of the various levels of government in regulating air quality.

PSD and Visibility Standards

The large number of Class I PSD areas in some parts of the eight-state region, the complexity of the PSD regulations, and uncertainty about future visibility regulations could block or slow the development of western energy resources.

The concept of PSD was first included in the Air Quality Act (1967) and was continued by Congress in the CAA Amendments (1970). However, the administrator of EPA did not include PSD in his 1971 implementation guidelines. His failure to do so was successfully challenged by the Sierra Club on the basis that the 1967 Act, the 1970 Amendments, and the legislative history of air-quality control required a policy of preventing the significant deterioration of air that was of better quality than that required by NAAQS (*Sierra Club* v. *Ruckelshaus,* 1972, 1973). PSD regulations were established by EPA in December 1974 and were immediately challenged by both industry and environmental groups (*U.S.C. Cong. and Admin. News,* 1978:1183; Ayers, 1975:460–64).

Congress actively considered PSD amendments to the CAA beginning in 1975, but it was not until 1977 that the PSD requirements (see Table 5-7) were actually established (CAA Amendments, 1977:§127).

Congress stated that the purpose of the PSD requirements was to protect public health and welfare; to preserve, protect, and enhance air quality in unique national lands; to ensure that emissions in one state would not interfere with another state's SIP; and to ensure that decisions to permit increased air pollution would be made only after careful evaluation and informed public participation (CAA Amendments, 1977).

The PSD requirements of the 1977 Amendments create a number of problems and questions that could significantly affect western energy development. Under the PSD regulations, a facility is given a permit to begin construction if predictions from an atmospheric diffusion model[4] indicate that the facility will not violate the PSD standard. This raises concerns because the current air-quality models generally can estimate the impact on air quality only to within a factor of two of the actual impact that will occur; and in areas of complex terrain or unusual meteorology these models are even less accurate.

Because of this limitation, the 1977 Amendments require EPA to hold a conference on air-dispersion modeling within six months of passage of the amendments and at least every three years thereafter. In addition, as noted previously, the amendments call for EPA to

study the four other criteria pollutants that are not now covered by PSD (NOx, HC, CO, and oxidants) and to set PSD regulations for these pollutants within two years. However, the interrelationships among all six criteria pollutants and their specific effect on the deterioration of air quality (for example, the formation of oxidants) are not well understood and with existing methods cannot be determined accurately. EPA has contended that it has neither the technology nor the modeling capability to regulate these pollutants on a case-by-case basis (*U.S.C. Cong. and Admin. News,* 1978:1485–86). The net effect of these inadequacies, therefore, is to increase uncertainty about future regulations and to delay—and perhaps constrain—some energy development.

Another potential problem has to do with the process of reclassifying PSD areas as specified in the 1977 Amendments. The process, which can be initiated by a governor or an Indian tribal council, requires a public hearing in the locale affected, approval by the state, and, when federal lands are affected, consultation with the appropriate federal land manager.

The attempt of Dunn County, North Dakota, to be redesignated as Class I illustrates some of the problems in this process. A petition requesting the required approval of local government was circulated, the petition was signed by a large number of people, and the county government passed an appropriate resolution. But the process then stalled because the county lacked sufficient funds for the required environmental assessment. Requests to the state and to EPA for funds were unsuccessful, so no further action was taken (Metzger, 1977).

Another example of the reclassification process in action is the redesignation of the Northern Cheyenne Indian Reservation in eastern Montana (see box, "Reclassification Conflict") to Class I (*Old West Regional Commission Bulletin,* 1977). According to the Northern Cheyenne, the request was made to protect their values and lifestyle and to prevent any

---

[4]These models are usually computer simulations which predict changes in ambient air quality based on the volume, rate of discharge, and temperature of emissions from the facility, its stack height, and the meteorology and terrain of the surrounding area.

further degradation in air quality (ES&T, 1977). On the other hand, the tribal chairman of the adjacent Crow Reservation, Patrick Stands Over Bull, wrote to EPA arguing that the Northern Cheyenne's request for redesignation of the reservation as a Class I area "has serious ramifications on coal development on the Crow Reservation" (*Business Week*, 1977). The state of Wyoming also filed an objection to the redesignation request by the Cheyenne Indians on the grounds that it could limit energy development in northern Wyoming (*Business Week*, 1977). Another objection came from Montana Power Company, concerning construction of Colstrip 3 and 4 near the reservation. EPA attempted to require a PSD permit for the facilities, but Montana Power successfully sued EPA on the grounds that these units were already under construction when the classification rules were passed and were, therefore, exempt. The case is currently under appeal. Montana Power attempted to begin site preparation in late 1977 but was blocked by EPA because the company did not have a permit, even though the court had ruled that one was not needed. In order to save time while the case was being appealed, Montana Power filed for a PSD permit even though the company believed that one was not required. On June 12, 1978, EPA issued its final decision not to grant a construction permit for Colstrip 3 and 4 because the plant would violate the Class I PSD standard on the nearby Northern Cheyenne reservation. More recently, however, EPA did agree to grant a PSD permit on the condition that more stringent SO_2 emission controls be instituted.

Another potential problem with the PSD requirements involves wind-blown dust from surface mining operations. If dust is included in the allowable PSD increment, it is likely to cause the particulate PSD Standard to be exceeded in a great many cases. Currently, EPA is planning to exclude these fugitive dust emissions from the PSD increment, but this policy may be challenged in the courts by environmental groups.

Reclassification Conflict

The Northern Cheyenne tribal chairman, Allan Rowland, has stated that "we are not requesting this redesignation because we are against progress. For us progress means developing our environmental resources in renewable and compatible manners such as timber and agricultural products. They are the cores of our value system as people."

When the State of Wyoming objected to the redesignation citing possible restrictions on development in some parts of the state, Eric Metcalf, a Cheyenne spokesman, replied, "They're saying: 'We want to make decisions in our area, but you can't make those decisions for yours.'"—Business Week, 1977.

Another major point of conflict concerning PSD occurs because of the large number of mandatory Class I areas in the eight-state region, especially in southern Utah and western Colorado (see Figure 5-3). This means that the PSD regulations could limit the region's economic development: the situation with the Intermountain Power Plant in Utah is a good example. This facility will utilize water from the Fremont River which is too low in quality to be suitable for any other beneficial use. However, the site is only eight miles from a mandatory Class I area, Capitol Reef National Park, and the plant would probably violate Class I PSD standards in the park. State Sen.

Table 5-8: *Estimates of Required Separation Distances from Class I PSD Areas*

Facility	Size	Separation Distance (miles)
Synthoil—Farmington	100,000 bbl/day	16
Power Plant—Rifle	1,000 MWe	14
Lurgi—Gillette	250 MMcfd	7
Synthoil—Colstrip	100,000 bbl/day	13
Power Plant—Beulah	3,000 MWe	75

Colorado Particulate Study

The Colorado Senate Committee on Health, Environment, Welfare and Institutions passed a resolution (subject to Senate approval) asking the Colorado Air Pollution Commission to study the health effects of air borne particles and suggest air quality standards. The goal is to find a way in which large dust particles that naturally occur could be discounted when air quality is measured and perhaps allow more industrial development in the southern part of the state.

State Senator Harvey Phelps argued that if naturally occurring dust particles, common to an arid climate, are included in the measurement of air quality, some areas of the state will always exceed air pollution standards. "Then we always come out high and aren't allowed to grow, while growth goes on in dangerously polluted areas."—Denver Post, 1978.

Ernest H. Dean has pointed out that this is very frustrating for the people of Utah, largely because the park's public facilities are of poor quality and the park is seldom used; on this basis, Senator Dean says, concern about its air quality does not seem to be justified (Dean, 1977).

To get an idea of the effects that PSD Class I areas have on energy development, Table 5-8 shows estimated separation distances that would be required for individual facilities of various types and at different locations in the West. As we have already indicated, certainly Class I increments pose the most serious problem for large power plants—a separation of 75 miles is estimated for a large power plant at Beulah. But notice, for example, that the Synthoil plant at Colstrip would require a separation distance of 13 miles—just slightly less than the distance from the town of Colstrip to the Northern Cheyenne Reservation which has been redesignated to a Class I status. This 13-mile separation is to allow for dispersion of the SO_2 emissions, and is based on the emissions from this one facility only—in fact, the power plant that is already at Colstrip and the additional units being built there will

consume most of the Class I SO_2 increment; thus, any additional facilities would have to be located far enough away so as to add, in effect, no additional SO_2. Since the Colstrip area is a likely choice for several large coal conversion facilities, it is clear that the PSD standards could limit development in the area.

It should also be noted that the calculations on which Table 508 is based do not include either the question of visibility or future regulations covering other pollutants. Given these factors, the fact that additional areas can be redesignated as Class I, and the current inadequacy of the air-diffusion models used to predict PSD violations, it clearly is not possible at this time to make a reliable assessment of the extent to which PSD regulations could constrain energy development in the West.

Emissions Offset Policies

Some westerners resent the fact that the emissions offset policy allows development in areas already violating NAAQS while the PSD policies could hinder development in areas where the air quality is, at present, better than is required by NAAQS.

An emission offset policy to permit industry in areas where national air-quality standards have not been met ("nonattainment" areas) was formulated by EPA in 1976 and was included in the 1977 CAA Amendments. This policy was intended to permit the growth of clean industry in polluted areas. The alternative seemed to be to prohibit all industrial development in all nonattainment areas until air-quality standards were met. As described previously, under the offset policy, new emissions sources are permitted if existing sources reduce their emissions enough to more than compensate for emissions from the new facility. The new facility (or major addition to an existing facility) must also meet a "lowest achievable emission rate" requirement (CAA Amendments, 1977:§129).

In our eight-state study area (and in other areas as well), the emissions offset policy generates controversy because it is perceived as having inequitable effects on different regions of the country (McKee, 1978:602–3). There are several reasons for this feeling. One reason that is important for the West has to do with the high levels of some pollutants, especially particulates and HC, that occur naturally in some areas. If these "naturally occurring" levels are included in ambient air quality measurements and the NAAQS are exceeded, there could be no emission sources to use as an offset for emissions from new industries. Thus, economic development in these areas would effectively be blocked, while development could be allowed in industrialized nonattainment areas (see box). The key issue revolves around uncertainty as to how emissions from naturally occurring sources will be counted when ambient air-quality standards are measured (Palomba, 1978:575).

Another reason for the controversy is that the emissions offset policy is viewed by some westerners as favoring economic growth in the already highly polluted areas over the region's "cleaner" areas. In areas with large numbers of "dirty" sources, a considerable amount of new economic growth could be allowed by reducing emissions from existing sources. Some people believe that economic growth in clean air areas of the West will be constrained to a greater degree than in polluted areas by Class I or Class II PSD increments.

BACT Requirements for SO2

Although BACT requirements will help to ensure a high level of air-quality protection, they could sharply reduce the short- and mid-term demand for western coal. They are also controversial because the control technologies are of unknown reliability, are expensive, reduce power plant efficiency, increase water consumption, and create a large amount of sludge waste.

The previous NSPS for SO_2 of 1.2 pounds per 10^6 Btu's established in 1971 by EPA for large, coal-burning facilities, was a major contributor to the boom in demand for western coal. The reason was that much of the western coal has such a low sulfur content that it can be used without, or with only limited, sulfur controls and still meet the NSPS (see Table 5-2). Therefore, many utilities in the Midwest and Southwest made long-term contracts for western coal (primarily Northern Great Plains coal). They calculated that it was cheaper to buy this inexpensively mined coal and ship it very long distances than to buy the higher sulfur local coal and install FGD units.

By 1977, however, Congress decided that this unintended side effect of the NSPS was undesirable and in the CAA Amendments developed two policies intended to produce a more balanced demand for western, interior, and eastern coal. The first of these policies is the NSPS requirement for some degree of active emission control regardless of the original sulfur content of the coal (i.e., BACT). The second is a provision that allows a governor to require utilities in his state to use locally mined coal rather than importing it from another state or region (CAA Amendments, 1977:§122). Of course, the BACT requirement is also intended to help protect air quality.

The 1977 CAA Amendments contain several major points concerning the NSPS provisions (see Krohm, Dux, and Van Kuiken, 1977). EPA is to review and revise the NSPS every four years. The law requires a substantial improvement in the standards of performance by the major sources of pollution through mandatory use of the "best technological system of continuous emission reduction" (BACT). Designed to take costs, health impacts, environmental impacts, and energy requirements into consideration, BACT is defined as either a process which is inherently low-polluting or nonpolluting or a system for the continuous reduction of pollution before the pollutants are emitted, including precombustion treatment. EPA may or may not require any particular technological system of control such as FGD. However, the use of untreated, low-sulfur fuel and certain other strategies, such as intermittent control systems, are precluded as a means of compliance. The new standards must include both an emission limitation and a percentage reduction of SO2, particulates, and NOx. Congress left the exact quantitative expression of these standards to EPA.

After considerable debate and controversy, EPA issued the final BACT regulations for large coal-fired power plants in May 1979 *(EPA Environmental News, 1979)*. The standards require SO2 emissions to be reduced by 70 to 90 percent, with a maximum allowable emission rate of 1.2 pounds of SO2 per million Btu's. The 90 percent level applies to any plant where controlled emissions are more than 0.6 pounds of SO2 per million Btu's, while plants with a lower level of controlled emissions are permitted to reduce sulfur by amounts between 70 and 90 percent.

Although several potential emission control processes under development may be used to burn coal cleanly, FGD is the most fully developed technology, and these scrubbers are now capable of removing up to 95 percent of the sulfur emissions. However, the scrubbers that are currently available produce an inert sludge that must be disposed of, usually in on-site evaporative holding ponds.

Some spokesmen for the electric utility industry have argued that SO2 scrubbers aren't reliable, are too expensive, and produce a significant problem of solid-waste disposal. One projection indicates that FGD units on coal-fired boilers could produce 55 million metric tons (dry basis) of sludge annually by the year 1955 (Teknekron, 1978a). FGD units also substantially reduce the efficiency of power plants; typical estimates suggest reductions from 38 percent to 35 percent, a net loss of around 8 percent efficiency. This means that in plants with scrubbers approximately 8 percent more fuel will be required to generate the same amount of electricity.

On the other hand, EPA argues that FGD units have been proven reliable, that their

costs are justified in order to protect human health and the environment, and that strict standards will encourage the development of improved sulfur-control technologies. (These might include "dry" FGD systems, fluidized bed boilers, or regenerable FGD processes that would virtually eliminate the problem of solid-waste disposal.) In fact, in its news release on the BACT standards, EPA emphasized that the requirement for lower SO2 reduction (as low as 70 percent) was designed to encourage the development of "dry" scrubbers. Although dry systems create a larger amount of waste, the disposal problems are considerably less than with wet sludge, and water consumption is also reduced.

Several studies have evaluated alternative BACT definitions and have concluded that the requirement of the 1977 amendments for percentage reductions will achieve the objective of reducing the demand for low-sulfur western coal. For example, one study (Krohm, Dux, and Van Kuiken, 1977) estimates that in 1990 coal production in the Northern Great Plains, under an across-the-board requirement of 90 percent sulfur reduction, will only be 52 percent of the amount that would be produced if the current NSPS were continued—a drop in 1990 from 388 MMtpy to 202 MMtpy. This compares to a 1973 production level of 46 MMtpy. Most of the change would be the result of decreased shipments from the Northern Great Plains to the east central region of the county.

Of course supply/demand projections based on econometric models are laced with uncertainties (e.g., future mining costs in different regions and future transportation costs); and at least one factor that may tend to reduce the magnitude of this potential decrease in demand is impossible to quantify. This is the apparent greater reliability of coal supplies from the predominantly nonunionized surface mines in the West as compared to the mainly unionized, underground mines in the central and eastern coal-producing regions. This is of special concern now after the protracted 1977–78 coal miners' strike.

Governmental Conflicts
in Air-Quality Regulation

Intergovernmental conflicts over air-quality control already have affected relations among the West and other coal-producing regions and among western states, the federal government, and Indian tribes.

This fourth area of conflict differs from the previous three in that it deals not with any particular regulation but rather with the more pervasive problem of the appropriate role of the various levels of government in regulating air quality. Although the question of federal versus state/local control is national in scope, the matter is more complex in the West because of the large areas of federally owned land and the unique political status of Indian tribes.

The key issue centers on whether the federally established national regulations allow enough flexibility to account for regional differences in the natural environment and for the goals of citizens in a particular area. As noted previously, the states are given primary responsibility for monitoring and enforcing air-quality regulations, but generally only within the fairly narrowly defined bounds set by the federal government. The states may set their own emission and air-quality standards but only if they are at least as strict as the federal ones. Given the relatively strict federal standards, such as the PSD and BACT regulations, this results in very little real state control. It should be noted too that the representatives of various states are not always on the same side of the argument; for example, Sen. Jake Garn (R-Utah) feels federal regulations have unduly restricted growth in energy production in his state, while Sen. Gary Hart (D-Colorado) believes that there needs to be greater federal leadership in pollution control (see box).

As noted earlier, the 1977 CAA Amendments give Indian tribes the right to redesignate their reservations as Class I PSD areas. Thus far only one tribe, the Northern Chey-

---

Air of a Different Quality

Senator Jake Garn feels that Utah is ideal for the development of huge "mine-mouth" generating plants at underground mines for the export of electricity to the growing southwest. He feels that using Utah's low sulfur coal and modern control equipment would essentially make such plants nonpollutors and that these developers (namely the $3.5 billion generating complex at Kaiparowits) would have created a new city of 15,000 residents and a 10 percent increase in Utah's tax base. The Senator states, ". . . we should . . . leave it up to the states to meet federal ambient air standards as they see fit. If we don't meet them then the feds can step in and enforce them."

Senator Gary Hart is particularly concerned about the health hazards of increased levels of nitrogen oxides (18 percent by 1985) that would result even with maximum use of scrubbers to reduce sulfur oxide emissions. He does not want Colorado to supply energy for people outside the state while people in his state will be exposed to the public health consequences. Senator Hart feels ". . . the leadership should come from the White House. They've got to evaluate pollution control to the same level of seriousness as the supply side issues . . . California can figure out how to generate its own electricity."— Kirschten, 1977:783–84.

---

enne, has opted for this classification, and we have previously discussed the conflicts this action has created. On the other hand, to date there is no case on record of an Indian tribe formally participating in the formulation or revision of a SIP. Given the legal status of the relationship between a state and an Indian tribe, it is doubtful that existing SIPs can be applied to new or existing sources of air pollution on trust lands.

Summary of Problems and Issues

The impact on air quality due to energy conversion facilities depends on several critical technological and locational factors: type of energy facility and the levels of pollution control, labor requirements, resource characteristics, background levels of pollutants, terrain, proximity to pristine air-quality areas, and meteorological conditions. In general, coal-fired power plants emit more pollutants than do any other energy conversion technology, and thus they give rise to more air-quality problems and issues. Within our eight-state study area, facilities located in the Rocky Mountain region are expected to have the most site-specific problems, especially when sited in complex terrain and close to Class I PSD areas.

The issues concerning air quality are potentially one of the most limiting factors on the development of western energy resources. The national goal of protecting and enhancing air quality comes in conflict in a variety of ways with the goal of increasing domestic energy production and continuing economic growth. Such conflicts have become especially prevalent in the West where vast, readily accessible energy resources can support large-scale energy development, but the development of these resources poses a threat to the region's pristine air.

Table 5-9: *Alternative Policies for Air Quality*

General Alternative	Specific Alternative
Modify standards and regulations	Define BACT for SO_2 emissions from coal combustion as 90 percent sulfur removal.
	Make PSD standards more flexible.
	Reserve a portion of PSD increments for energy production.
Alter technological and siting choices	Construct smaller, dispersed energy conversion facilities.
	Concentrate facilities in energy parks.
	Increase commercialization programs for new, less polluting energy technologies.
Improve procedural mechanisms for approving new facilities	Establish a task force for identifying future sites for energy facilities.
	Increase assistance to state and local government planners.
	Establish a regional air-quality council.

Policy Alternatives for Air Quality

As the two previous sections have shown, western energy development will both affect air quality and be affected by air-quality regulations. And this gives rise to several problems and issues which policymakers must attempt to resolve. In dealing with these problems and issues, policymakers will have to attempt to reconcile energy production and environmental protection; that is, the overall objective will be to safeguard air quality in the West while proceeding with the development of western energy resources.

A variety of alternative responses are available to policymakers to achieve this objective. Three general alternatives and several specific alternatives are listed in Table 5-9. The general alternatives suggest that policymakers can take approaches which emphasize standards and regulations, technology and siting

choices, and/or institutions and procedures. These alternatives are described briefly below, followed by a more detailed description and evaluation of four of the specific alternatives.

The first set of specific alternatives focuses on modifying existing regulations or adding new ones. One alternative would be to set the BACT standard for SO_2 from large coal-burning facilities mandated by the 1977 CAA Amendments at 90 percent reduction with an emission maximum of 1.2 pounds of SO_2 per million Btu's of input. Essentially, this requirement would mean that SO_2 emissions would be as low as is practically feasible, thus assuring a large degree of environmental protection while still allowing for energy conversion plants. The other two specific alternatives in this category are aimed at resolving some of the problems created by current PSD

Scrubber on Coal-Fired Power Plant

regulations. The first would allow higher levels of energy development by relaxing the PSD standards slightly. For example, modifying the PSD regulations to allow a greater number of times that the 3-hour and 24-hour increments could be exceeded in a year would lessen the possible constraint of PSD regulations on energy facilities. Implementation of this policy would require amending the CAA, and it is likely that such a move would be opposed by environmental groups which support strict PSD standards. A related alternative would reserve a percentage of allowable PSD increments for energy conversion facilities. This would ensure that energy development is not blocked but would not relax current PSD standards. Since the need for such "reserved increments" would vary considerably from region to region, percentage levels could be set by the states, possibly as part of their SIPs. This alternative, however, would probably require EPA approval and possibly congressional action and would probably be opposed by other industries.

The second category of specific alternatives listed in Table 5-9 includes policies which would alter choices of technologies and/or sites. Under the proposal for smaller, dispersed energy facilities, for example, instead of building one 1,500 MWe power plant, industry would build three separate 500 MWe power plants sited far enough apart to avoid significant air-emission interactions. By siting smaller facilities over a wide area, the ground-level concentrations of air pollution at any one point would be reduced. This would help reduce the constraints on energy development that result from ambient air-quality regulations. An opposite approach would be to concentrate energy facilities in "energy parks." This alternative would almost certainly sacrifice the quality of the air locally, but such parks could be located away from both major population centers and Class I PSD areas so as to minimize the adverse environmental and health effects. Federal legislation might be required to relax the current PSD regulations for these areas, and states would have to include provisions for energy parks in their SIPs. In the absence of federal action, the states could implement this alternative on a limited scale by redesignating some areas to a PSD Class III status.

A third specific alternative in this category would be to increase commercialization programs for new, less polluting energy resources (e.g., geothermal energy) or conversion technologies (e.g., coal-based synthetic fuels). As indicated in Table 5-1, some of the energy development technologies have fewer air pollution problems than others. Currently, most of these cleaner technological alternatives are quite new and are either still in the research-and-development stage or are considered by industry to be too economically risky for commercialization. The aim of this alternative policy is to make the newer, less polluting technologies commercially attractive earlier than they might otherwise be.

The last set of specific policy alternatives listed in Table 5-9 focuses on the procedural mechanisms by which approval is given for the construction of new facilities. As noted earlier, air-quality regulations and the process for assuring compliance with these regulations is very complex and can often lead to long delays in siting. Under the first alternative in this category, each state in our eight-state study area would establish a task force which would include among its members representatives of the federal government, the state, industry, and other interested parties (e.g., environmental groups, farmers and ranchers, etc.). This task force would be charged with identifying areas which could be used as future sites for energy and other industrial facilities.

The second alternative would increase financial and technical support for interested groups, primarily using resources from the federal government. For example, technical experts could be provided by the EPA regional offices to serve as consultants to state and local governments and to public interest groups, or financial support could be provided to these groups so that they could hire their own expert consultants. This kind of activity has already begun to some limited extent. For example, EPA jointly planned and sponsored a series of meetings in the West with the Friends of the Earth to provide technical assistance, to disseminate information on the 1977 CAA Amendments, and to give EPA a sense of how environmentalists saw the proposed regulations. (For a more general discussion of siting procedures, see Chapter 11.)

A third specific alternative would establish a regional air-quality council to support interstate planning agreements. Federal and state funding could be provided for a western interstate air-quality board for the eight states in our study area. Such a board could be patterned on major existing multistate organizations such as the Western States Water Council or the Western Interstate Energy Board. This council would provide advice on air quality to federal, state, and local governments and would help coordinate multistate air-quality planning and impact assessment.

Together, these nine specific alternatives provide a range of options that might be used to help resolve some of the issues involving energy development and air quality that have been previously identified. In the following sections, four of these alternatives are considered further. These are to:

- Define BACT for SO_2 emissions from coal combustion as 90 percent sulfur removal;
- Construct smaller, dispersed energy conversion facilities;
- Increase commercialization programs for new "clean" energy technologies; and,
- Establish a task force for identifying future sites for energy facilities.

Each of these options will be described in more detail and evaluated in terms of the five basic criteria we have identified and defined in Chapter 2: effectiveness, efficiency, equity, flexibility, and implementability.

BACT for SO₂ Emissions

As indicated above, this alternative involves setting a standard that would require 90 percent sulfur removal and a maximum emission rate of 1.2 pounds of SO_2 per million Btu's of fuel input for large, coal-fired power plants. The general purpose of our evaluation of this option is to identify the effects of one particular specification of BACT, as required by the 1977 CAA Amendments. At the time of our research, the final BACT standard had not yet been promulgated, and the 90 percent level was chosen as a likely candidate. In May 1979, the standard was finalized, and as described previously, it requires between 70 and 90 percent removal depending on the sulfur content of the coal. Since this final standard is not too different from the 90 percent level of control we projected, many of our original conclusions will remain unchanged.

The following evaluation will examine the incremental effects of the 90 percent sulfur-control requirement as compared to what the situation would be without it (i.e., the "baseline"). One of the difficult aspects of this evaluation is determining what the "baseline"

should be. There are numerous state and federal emission and air-quality standards, and the controlling standard can vary from case to case. For example, Colorado currently has a stringent SO₂ emission limit for new plants of 0.2 pounds per million Btu, but whether this or the 90 percent reduction requirement would be the limiting standard depends on both the sulfur and the heat content of the coal. In other cases, the emission limits required to meet Class I PSD increments may be the controlling factor. We have attempted to clarify what our "baseline" assumptions are, but in several of the studies we consulted in making this evaluation, the "baseline" conditions are not always completely compatible.

Another general point concerning this specific alternative is that the BACT standard will be applied only to those plants whose construction commences after publication of the proposed BACT definition—that is, after June 1979; therefore, in general, the BACT standard will only apply to plants that begin to operate sometime during or after the 1984–86 time period.

How Will Demand for Western Coal Be Affected? The Regional Studies Program at Argonne National Laboratory has analyzed the impact of this policy alternative on the demand for western coal (Krohm, Dux, and Van Kuiken, 1977). Argonne used a computer model which simulates the behavior of competitive coal markets and allocates coal

Table 5-10: *Regional Steam-Coal Production Estimates (millions of tons per year)*

Coal-Producing Region	1975 Steam-Coal Production	1985		1990	
		Baseline Case	90% Sulfur Removal	Baseline Case	90% Sulfur Removal
Northern Great Plains	46.0	249.3	188.6	387.8	201.6
Other Western[a]	60.0	162.5	164.2	204.7	216.9
Total National	499.7	925.1	922.9	1,235.0	1,211.3

Source: Krohm, Dux, and Van Kuiken, 1977.

[a]This production comes primarily from Colorado, Utah, Arizona, and New Mexico, but it also includes Texas lignite and middle-sulfur coal from Oklahoma, Kansas, Arkansas, and Texas.

from each area of supply to each area of demand (both utilities' and industries') in such a way as to minimize total market costs. The model considers such factors as coal availability; coal quality (Btu and sulfur content); regionally disaggregated coal supply and demand; the costs of mining, transportation, and pollution control; and others. Some of the results of this study are summarized in Table 5-10.

The baseline used by Argonne assumes a continuation of the current NSPS for SO_2 of 1.2 pounds per million. As the table shows, a requirement for a 90 percent sulfur reduction would cause the demand for Northern Great Plains coal to drop sharply to about 52 percent of the baseline case by 1990. Almost all of this drop would occur in the shipments from the Northern Great Plains to the east central region (Ohio, Michigan, Indiana, Illinois, and Wisconsin); in 1990 there would be a drop in these shipments from 180 million tons per year in the baseline case to 40 million tons per year under the requirement for 90 percent sulfur removal. This shift is attributable to the fact that once scrubbers are required uniformly, it will be less expensive for utilities and industries in the east central region to use local coal. In fact, the continued transporting of coal from the Northern Great Plains to the east central region after 1985 will be largely the result of contractual agreements made before the BACT policy took effect. In sum, the Argonne study indicates that BACT will dramatically shift new contracts for the coal needed in the nation's east central region from western to local sources.

Nevertheless, it should be noted that the data in Table 5-10 still indicate a large growth in the demand for western coal even with the 90 percent sulfur-control requirements: the total demand for western coal is projected to be 418.5 million tons per year in 1990—a 294 percent increase above the 1975 level.

Two factors could reduce this dramatic shift from western to east central coal. First, the coal industry may not be able to shift production capacity this rapidly from western

mining areas back to the Midwest and Appalachian areas. And, second, most western coal is currently produced from nonunionized surface mines, while most midwestern and eastern coal is produced by unionized miners. The protracted coal strike of 1977–78 has heightened the concern of industrial and utility coal users about the reliability of their supplies from unionized mines. This concern may result in a greater demand for western coal than would be expected on the basis of economics alone.

Another important factor which makes these calculations uncertain is the degree to which the increased cost of pollution controls for coal-fired power plants will lead utilities to choose nuclear power over coal. As indicated in Table 5-10 the difference in the nation's total demand for coal between the baseline case and that of 90 percent sulfur removal is not very large. However, recent studies (Rudasill, 1977) have reaffirmed the economic attractiveness of nuclear power for baseload generation; and since the BACT requirement should enhance the cost advantage of nuclear power, it could lead to a substantial overall reduction in coal demand (Krohm, Dux, and Van Kuiken, 1977:49–51).

What Will Be the Effect on SO_2 Emissions?
Another important question concerning this alternative is the impact it would have on the levels of air pollution both within and outside the eight-state study area. This question has been examined in a study of alternative NSPS definitions performed under contract for EPA (Teknekron, 1978a). Table 5-11 presents some of the results of this study which was based on a "moderate" rate of growth in the demand for energy (5.8 percent annually through 1985 and 3.4 percent annually thereafter). Unfortunately, the regions used in this report do not correspond precisely with our eight-state study area.

As indicated in Table 5-11, Teknekron's study shows that a requirement of 90 percent SO_2 control would have a significant effect on the emission levels in the mountain re-

Table 5-11: *Regional SO$_2$ Emission Estimates from Power Plants* [a] *(million metric tons per year)*

Region	1976	1985 Baseline	1985 90% Control	1995 Baseline	1995 90% Control
North Mountain (Idaho, Wyoming and Montana)	0.12	0.09	0.06	0.18	0.06
South Mountain (Nevada, Utah, Colorado, Arizona, and New Mexico)	0.34	0.23	0.19	0.28	0.16
U.S. (Total)	13.6	15.2	15.2	15.8	13.6

Source: Teknekron, 1978a:3-12 to 3-15.

[a] Based on "moderate" growth in electricity demand: 5.8 percent per year through 1985 and 3.4 percent per year between 1985 and 2000.

gions: by 1995, for example, emission levels would be reduced by 43 percent in Nevada, Utah, Colorado, Arizona, and New Mexico, and by 67 percent in Idaho, Wyoming, and Montana. However, equally important are the extremely low levels of emissions for the mountain regions in the baseline case as compared to the U.S. total. For example, the mountain regions in the baseline case for 1995 account for only about 3 percent of the nation's total SO2 emissions, and yet the capacity of coal-fired power plants in the mountain region represents 7.7 percent of the U.S. total.

The reasons for this low baseline rate of emissions in the mountain region are the availability of low-sulfur coal, the already strict emissions standards set by the states, and the federal ambient air-quality standards (e.g., PSD) that result in relatively tight SO2 controls for many plants even in the baseline case. The states' strict controls, required even without BACT, are indicated in Table 5-12. The large percentage difference between the baseline case and that of 90 percent SO2 removal for the mountain region can be explained by the fact that given stringent controls, tightening them even slightly results in a relatively large percentage of reduction in emissions; for example, tightening SO2 removal rates from 80 to 90 percent means a 50 percent reduction in emissions. In sum, the 90 percent SO2 control alternative shows a large relative drop in total SO2 emissions in the

West; but in fact the standard would not, in general, be significantly more strict than the emission limits already imposed by many western states or than the emission limits likely to be required on many plants in order to meet the PSD or other new standards.

How Will Electricity Prices be Affected?
The Teknekron study (1978b) cited earlier also examined the effects of BACT on electricity prices. Table 5-13 presents projected real prices (in 1975 dollars) for the baseline and 90 percent SO2 control cases. Tighter SO2 and particulate controls are expected to increase electricity prices by 0.05 ¢/kWh nationally (a 1.7 percent increase) in 1995. For the mountain regions, the real price increases in 1995 will be slightly larger—0.07¢/kWh for Idaho, Wyoming, and Utah (a 3.5 percent increase) and 0.1¢/kWh for Nevada, Utah, Colorado, Arizona, and New Mexico (a 3.75 percent increase).

One important disadvantage is that the requirement for 90 percent sulfur reduction provides no incentive for using low-sulfur coals. Therefore, this approach is inherently inefficient since the same emissions levels could be achieved at a lower economic cost by using low-sulfur coals in conjunction with a lower percentage of SO2 removal.

Are There Other Costs or Benefits? The use of currently available nonregenerable FGD

Table 5-12: *Sulfur Removal Efficiencies Required for Coal-Fired Power Plants to Meet All Federal and State SO$_2$ Standards (excluding BACT)*

State	Removal Required (%)	Governing Regulation
Colorado Rifle 1,000 MW and mine	96.2	State Category II Ambient
Montana Colstrip 3,000 MW and mine	Maximum control capability. State NSPS technically practicable and economically feasible as determined by Air Quality Bureau 80	 Federal 24 hr. Class II PSD
North Dakota Beulah 3,000 MW and mine	85	Federal 24 hr. Class II PSD
New Mexico Farmington 3,000 MW and mine	78	Federal 24 hr. Class II PSD
Utah Escalante 3,000 MW and mine	94	Federal 24 hr. Class II PSD
Wyoming Gillette 3,000 MW and mine	82	State NSPS

processes, while reducing SO$_2$ emissions, creates a number of other environmental problems, including the disposal of large amounts of sludge wastes, lowered efficiency due to the energy needed for the scrubbers, and increased consumption of water. The requirements for land on which to dispose of sludge should not be a significant problem for the West, but these wastes do create the threat of water pollution, as discussed in Chapter 4. The lower energy efficiency means increased coal and water consumption (for both cooling and for the FGD system) per unit of electricity produced.

Table 5-13: *Electricity Price Impacts of 90 Percent SO$_2$ Control: Moderate-Growth Scenario[a] (cents per kilowatt hours, in 1975 dollars)*

Region	1985		1995	
	Baseline	90% Control[b]	Baseline	90% Control[b]
North Mountain (Idaho, Wyoming, and Montana)	2.30	2.31	2.02	2.09
South Mountain (Nevada, Utah, Colorado, Arizona and New Mexico)	2.67	2.82	2.67	2.77
U.S. Average	2.81	2.85	2.93	2.90

Source: Teknekron, 1978b:3-20.

[a]Assumes a 5.8 percent annual increase in electricity demand through 1985 and 3.4 percent annually between 1985 and 2000.

[b]The 90 percent SO$_2$ control scenario also includes a tightening of the particulate standard from 0.1 lbs/10^6 Btu to 0.03 lbs/10^6 Btu. The tighter particulate standard also contributes to the price increases shown in this table, but the great bulk of the increase is due to the SO$_2$ removal requirement.

How Are Costs and Benefits Distributed Regionally?

The legislative history of the 1977 CAA Amendments makes it clear that one of the purposes of the BACT requirement was to redress an inequity caused by the existing NSPS which gave low-sulfur western coal a competitive advantage. The intent was to put western coal on a more equal footing with midwestern and eastern coal. As indicated previously, the BACT requirement does this by substantially shifting demand from western to east central coal.

In looking at the impact on the eight states in our study area, the question of equity needs to be considered in both environmental and economic terms. Some states in the West have much stricter emission and ambient air-quality standards than do others. Since air pollutants do not respect state borders, this variation can become the basis for conflict. The states with less stringent standards have some economic advantage in attracting industry and yet the "clean-air" states could suffer a reduction in air quality from energy facilities near their borders or from the long-range transport of pollutants. The mandatory requirement for 90 percent SO_2 removal would remove the environmental and economic inequities by requiring all states to control SO_2 emissions to approximately the same degree.

Are Utilities or Investors Affected Financially?

Another aspect of "equity" to be considered is the economic impact of the 90 percent SO_2-control requirement on the utility industry. In principle, regulators allow utilities to set prices just high enough so that the revenues collected will equal the costs of providing service and an appropriate return on investment. In practice, ratemaking is subject to a regulatory lag—that is, rates are adjusted only once every few years. And although the full cost of such equipment as FGD units is supposed to be borne by a utility's customers, the lag in setting rates often results in a utility's investors receiving less than the allowed rate of return as long as intensive construction schedules persist.

Table 5-14: *Percentage of Return on Equity for Electric Utilities under Alternative Scenarios, 1985-1995 (moderate-growth scenario)*

Region	Baseline	90% Standard	Difference
North Mountain[1]	16.5	13.9	2.6
South Mountain[2]	13.2	12.7	.5
U.S. Average	12.4	11.9	.5

Source: Teknekron, 1978b:3-37.

[1] Idaho, Wyoming, and Montana.

[2] Nevada, Utah, Colorado, Arizona, and New Mexico.

Teknekron has estimated the investor-borne costs associated with a 90 percent sulfur-removal standard. The two measures which were employed, both considered important by participants in capital markets, are summarized in Tables 5-14 and 5-15. It was assumed that the return on equity—the annual profit per dollar invested by the shareholder—is set at 13 percent by the regulatory authorities. As can be seen in Table 5-14, the average utility will probably not achieve the allowed rate of return even in the baseline scenario. Imposition of the 90 percent sulfur-removal standard would double that discrepancy for

Table 5-15: *Interest-Coverage Ratios for Electric Utilities under Alternative Scenarios, 1985-1995 (moderate-growth scenario)*

Region	Baseline	90% Standard	Difference
North Mountain[1]	4.38	4.36	.02
South Mountain[2]	3.42	3.31	.11
U.S. Average	3.14	3.14	0

Source: Teknekron, 1978b:3-39.

[1] Idaho, Wyoming, and Montana.

[2] Nevada, Utah, Colorado, Arizona, and New Mexico.

the nation as a whole (1.1 percent short versus 0.6 percent short). The north mountain states (Idaho, Wyoming, and Montana) would experience a much larger drop in return on equity under the 90 percent sulfur standard, but these utilities would still end up with a higher rate of return than is allowed.

Table 5-15 expresses the returns to capital in terms which highlight the risks to lenders. The interest-coverage ratio

$$\frac{\text{earnings} + \text{interest payments}}{\text{interest payments}}$$

shows the safety margin which protects lenders in the face of a utility's unpredictable earnings. Utilities often enter into agreements with bond buyers not to extend their debt if the interest-coverage ratio falls below the 2.0 to 1.75 range. Table 5-15 shows that the proposed sulfur-removal standard would not put utilities very much closer to ratios that would be considered financially risky. There would be essentially no effect nationally or in the north mountain region.

In sum, Teknekron's study indicates that in the face of an inevitable regulatory lag, the utilities' shareholders will bear a measurable amount of the cost of implementing a sulfur-control standard. The actual effect on shareholders will probably exceed that indicated by the Teknekron model because it assumes annual ratemaking reviews, whereas in practice comprehensive reviews and adjustments normally occur only once every two or three years. Nevertheless, the extra cost borne by investors will not be large enough to have much effect on the credit rating of the average utility.

Are There Implementation Difficulties?
As discussed, a BACT standard has already been established by EPA under the provisions of the 1977 CAA Amendments. However, the standard could result in some practical difficulties for the agency's monitoring activities. The BACT standard would require EPA to monitor both emission rates and the effectiveness of sulfur-removal systems. In measuring the percentage of sulfur removal, EPA would have to monitor both coal-cleaning plants and FGD systems. During the course of our study no information was available on the economic cost of these increased monitoring tasks.

Smaller, Dispersed Energy Facilities

In order to evaluate the option of building smaller, dispersed energy facilities, we have assumed that the basic hardware would be the same in either a single, large plant or in a number of smaller, dispersed plants. For example, with coal-fired power plants a single large plant of, say, 1,500 MWe capacity might consist of three 500 MWe units, with each unit made up of a boiler, a turbine, a generator, and a cooling tower.[5] With smaller, dispersed facilities, these basic units would simply be dispersed rather than built side by side.

The idea of smaller facilities is receiving considerable attention from both utilities and state governments. For example, the New York Public Service Commission has required a utility to examine the pros and cons of building 300 to 600 MWe facilities instead of the more conventional strategy of relying on 900 to 1,200 MWe plants. The commission stated:

> It has been conventional system planning wisdom for a long time that construction of plants of ever increasing capacity is desirable (1) to achieve scale economies, and (2) to reduce the environmental impact associated with siting a number of smaller generating plants. We believe these arguments must now be reexamined in light of current conditions and future prospects [N.Y. Public Service Commission, 1978].

---

[5] With other types of cooling systems, such as once-through cooling or cooling ponds, the three units might actually share the same cooling systems rather than having three distinct systems.

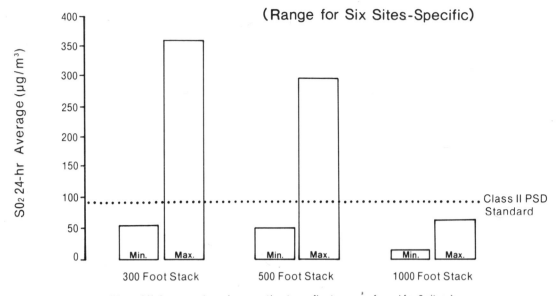

(Note: Minimum and maximum estimates reflect a range found for 6 sites.)
Figure 5-4: *Effect of Stack Height on Peak Sulfur Dioxide Levels for Power Plants*

There are also two research projects currently underway which are comparing the environmental and economic trade-offs between large and small power plants; one is being conducted at Los Alamos Scientific Laboratory in New Mexico (sponsored by EPA) and the other at the Electric Power Research Institute in Palo Alto, California (Ford, 1978a; Wyzga, 1978).

Will Increased Western Energy Development Be Allowed? As was indicated earlier in this chapter, only coal conversion facilities (especially power plants) and oil shale plants face serious restrictions on size due to ambient air-quality standards. However, there is very little flexibility in siting oil shale facilities due to the concentration of the resource in a relatively small geographical area and the impracticality of shipping raw shale for retorting at a distant location. Therefore, the option of building smaller, dispersed facilities is primarily of interest for coal. Because coal resources are widely distributed throughout our eight-state study area and because it is economically feasible to ship coal some distance from the mine for conversion, there is con-

siderable flexibility in siting coal conversion facilities.

Even if plants meet strict BACT emission restrictions, air-quality regulations (NAAQS, PSD, or state standards) can restrict the location of large energy facilities. The maximum size allowed will depend on such factors as the emission rates of various air pollutants, the terrain, meteorological conditions in the vicinity of the site, and its proximity to PSD Class I areas. Where large facilities are prohibited because of air-quality regulations, the approach suggested by this policy alternative is to build the same total conversion capacity in smaller units, dispersed over a wide enough area to minimize significant air pollution. This plan would reduce the concentration levels of pollutants and thus help avoid restrictions on the increased development of western energy resources.

Because we have assumed that the same basic hardware is being used in both cases, total emissions will be the same regardless of the plant-siting strategy. Thus, the regional effect of the long-range transport of pollutants would not be significantly different under this alternative. In fact, the impact on air quality of smaller facilities is really no different from that of larger plants using taller stacks to disperse pollutants. But, it is unclear whether or not current air regulations will allow tall stacks as a dispersion technique, whereas smaller facilities are permissible. Figure 5-4 shows the effects of taller stacks on the concentration levels of air pollution. Although not shown in Figure 5-4, for three of six baseline cases considered (with 500 foot stacks), Class II PSD increments for SO2 were not met; however, with a 1,000 foot stack, all six plants met the standard.

In addition to considerations of air quality, smaller facilities could also alleviate siting difficulties caused by limited water resources because each smaller plant would consume proportionally less water than a large facility. This means, for example, that smaller facilities could be sited near small rivers where it might not be feasible to build a plant three or four times larger.

One potential drawback of this alternative in terms of its effect on the levels of energy development has to do with the siting process. As discussed in Chapter 11, obtaining permits and meeting other regulatory requirements can be a very time-consuming, expensive, and frustrating process. The plan to build smaller facilities increases the number of siting applications a developer would need to make, more sites would need to be acquired, possibly more environmental impact statements would need to be prepared and more permits would need to be obtained. This proliferation of activity could potentially clog the process of approving sites and thereby slow energy development. However, this might be avoided, particularly since the concentration of adverse environmental impacts would generally be far less for the smaller facilities. For this reason, they might face less opposition, thereby easing the siting process for each small facility as compared to a larger one.

Are There Economic Costs or Benefits Associated with This Option? The economic trade-offs between the strategies of

building large, and small facilities depend on a number of factors which can be dealt with in detail only on a specific, case-by-case basis. For power plants, it is clear that utilities until quite recently believed that very large facilities offered an economy of scale. However, this thinking has begun to be questioned for a variety of reasons, some of which will be discussed in this section, focusing in particular on the general trade-offs involved for electric power plants. Similar considerations apply to other types of energy conversion facilities, although the specifics would, of course, vary.

There is little question that the construction of large power plants offers advantages on the basis of dollars per kilowatt-hour capacity. Assuming that the basic hardware (i.e., boilers, turbines, and generators) are the same regardless of the size of a facility, the capital costs for these items would be approximately the same (the labor costs for construction might be lower for a single large plant). However, the construction of smaller facilities would have some diseconomies associated with coal-handling equipment and land. For example, three small, separate facilities would require from 2 to 2.5 times as much land (depending on the type of cooling system) as a single large facility (Wyzga, 1977). The estimates of cost differentials vary considerably. The results of one study indicate that overall, three separate 700 MWe units would have a 20 percent higher capital cost than a single 2,100 MWe plant (Wyzga, 1977); another study, however, concludes that five separate 500 MWe plants (single units) would have a total capital cost of only about 1.3 percent higher than one single plant consisting of five 500 MWe units (Gassemi, 1978).

The smaller, dispersed facilities would also have some disadvantages in operating costs since single, large plants require fewer personnel on a per kilowatt basis. The cost of coal (including transportation) could also be higher for smaller facilities, although this would depend on many factors, such as the location of supply sources relative to the plant sites and the availability of existing rail lines.

Another possible diseconomy associated with smaller, dispersed facilities would involve the delivery of energy (by electric transmission lines or gas pipelines). These energy transportation systems offer significant economies of scale. For example, an electric transmission line capable of handling 500 MWe would cost approximately 30 to 40 percent more on a per kilowatt-hour per mile basis than a line capable of handling 1,500 MWe (see FPC, 1971: Vol. I, p. 13-9). The net effect on the cost of delivering energy would depend on a variety of considerations such as the location of a plant site in relation to demand and the availability of transportation rights-of-way. If building smaller facilities meant they could be located closer to where the demand is, this option might well have lower energy delivery costs than a single large plant.

Depending on how the construction is timed, building smaller facilities could offer certain economic advantages that would outweigh the higher capital and operating costs. If a large plant is constructed in such a way as to have all of its units ready for operation simultaneously, the total elapsed time from the decision to proceed to commercial power production may range from about 9.5 to 12 years. The equivalent time period for a small plant would be from 5.5 to 7.5 years. Furthermore, the shorter planning period for small plants reduces the uncertainty of forecasting the demand for electricity and thus reduces the likelihood that a utility will be caught with an excess or shortage of capacity. The shorter lead time for smaller facilities also means that the financial burden of carrying a large amount of "construction work in progress" would be reduced (Ford, 1978b).

In sum, it is not possible to state categorically that the option to build smaller, dispersed facilities is either more or less economically attractive than building a single, large plant. The economic trade-offs would depend on many factors that can only be evaluated precisely on a case-by-case basis. However, as reported above, one recent study estimated that by adding 2,500 MWe of capacity in five separate plants would have a total annual cost of only about 1.3 percent more than five units in a single plant case (Gassemi, 1978).

Are There Other Environmental or Social Effects? The construction of smaller, dispersed energy facilities could have an important effect on social and economic factors. Small facilities will generally have a larger peak labor force on a per unit of energy basis than a larger facility. For example, one source estimates that a 500 MWe power plant would have a peak construction requirement of 650 persons, or 1.3 persons per MWe, while a 3,000 MWe plant would only have a peak construction force of 2,850 persons, or .95 persons per MWe (Ford, 1978a). However, if the smaller facilities are built sequentially or are separated enough so that the impact of population growth falls on different communities, the "boomtown" effects discussed in chapters 7 and 8 could be reduced significantly.

Two potential effects of the smaller facilities that would have a negative impact on the environment are difficult to quantify. First, smaller facilities will mean the use of more land (as indicated previously, on the order of 2 to 2.5 times as much for the same total capacity in 700 MWe power plant units as opposed to a single 2,100 MWe facility). Second, smaller facilities could possibly lead to a greater proliferation of power lines which not only require more land for rights-of-way but also have a negative aesthetic impact. However, as mentioned earlier in connection with the economics of energy transportation, it is not possible to generalize since these factors will depend upon so many specific conditions.

Are There Equity Considerations? One disadvantage of building very large energy facilities is that they concentrate most of the adverse environmental and social effects in a very small area. In addition, once the facility is operating and population levels have stabilized, the overall tax benefits particularly for counties will often greatly exceed their needs. This can create serious problems of equity if many of the energy facility's employees live in other counties or in municipalities which do not receive these tax benefits (see chapters 7 and 8). Smaller, dispersed energy facilities could alleviate some of these inequities. The problems would be distributed more widely, and, as noted previously, the undesirable social and economic effects caused by the "boomtown" conditions would be reduced. Also, if the smaller facilities were sited in different towns or counties, the tax benefits would be more evenly distributed over the region.

How Could This Alternative Be Implemented? As indicated above, the advantages and disadvantages of smaller facilities can vary significantly from case to case. For this reason, any strategy for implementing this approach should be flexible enough to allow consideration on a case-by-case basis.

One strategy which gives flexibility without ignoring possible benefits would be for the state utility commissions to require industry to evaluate the option of smaller, dispersed facilities explicitly in their planning. (This is the tack taken by the New York Public Service Commission [1978].) Alternatively, federal agencies could require that this option be considered explicitly in the section on alternatives in environmental impact statements for new energy facilities. These approaches would not necessarily force industry to choose the option, but they would force them to look at it. In the process, more information would be made publicly available about the possible costs and benefits of smaller, dispersed facilities.

Commercialization of New Energy Technologies[6]

The threat to the quality of western air from large-scale energy development could be significantly reduced if the "cleaner" production technologies that are presently being devel-

oped prove successful and become available for commercial use. Thus, there are advantages for the West in a policy alternative which rapidly advances new, cleaner technologies through the research, development, and demonstration (RD&D) phases to a point where they are commercially available within the next ten years. This means efforts should focus on those processes that are currently ready or that will soon be ready for the construction of commercial-scale demonstration plants.

Such commercial-scale demonstrations represent the final developmental stage and should take place only after the technology is well understood. These demonstrations serve to determine what the performance characteristics will be (e.g., air emissions and efficiency) as well as to help identify demand and other important factors which go into the ultimate decision on whether to adopt the technology (Baer, Johnson, and Merrow, 1976; Kash et al., 1976). The questions of particular concern in the West should involve air quality, water consumption, and labor, or population, growth.

Given the high cost of commercial-scale demonstrations (as much as $1 billion for synfuel coal projects), major federal support might be essential. At the same time, the need to demonstrate the impact on air quality to western states and to various interest groups

[6] After the original drafting of this section, President Carter in July 1979 called for a major national program to produce approximately 2 million barrels per day of synthetic fuels by 1990, of which about 1.2 million barrels per day were to come from the West. If this program is implemented, many of the goals of the policy alternative discussed here will be achieved, but without the advantages of a demonstration/evaluation step. In fact, one of the major obstacles to President Carter's program is the lack of information on the impact of commercial-scale synthetic fuel facilities. Therefore, our evaluation of this alternative is instructive in that it provides insight into the advantages of a well-designed commercialization program as compared to a "crash" program. (See Kash et al. [1976] for a description of energy R&D needs, including needed demonstration programs, as they were viewed in 1976.)

requires that they be included in the planning and execution of any demonstration program that focuses on the West. Thus, such a demonstration program must involve at least four different categories of participants: the federal government, the western states, energy developers, and other affected interest groups such as environmentalists, consumers, farmers, and ranchers.

Several possible strategies could be used to implement such a program. It could be accomplished either by having the Department of Energy establish a continuing committee or commission that would involve these interests or by establishing a program within an organization of the western states such as the Western Governors' Policy Office. The committee approach would likely be easier to manage, but would not be as likely to build a regional base of support for the commercialization program as the approach linked with a regional organization.

Is the Alternative Effective? Table 5-1 lists the air emission rates for the various technologies considered in this study. As discussed above, many of these energy development systems have fewer air emissions per unit of energy produced than do conventional coal-fired power plants. In addition, there are other new conversion technologies which offer the potential for low emissions that are not considered in this study. These include fluidized bed combustion, low-Btu gasification/combined-cycle power plants, and solvent-refined coal plants.

Of course, technologies cannot be ranked solely on the basis of air emissions and impact on air quality. Economic costs are crucial in any situation, and water consumption and labor requirements are especially important in the West. Since considerable uncertainty surrounds the cost and characteristics of any new technology, none of the options are clearly superior to conventional coal-fired power plants in all respects. However, several offer potential advantages for the West; for example, coal-based synthetic fuel processes generally emit fewer air pollutants and require

less water than coal-fired power plants. The major uncertainty with regard to these technologies at the moment concerns their economics. A demonstration/commercialization effort which will improve the available data on these newer technologies and, thereby, possibly ensure or speed their commercialization would be an effective policy alternative for protecting the quality of western air while still allowing the region's energy resources to be developed.

What Are the Economic Costs and Risks?
The economic costs of energy RD&D, especially at the demonstration stage, are quite high. For example, a commercial-scale coal gasification plant (250 MMcfd) would cost on the order of $1 billion. In addition to the cost of the plant, total costs would include the research needed to determine the environmental and socioeconomic effects of the facility. Costs of a similar magnitude would apply for coal liquefaction plants producing 50,000 bbl/day, while oil shale plants producing 50,000 bbl/day would cost on the order of $600 million (see Chapter 9, Capital Availability). Commercial-scale geothermal demonstration plants can be much smaller—say, 25 to 50 MWe—and thus less costly: two demonstration power plants for liquid-dominated resources would cost on the order of $100 million.

The costs of these facilities would presumably be divided between the public sector (both federal and state governments) and industry. The proportion of costs assumed by each party and the distribution of future revenues (if any) would depend on the particular case. The initial costs and economic risks for the public sector could be minimized by offering certain incentives to industry (perhaps guaranteed prices for production) to construct and operate the demonstration plant. However, the associated research on the plant's impact must be paid for by the public sector to ensure credibility. The demonstration program also must be organized so that there is a realistic option of shutting down the demonstration plant or at least preventing further development if the impact of the plant is found to be unacceptable. (See Kash et al. [1976] and Baer, Johnson, and Merrow [1976] for a discussion of strategies for energy RD&D programs.)

How Are Costs and Benefits Distributed?
If an RD&D program were successful in developing new technologies for producing western energy resources while minimizing the impact on air quality, several groups would benefit. The nation as a whole would benefit from the increased domestic production of energy; the western states would enjoy the economic benefits that such development would bring while still maintaining a high standard of air quality; and industry would obtain profits from the commercial development. Thus, to be most equitable, the demonstration/commercialization program should be organized so as to share the costs and risks among these groups.

Is the Policy Flexible? To be most flexible, the demonstration programs should pursue a variety of technologies, subject, of course, to budgetary constraints. Then, the best option should be chosen for large-scale commercialization. Pursuing the widest range of technologies that funding allows will ensure that potentially attractive options are not prematurely dropped. Each project should be fully evaluated along the way using a broad set of criteria including technical and economic feasibility, environmental impact, socioeconomic impact, and institutional constraints. Both the planning and evaluation procedures should be open and allow for meaningful public participation so that legitimate concerns are addressed in the demonstration program and so that public confidence in the decisionmaking process in nurtured.

Are There Implementation Constraints?
This alternative would not be easy to implement because of several fundamental problems. Perhaps the most important is that RD&D requires large amounts of money from both the public and private sectors. It also requires a considerable amount of cooperation and coordination among industry, the federal government, and state governments; and depending on the implementation strategy pursued, it could require creation of a new institutional arrangement among the western states for the joint demonstration program.

Establishment of a Siting Task Force

The aim of establishing a siting task force is to create an overtly political process for reaching an accommodation between the conflicting interests of energy/economic development and air-quality protection. The specific goal of such a task force would be to reach a consensus on future sites for major energy facilities, and thus avoid the expensive and time-consuming delays often encountered under present siting procedures. Such an approach has been undertaken in Utah, where the governor's office formed a task force composed of representatives from federal land-management agencies, state and local governments, environmental groups, consumers, and industry (Utah Energy Office, 1978). Although the Interagency Task Force on Power Plant Siting was initially formed to resolve siting conflicts of the Intermountain Power Project near Capitol Reef National Park, Utah's utilities now consistently seek the advice of the task force, nominating six to twelve sites for evaluation. The result has been an agreement to avoid siting power plants in Utah's scenic and national park-filled southeast quadrant, with eight sites in the central part of the state designated for future development.

Is the Alternative Effective? Establishing a siting task force will not do anything to reduce air emissions from energy plants. What it will do, however, is site power plants in locations where a consensus has been reached that the air emissions will have the fewest adverse effects. Thus, the process could help eliminate or significantly reduce the debilitating delays that often accompany the siting of large energy facilities. For this reason, it could help increase energy development.

On the other hand, although industry would not be legally bound by the findings of the task force, it would probably be difficult to secure permission to locate energy facilities in areas the group found unacceptable. Thus, the conversion of substantial energy resources in some areas could be prevented or delayed. This effect has already been felt in Utah, where future energy export schemes are being scrutinized. The head of Utah's Energy Office stated that the Intermountain Power Project, backed by the Los Angeles city utility, may be the last export project in Utah "at least for a long time, since we can't afford to use up any of those other seven sites for outsiders until we're sure we won't need them for Utah's own future power" (Gill, 1978).

What Are the Economic Costs? The administrative cost of the task-force program would be relatively low and would include travel expenses, report preparation, and possible per diem expenses or subsidies for some participants. Costs would increase if doing research (such as conducting environmental studies) became a major part of the activities. In Utah, many of these costs have been absorbed by participating agencies or groups (Utah Energy Office, 1978).

The program could lessen energy development and production costs by avoiding lengthy delays in the siting process. On the other hand, the task force could pick sites that are more expensive than those industry would have chosen, when such items as land, cooling systems, and transportation costs are included.

What Are the Other Pros and Cons? While some groups may not agree with the final decisions of the task force, the whole concept of this policy alternative is to find an equitable way to reach an accommodation among conflicting interests. If the task force is properly structured so as to permit participation by any group that is potentially affected, then the approach should ensure a high degree of equity.

This alternative is also highly flexible because it is convened at the request of a governor or agency head. Its decisions attempt to introduce flexibility into a system that is often characterized by intransigence on the part of bureaucracies, industries, and environmental groups. This option can also be relatively easy to implement because advisory functions are usually attractive to most participants and the costs are low. However, because the effectiveness of the task force requires a specific response on the part of industry, there could be some resistance to implementation. In Utah though, industry has been a very willing participant (Utah Energy Office, 1978).

Summary and Comparison of Air-Quality Alternatives

We have described three categories of alternatives and nine specific alternatives for protecting the quality of the air in the West while proceeding with development of the region's energy resources. Four of the specific alternatives were considered in detail. Table 5-16 provides a brief summary and comparison of these four alternatives on the basis of the five criteria outlined in Chapter 2 and used throughout this study. While the net effect of any one alternative is uncertain, each has the potential for contributing to the achievement of the policy objective. The BACT definition of 90 percent sulfur removal is a regulatory program aimed at limiting SO_2 emissions for large coal-burning facilities, especially power plants. This standard is not expected to have a major effect on the number of power plants constructed in the western region, but it will dampen the demand for raw western coal. A complimentary approach is offered by the commercialization program, the goal of which is to bring into practical use new energy conversion technologies which are inherently less polluting than conventional coal-fired power plants. This could allow

Table 5-16: *Summary and Comparison of Air-Quality Policy Alternatives*

Criteria	BACT Definition	Smaller, Dispersed Facilities	Commercialization Programs	Siting Task Force
Effectiveness: Achievement of policy objective	Air emissions per unit of electricity would be reduced. Also, drop in exports of Northern Great Plains coal relative to continued current NSPS.	No direct effect on emission rates. Eases siting difficulties for individual plants by reducing lead times and concentration of adverse effects. However, increases the number of plants sited, which could slow development. Pollution dispersion similar to using tall stacks.	Reduces emissions per unit of energy produced by encouraging development of energy resources and conversion technologies with fewer air emissions than power plants.	No direct effect on emission rates, but sites selected to minimize adverse effects. Energy development encouraged by making siting easier, but level of energy development could possibly be constrained by limited number of sites agreed upon.
Efficiency: Costs, risks and benefits	Electricity prices increased by 3.5 percent for West. Inefficient since no incentives for using low-sulfur coal. Adverse side effects: large amounts of sludge waste, increased water consumption and lower energy efficiency.	Net effect is uncertain. Capital and operating costs increased. Land used for power plants increased, while land use for energy transportation depends on site-specific factors. Boom-town impact reduced.	Capital costs high, but if large-scale commercialization occurs, it represents only a small fraction of the economic value of all future production.	The economic costs of the task force itself are small. Energy production costs could be reduced because of fewer delays in siting. However, these cost reductions could be offset if the site chosen requires higher development costs, e.g., stricter pollution control, transport of water, etc.
Equity: Distribution of costs, risks, and benefits	Achieves national objective of putting western coal on more equal economic footing with coal from other regions. Electricity price increases highest for the West. Emission rates among states equalized.	Air pollution, tax benefits, and other economic benefits distributed more evenly across the region.	Costs and benefits shared throughout the nation, western region, and industry.	Would provide a procedural mechanism for reaching an equitable decision, balancing widely differing viewpoints and values.
Flexibility: Adaptability to locational needs and changes over time	Not very flexible — locational variables and state and local controls not taken into account.	Flexibility depends on specific implementation strategy. Not advisable to force this approach; rather it could be evaluated by state utility commissions and federal agencies in industry planning requirements.	Planning and evaluation should include public participation. The option to preclude full-scale commercialization should be retained.	Highly flexible; ad hoc in nature with public participation key feature.
Implementability: Institutional constraints and acceptability	Could be established under 1977 Clean Air Act Amendments. Creates practical difficulties in monitoring for EPA.	Encouragement by state utility commissions and/or federal agencies straightforward. Industry likely to strongly oppose any mandatory requirements for smaller, dispersed facilities.	Difficult to implement. Dollar costs are high and requires cooperation among federal government, western states, and industry.	Dollar costs for administration low; no new institutions or laws required.

greater levels of western energy development while still protecting the quality of the air. Unlike these two policies, the options of building smaller, dispersed facilities and of creating a siting task force do nothing to reduce emission rates. Rather, the plan for smaller, dispersed facilities attempts to minimize adverse effects on the air by reducing the peak concentration levels of pollution that would otherwise occur. In general, its effect on the dispersion of air pollution is similar to that of using taller stacks. The plan's effect on the overall level of energy development is uncertain. The creation of a siting task force primarily aims at eliminating some of the procedural problems which hinder western energy development; it contributes to the protection of air quality by identifying those areas where it is agreed that the air emissions would have the fewest adverse consequences.

The other four criteria (efficiency, equity, flexibility, and implementability) reflect other pros and cons of the policies which can be measured against their effectiveness. Since the specific goals of each policy vary widely and the net effects are uncertain, as we have just described, it is difficult to make direct comparisons among the alternatives. Nevertheless, certain important conclusions stand out. The BACT option could be easily implemented under the provisions of the 1977 CAA Amendments and would achieve equity. However, it is not a very flexible plan since local variables are not taken into account, and it is not a very efficient plan because there are no incentives for using low-sulfur coals. In some instances, lower or equal SO_2 emission rates could be achieved by requiring a smaller percentage of sulfur removal in conjunction with the use of a lower sulfur coal. This would also lessen the problems of sludge disposal, water consumption, and energy inefficiency associated with the FGD processes that are currently available.

While smaller, dispersed facilities offer some potentially attractive features, the pros and cons can vary greatly depending on the particular case. Given the uncertainty, an attractive implementation strategy would be for utility commissions and federal agencies to require industry to consider this approach in its planning activities. This would give recognition to the idea and generate better information on its possible benefits, but it would not force an across-the-board adoption of the alternative.

Over the long term, the commercialization program offers the greatest possible benefits of the four specific alternatives we have considered. However, there are a number of major obstacles to this approach. The economic costs of demonstration plants are high. Even though these costs would be small relative to the value of future energy production, the possibility of a negative decision based on the demonstration makes the economic risks substantial. These high initial costs and the need for cooperation and cost-sharing among the federal government, western state governments, and industry combine to make implementation difficult.

The establishment of a siting task force represents an attractive option for policymakers. Economic costs are low, flexibility is high, and this alternative should be relatively easy to implement. Thus, although the effectiveness is uncertain, it does offer potential benefits with very little cost or risk.

Of course, these four specific alternatives are not mutually exclusive. These (and others) could be implemented simultaneously, with each making a different contribution to the resolution of the problems and issues surrounding energy development and the protection of air quality in the West.

Conclusion

Along with issues involving water availability, those of air quality have a pervasive influence on the development of western energy resources. Concerns about the quality of the air will continue to directly affect which resources will be developed, the timing of that development, and the cost of the energy produced.

The issue of air quality is also similar to that of water availability because of the complex regulatory system which has attempted to reflect multiple goals and values. The intricate and constantly changing regulations and standards greatly increase the risks to energy developers and make planning at all levels highly uncertain. A number of recent or potential legal changes, regulatory interpretations, or court decisions could substantially change the future role of western coal and other energy resources. For example, the BACT requirement in the 1977 CAA Amendments, in conjunction with regulations required by the 1976 RCRA which, when promulgated, may classify FGD sludge as a hazardous waste, will put new burdens on coal users. It is generally recognized that more stringent environmental regulations will require improved technologies, which will, in turn, increase the price consumers must pay for energy. However, there is little evidence that policymakers have recognized or attempted to deal with the problems created by the myriad procedural requirements and uncertainties that now characterize the policies governing and protecting the quality of the air.

REFERENCES

Air Quality Act of 1967, Pub. L. 90-148, 81 Stat. 485.

Ayers, Richard E. 1975. "Enforcement of Air Pollution Controls on Stationary Sources Under the Clean Air Amendments of 1970." *Ecology Law Quarterly* 4 (No. 3):452, n. 28.

Baer, Walter S., Leland L. Johnson, and Edward W. Merrow. 1976. *Analysis of Federally Funded Demonstration Projects: Final Report,* prepared for U.S. Department of Commerce, Experimental Technology Incentive Program, Santa Monica, Calif.: Rand Corporation.

Ball, Richard H., Project Officer. 1977. *A Description of the SEAS Model.* Washington, D.C.: Environmental Protection Agency.

Berry, W. T., and R. E. Amber. 1977. "Report on the Fifth APCA Government Affairs Seminar: 'A New Look at the Old Clean Air Act.'" *Journal of the Air Pollution Control Association* 27 (June):529.

Business Week. 1977. "The Cheyennes Drive for Clean Air Rights." April 4, p. 29.

Clean Air ACT (CAA) Amendments of 1970, Pub. L. 91-604, 84 Stat. 1676.

Clean Air Act (CAA) Amendments of 1977, Pub. L. 95-95, 91 Stat. 685.

Code of Federal Regulations. 1978. Title 40, §50.

Crenshaw, John, et al. 1976. *Alternatives for Revising the SO₂ New Source Performance Standard for Coal-Fired Steam Generators,* Staff Study. Research Triangle Park, N.C.: Environmental Protection Agency, Office of Air Quality Planning & Standards Division, Energy Strategies Branch.

Ctvrtnicek, T. E., S. J. Rusek, and C. W. Sandy. 1975. *Evaluation of Low-Sulfur Western Coal Characteristics, Utilization, and Combustion Experience,* EPA-650/2-75-046. Dayton, Ohio: Monsanto Research Corporation.

Dean, Ernest H., Utah State Senator. 1977. Comments at National Conference of State Legislators, Energy Policy Planning Workshop, Denver, Colorado, December 9.

Denver Post. 1978. "Study May Suggest New Air Standards." January 19, p. 45.

Environmental Science and Technology (ES&T). 1977. "Currents." 2 (October):946.

EPA Environmental News, May 25, 1979.

Federal Power Commission (FPC). 1971. *The 1970 National Power Survey,* part 1: *Guidelines for Growth of the Electric Power Industry.* Washington, D.C.: Government Printing Office.

Federal Register 43 (October 5, 1978):46258.

Ford, Andrew, Los Alamos Scientific Laboratory. May 1978a. Personal communication.

Ford, Andrew. 1978b. "Expanding Generating Capacity for an Uncertain Future: The Advantage of Small Power Plants." Paper presented at the Conference on Simulation, Modelling, and Decision in Energy Systems,

Montreal, Canada, June 1.

Garvey, Doris B., et al. 1978. *The Prevention of Significant Deterioration: Implications for Energy Research and Development.* Argonne, Ill.: Argonne National Laboratory, Office of Environmental Policy Analysis.

Gassemi, Esmaeil. 1978. "Economic Analysis of Large and Small Coal Fired Power Plants." M.S. thesis, University of Oklahoma.

Gill, Douglas. 1978. "Deal Spares Utah Parks Area." *Denver Post,* May 28.

Jones, Charles O. 1975. *Clean Air: The Policies and Politics of Pollution Control.* Pittsburgh, Pa.: University of Pittsburgh Press.

Kash, Don E., et al. 1976. *Our Energy Future: The Role of Research, Development, and Demonstration in Reaching a National Consensus on Energy Supply.* Norman: University of Oklahoma Press.

Kirschten, J. Dicken. 1977. "Converting to Coal—Can It Be Done Cleanly?" *National Journal* 9 (May 21):781–84.

Krohm, G. C., C. D. Dux, and J. C. Van Kuiken. 1977. *Effect on Regional Coal Markets of the "Best Available Control Technology" Policy for Sulfur Emission,* National Coal Utilization Assessment. Argonne, Ill.: Argonne National Laboratory.

McKee, Herbert C. 1978. "The Problem of Equity in Emissions Offset." *Journal of the Air Pollution Control Association* 28 (June).

Metzger, Charles. December 9, 1977. Personal communication.

New York Public Service Commission. 1978. *Opinion Analyzing Plans Pursuant to Section 149-b of the Public Service Law and Order Directing Additional Studies,* Opinion No. 78-3, Case 27154—Long Range Electric Plans. Albany: New York Public Service Commission, March 6.

Old West Regional Commission Bulletin 4 (September 1, 1977):4.

Palomba, Joseph, Jr., Colorado Air Pollution Control Commission. 1978. Comments in "Report on Sixth APCA Government Affairs Seminar." *Journal of the Air Pollution Control Association* 28 (June).

Rasmussen, Reinhold A. 1972. "What Do the Hydrocarbons From Trees Contribute to Air Pollution?" *Journal of the Air Pollution Control Association* 22 (July):537–43.

Resource Conservation and Recovery Act of 1976, Pub. L. 94-580, 90 Stat. 2795.

Rudasill, Charles L. 1977. "Comparing Coal and Nuclear Generating Costs." *EPRI Journal* 2 (October):14–17.

Sierra Club v. *Ruckelshaus,* 344 F. Supp. 253 (D.D.C.1972), *affirmed sub nom., Fri* v. *Sierra Club,* 412 U.S. 541 (1973).

Teknekron, Inc. 1978a. *Review of New Source Performance Standards for Coal-Fired Utility Boilers,* vol. 1: *Emissions and Non-Air Quality Environmental Impacts.* Berkeley, Calif.: Teknekron.

Teknekron, Inc. 1978b. *Review of New Source Performance Standards for Coal-Fired Utility Boilers,* vol. 2: *Economic and Financial Impacts.* Berkeley, Calif.: Teknekron.

U.S. Department of the Interior (DOI), Bureau of Land Management (BLM). 1976. *Final Environmental Impact Statement: Proposed Kaiparowits Project,* 6 vols. Salt Lake City, Utah: BLM.

U.S. Department of the Interior (DOI), Bureau of Reclamation (BuRec). 1976. *Western Gasification Company (WESTCO) Coal Gasification Project and Expansion of Navajo Mine by Utah International Inc., San Juan County, New Mexico: Final Environmental Impact Statement,* 2 vols. Salt Lake City, Utah: BuRec.

U.S. Code (U.S.C.) Congressional and Administrative News, vol. 2: Legislative History. 1978. "Clean Air Act Amendments of 1977: Legislative History." St. Paul, Minn.: West.

Utah Energy Office, Staff. June 1978. Personal communication.

White, Irvin L., et al. 1979a. *Energy From the West: Impact Analysis Report,* 2 vols. Washington, D.C.: Environmental Protection Agency.

White, Irvin L., et al. 1979b. *Energy From the West: Energy Resource Development Systems Report.* Washington, D.C.: Environmental Protection Agency.

Wyzga, Ronald E. 1977. "Concentrated vs. Large Dispersed vs. Dispersed vs. Atomized Power Plants-Overview." Paper presented at Engineering Foundation Conference on Non-Conventional Siting of Power Plants, Henniker, New Hampshire, July 10–15.

Wyzga, Ronald E., Electric Power Research Institute. May 1978. Personal communication.

CHAPTER 6

Land Use and Reclamation

Energy development can lead to three critical land-use issues in the West. First, energy development will often produce or intensify conflicts between energy developers and others such as farmers, ranchers, environmentalists, and recreationists. Second, although only a small percentage of the region's lands will be affected, the percentage of land involved in a particular locale can be quite large. Third, it is not clear that lands disturbed by the development of energy resources can be successfully reclaimed—that is, returned to predevelopment uses—in all parts of the region.

Land-use and reclamation issues are already challenging the planning, monitoring, and enforcement capabilities of all levels of government. In part this is because demand for the land needed for energy development comes at a time when the public interest in land use is being substantially redefined, in terms of both broad environmental goals and the policies and institutions which control the way land is used. Increasingly, the proposition that land, like air and water, is a basic national resource in which the public has a substantial interest is being accepted and codified (Clawson and Held, 1957; Bosselman and Callies, 1972; Healy, 1976). Thus, today the need is to strike a balance between energy goals and environmental values. It is crucial to find effective and equitable strategies for dealing with western land-use issues because a range of interests is at stake and because

competing claims to the benefits of our finite land resources have to be accommodated.

Consequences of Energy Development

The extent and type of effects on land use can vary depending on both the energy technology and the location chosen. The significant technological factors include the method of mineral extraction, the type of conversion facility, and the requirements for labor associated with that technology. Critical local factors include the characteristics of the overburden, the thickness of the coal seam, the heat content of the resource, the characteristics of the climate and soil, the nature of the terrain, and the existence of biological communities in an area.

Variations in Land Use by Technology

Variations in the amount of land used by energy technologies are shown in Table 6-1, both for the size of facilities projected in our scenarios and on an equivalent-energy basis. When compared on an equivalent-energy basis, the land requirements for various conversion facilities do not differ significantly, and the land required for an underground coal mine is similar to that needed for a conversion facility. However, the land disturbed by surface coal mines can be ten times greater than

155

Uranium Mining in Wyoming

either conversion facilities or underground mines depending on the thickness of the seam and on the heating value of the coal (see the discussion of local characteristics below).

By the year 2000, depending on the level of development, between 1,000 and 1,500 square miles of land could be distrubed in our study by coal mining alone. As shown in Table 6-2, the magnitude of this surface land disturbance is projected to be greater for the Northern Great Plains than for any other area. Based on our scenarios, at some locations up to 18 percent of some entire counties could be disrupted. Although the consequences we have assessed for the eight-state area indicate that coal mining will disturb the largest amount of land regionally, the development of uranium, oil shale, oil and gas, and geothermal energy will also produce considerable land disruption, which can be especially critical in some regions; for example, extensive uranium mining in the Grants Mineral Belt of New Mexico is a major concern.

Direct use of the land produces a range of effects. The topographic alteration during mining can be aesthetically shocking. Protective vegetation is destroyed, and the soil and rocks above the coal or other mineral deposits are scraped away, resulting in a massive, visible alteration of the land. Such activities can also lead to water pollution problems, can disrupt the normal flow of surface water and of groundwater,[1] can cause soil erosion and related ecological consequences, and ultimately can restrict future uses of the land.

Generally, the ecological impact will be local, although if many facilities are sited in

[1] See Chapter 4.

Table 6-1: *Land Use by Technology*

Facility	Typical Size (acres per 30-year life of facility)	Equivalent Energy (acres per 10^{12} Btu produced)	Product
Conversion Facilities[a]			
Power Plant (3,000 MWe)	2,400	1.2[b]	Electricity
		0.4[b]	Electricity
Lurgi Gas (250 MMcfd)	805	0.3	Gas
Synthane Gas (250 MMcfd)	805	0.3	Gas
Synthoil Oil (100,000 bbl/day)	2,060	0.4	Oil
Oil Shale Retort (50,000 bbl/day)[c]	2,150	0.8	Oil
Uranium Mill (1,000 tpd)	280	0.03	Yellowcake
Mines			
Underground Oil Shale (26 MMtpy)	500	0.1	Oil Shale
In Situ Oil Shale (57,000 bbl/day)[d]	180	0.06	Oil
Underground Coal[e] (12 MMtpy)	1,760	0.3	Coal
Surface Coal[f] (12 MMtpy)	3,300-25,200	0.6-4.4	Coal
Surface Uranium[g] (1,100 tpy)	3,450	NC	Uranium ore
Underground Uranium (1,100 tpd)	70	0.01	Uranium ore
In Situ Uranium (1,000 tpd)	U	U	Uranium ore

NC = not considered.   U = unknown.

[a] Land required for above-ground energy conversion facilities; this number remains unchanged through the lifetime of the facility.

[b] The higher figure is based on assessing the electric output of a power plant only in terms of its heating value (i.e., 1 kWh = 3,415 Btu's); the lower figure is based on assessing electricity as being roughly 3 times more valuable than that (i.e., 1 kWh = 10,000 Btu's).

[c] Includes 650 acres for the retort site and 1,500 acres for the disposal of spent shale over 30 years.

[d] Includes land for surface structure associated with the mine (100 acres) and *in situ* process (80 acres).

[e] Includes only that portion of the mine site to be permanently occupied. At Kaiparowits, 30,000 acres will be subject to subsidence over the 30-year life of the mine; it is expected that nearly all of this land will be reclaimed and returned to productive use.

[f] The range of values reflects that found at the four sites using surface coal mining and is calculated on 12 MMtpy production rates; ranges are attributable to variation in seam thicknesses and the coal's heating value.

[g] Assumes annual acreage requirement for mining of 115 acres per year and ore waste disposal in mined-out areas.

a single area, the result will be the elimination or reduction of a large percentage of habitats and significant area-wide reductions in the land's ecological carrying capacity. Also, the impact from erosion and water pollutants, as well as air pollutants, from mines and power plants can cause widespread ecological effects such as decreasing crop yields and damaging wildlife, plants, and croplands.

Underground coal, oil shale, and uranium mining pose surface problems in preventing and controlling land subsidence (see box). Changes range from surface irregularities (uneven ground and/or cracks) to drastic alterations (gaping holes). Like strip mining, underground mining can also lead to water pollution problems, intercept groundwater aquifers, require mine dewatering, deplete aquifers, and ultimately change land uses.

Another land-use problem concerns the availability and location of sites for waste disposal. Land is required for evaporative

Table 6-2: *Acreage Disturbed by Surface Coal Mining for the Low-Demand Scenario[a] by the Year 2000*

Area	Acres Disturbed (thousands of acres)
Northern Great Plains	
North Dakota Lignite[b]	447
Powder River[c]	396
Intermountain Area[d]	39
Four Corners Area[e]	69

[a]The Low-Demand case is described in Chapter 2.

[b]Includes Billings, Bowman, Dunn, Hettinger, McKenzie, McLean, Mercer, Oliver, Slope, Stark, and Williams counties, North Dakota (assumed average seam thickness of 12.5 feet).

[c]Includes Powder River, Bighorn, and Rosebud counties, Montana (assumed average seam thickness of 27 feet); and Campbell, Johnson, and Sheridan counties, Wyoming (assumed average seam thickness of 64 feet).

[d]Includes Rio Blanco, Garfield, and Huerfano counties, Colorado (assuming one-third of the projected mines are underground and an average seam thickness of 7 feet).

[e]Includes San Juan County, New Mexico (assumed average seam thickness of 10.3 feet); and Kane and Garfield counties, Utah (assuming half the projected mines are underground and an average seam thickness of 10 feet).

holding ponds to dispose of the fly ash and scrubber sludge associated with coal conversion facilities, for mine settling ponds to contain toxic mine wastes, and for ponds to dump uranium mill tailings. Because of their potentially hazardous nature, these waste-disposal sites require continual maintenance and monitoring, even after energy sites and facilities are abandoned.

Finally, a network of unpaved service and maintenance roads contributes to area-wide erosion, increases water and air pollution, and fragments wildlife habitats. Wells in oil and gas fields are linked by a web of eroding roads that reduce both the aesthetic and the biological value of the land. Procedures to stabilize these roads may not be adequate, and new roads may have to be added when washouts occur (U.S. DOI, Geological Survey Staff, 1977).

Energy-Related Population Increases

Variations in the labor intensity of technologies can be another important land-use factor. Some technologies require a larger work force to produce the same quantity of energy than do others (see chapters 7 and 8). As illustrated in Table 6-3, for the Low-Demand Scenario, operational, or permanent population increases alone would result in 157,200 new people coming to the region by 1990 and another 299,400 by 2000.[2]

The ecological stresses brought on by energy development are closely related to the increase in human populations. Anticipated population growth will occur disproportionately near areas highly valued for backcountry recreation. Estimates made for the Missouri River Basin Comprehensive Framework Study indicate that it is reasonable to expect recreation demand (not just from energy development but from all sources) to double or triple by the year 2000 (Missouri Basin Interagency Committee, 1971: Vol. 1, p. 137). Since the early 1970s, backcountry recreational activi-

[2]The large increase during the 1990s is due primarily to assumptions about large-scale development in Colorado and extensive coal production in the Northern Great Plains.

Table 6-3: *Permanent Population Increases after 1975[a] Due to Energy Development in Eight-State Region (Low-Demand Scenario)*

Year	Permanent Population Increase
1980	45,000
1985	118,200
1990	157,200
2000	456,600

[a]The 1975 estimated population for the eight states was 9,551,000 (see U.S. Dept. of Commerce. Bureau of Census, 1976).

Population Effects

Grand Teton National Park is located 100 miles north of energy boomtown Rock Springs, Wyoming. Three million residents and out-of-state tourists visited the park in 1976. "Wyoming is feeling the effects of energy development," says State Senator John Turner. "It's doubled the population in some communities, brought in a rough crowd, caused an increase in mental health problems, divorce, alcoholism and crime." Due to these increases 70 percent of the respondents in a public opinion survey of residents of Jackson Hole (in the park) favored strong land-use controls and very limited growth.—Leydet, 1976:771.

---

ties such as hiking, backpacking, snowmobiling,[3] and trail riding with horses or jeeps have been rising in popularity, especially in the West; these activities account for 5 to 15 percent of the total use of certain areas such as national forests.

The locations for recreational activities generally fall into three categories: major established tourist attractions (e.g., Yellowstone and Grand Teton national parks); areas near population centers (e.g., Grand Mesa National Forest, near Grand Junction, Colorado); and isolated areas or areas with limited recreational opportunities (e.g., Black Hills National Forest in Wyoming and South Dakota). If access to such areas is limited or controlled, the bulk of the growing demand will fall on adjacent, nondesignated areas which may have a strong historical or aesthetic appeal.[4]

Furthermore, increases in population have both aquatic and terrestrial consequences.

More housing, roads, and recreational activities fragment wildlife habitats into small parcels that are less usable by either resident or migratory species, causing some species to desert an area. Easier human access to recreational areas can lead to increasing erosion; vegetation damage; damage to fish, birds, and other wildlife; and diminished enjoyment of resources. Based on current and projected energy development patterns, these effects are likely to occur primarily in four geographic areas: western Colorado and eastern Utah, the coal regions in Wyoming, the Four Corners area,[5] and the coal fields of Montana and North Dakota. In western Colorado, the large population influx is expected to use nearby prime outdoor recreation areas (including national forests, parks, monuments, and wilderness/primitive areas).[6] Wy-

---

[3]In Hayden, Colorado (pop. 10,000), which is located in the middle of the Yampa coal field, snowmobiling is the most popular form of recreation. At least one, and often two, snowmobiles can be found in every yard: "At one residence, there were four, lined up like bicycles" (Gits, 1978:32).

[4]A limitation in projecting recreational demands is the difficulty of anticipating trends in recreational styles. For example, a technological innovation such as snowmobiling is a recent phenomenon. Hydrofoil and shallow-draft boats make many western rivers available for recreational use. Similarly, uncertainty exists in land-management practices: the current trend is to increase restrictions on wilderness and backcoun-

try areas, but economic factors encourage the Forest Service to promote dispersed recreation activities by building trails and improving access.

[5]The energy-related population growth in Utah and New Mexico may affect such areas as San Juan National Forest (Colorado), Chaco Canyon National Monument (New Mexico), and the Dixie National Forest and Glen Canyon National Recreation Area (Utah).

[6]For example, Rio Grande National Forest, San Juan National Forest, Rio Grande National Wilderness Area, Upper Rio Grande Primitive Area, Gore Range/Eagles Nest Wilderness Area, and Mesa Verde National Park.

oming and North Dakota will also experience substantial population increases; however, in these places, the choice of recreation areas will be more limited with some receiving substantially heavier use (see box).

Variations in Land Use by Location

The impact of energy development on land may also vary according to several locational factors. For example, in the case of coal, the amount of land disturbed for each unit of energy extracted is highly dependent on the Btu content of the coal and on its seam thickness. Table 6-4 shows that the land requirements for mines at different sites vary considerably.

Climate, soil, and topography at a site are also important locational features. Rainfall, including its seasonal distribution, is one of the most significant climatic variables because it affects the cost and potential success of reclaiming mined lands. Rainfall data for three sites are shown in Table 6-5. Generally, the Southwest receives the least amount of rainfall overall; the Rocky Mountain region and the Northern Great Plains have similar annual totals. However, during the summer growing season, rainfall in the Northern Great Plains is twice that of the Rocky Moun-

Table 6-5: *Precipitation Patterns in the West*

Location	Annual Average	Record Low
Northern Great Plains (Cheyenne, Wyoming)	20.25	5.9
Western Rocky Mountains (Grand Junction, Colorado)	17.07	4.4
Southwest (Albuquerque, New Mexico)	9.1	4.0

Source: Ruffner and Blair, 1977.

tains. For a mine supplying a 3,000 MWe power plant, the amount of land requiring reclamation is generally largest in the Southwest; that also happens to be the area with the lowest average rainfall.

The semiarid climate and good topsoil of the Northern Great Plains create an ecosystem that can be restored with less difficulty than in the Southwest (Hodder, 1978; Weissenborn, 1977). Forest areas of the Rocky Mountains are more difficult to restore due to difficult topography and the amount of time required to establish a mature forest. In deserts, infrequent rainfall and overgrazing by livestock limit the success of reclamation efforts. For example, attempts at reclamation in the arid Black Mesa area, which is over-

Table 6-4: *Land Use for Mines at Six Specific Locations*[a]

Type	Site[b]	Typical Size Mine[c] (acres over 30 years)	Equivalent Energy (acres per 10^{12} Btu)
Surface Coal	Navajo/Farmington	4,000-27,820	.5-3.2
Surface Coal	Gillette	4,030	0.5
Surface Coal	Colstrip	9,680	1.2
Surface Coal	Beulah	24,210	3.0
Underground Coal	Kaiparowits	1,700	0.2
Underground Oil Shale	Rifle	4,200	0.5

[a]Includes all surface lands disturbed, including disposal of wastes (such as spent shale).

[b]Seam thickness and heating values assumed: Navajo/Farmington, seams 10.3 feet thick with between 1 and 7 multiple seams, 8,600 Btu per pound; Gillette, 64 feet, 8,000 Btu per pound; Colstrip, 28 feet, 8,600 Btu per pound; Beulah, 13 feet, 6,950 Btu per pound.

[c]In all cases the assumed mine size supports a 3,000 MWe power plant.

grazed and does not have a source of irrigation, have been ineffective (Green, 1977). But through irrigation at the Navajo and San Juan mines near Farmington in northwest New Mexico, vegetation has been established in some areas with greater cover than exists in adjacent unmined areas.[7] Still unanswered, however, is the question of whether reclaimed areas will remain productive. In part, this is because problems concerning the effects of phasing out water supplements on arid sites are still unresolved (Farmer et al., 1974; U.S. DOI, BLM, 1975:172). Even in semiarid places, such as the coal mining areas of Decker, Montana, and Carbon County, Wyoming, withdrawing irrigation and fertilizer supplements may be difficult because of unfavorable spoil and soil properties.

Generally speaking, using present reclamation techniques, areas with six to ten inches of rainfall per year can usually support plant regrowth without supplemental irrigation (NAS, 1974:2). Annual precipitation levels on strippable land in northeastern New Mexico, southwestern Wyoming, and south-central Utah often fall below ten inches. Based on the results of several years of investigation, however, Grant Davis of the U.S. Forest Service's Surface Environment and Mining Program, reports that his group has found that when "mined and graded with care, and planted properly to suitable species, areas receiving as little as six inches average annual precipitation have been revegetated." The key phrases here are "with care," "properly," and "suitable species" (Davis, 1976). This requires close consideration of site-specific characteristics (LaFevers, 1977; Riess, 1977). It should be noted here that much progress has been made over the last five years—since publication of a National Academy of Sciences (NAS) report—in understanding the factors

that affect the success of reclamation in the West. Moreover, extensive research in this field is being carried out by the federal government (for example, the U.S. Forest Service, DOE, BLM, EPA, USGS, and the Bureau of Mines); at Argonne, Oak Ridge, and Pacific Northwest laboratories; at universities; by regional commissions and organizations; and within state governments and industries (Hodder, 1978; Weissenborn, 1977; Davis, 1977; McKell, 1977; Grogan, 1977).

As noted above, the timing of precipitation appears to be much more critical than the average annual amount. This is the conclusion of an NAS study and is noted in Grant Davis's personal communication. In commenting on the NAS report, Robert R. Curry, associate professor of environmental geology at the University of Montana, has stated:

> Both soil formation and plant growth rely upon *soil moisture,* not rainfall, and in the arid west rainfall bears relatively little relationship to soil moisture. In addition to summer rainfall, concomitant summer evaporation which reduces soil moisture must be considered [Reviewer comments in NAS, 1974: 168].

If most of an area's moisture occurs during the winter and practically none occurs during the growing season, reclamation practices may be thwarted by the need for (and cost of) irrigation, by wind erosion, and by other factors.[8] In addition, periodic dry spells lasting up to several years will curtail successful revegetation unless irrigation can be provided (NAS, 1974; Packer, 1974).

Problems and Issues

As the previous discussion demonstrates, the

---

[7] Annual precipitation rates at the Navajo mine of approximately six inches have called for the use of supplemental water for at least 1.5 years (EPA, Office of Energy Activities, Region VIII et al., 1976).

[8] The Forest Service has been working with supplementary irrigation equipment and finds that it should be available as insurance against long dry periods, regardless of average annual precipitation in the West (Davis, 1976).

impact on land use of developing energy resources occurs as a consequence of two kinds
of land-use requirements: first, the land required for the development itself (such as for
mines, for conversion and support facilities,
and for waste disposal); and second, the land
required to meet the needs of energy-related
increases in the population (such as for housing and recreation). These land-use requirements affect both competition for public and
private lands and efforts to protect environmental quality and aesthetics.

Although the problems and issues addressed
here are likely to intensify as western energy
development proceeds, they are by no means
entirely new. In several instances, these questions have been before the public for most
of this century—for example, developing appropriate mining laws or making public lands
available for other private or public uses. In
certain other instances, the policy debates
associated with land use are relatively new,
such as those on the recent federal regulations
for developing and implementing state reclamation programs (U.S. DOI, OSMRE, 1979).

In the following section, we will summarize
selected elements of the land-use policies and
issues that are affected by energy development. Our discussion is organized around
three general issue themes:

- The role of government in planning and
  controlling land use;
- The objectives and requirements of postmining land use; and
- The accommodation and control of recreational and environmental uses of public
  lands.

Land-Use Planning and Management

*The development of energy resources must
compete with other economically productive land uses such as grazing and row
crops; and it must compete as well with
preservation, conservation, and recreational
uses. Competition already exists within and
among these three categories of land use
in the West. The large-scale development
of energy resources and the associated increase in population that can be expected
will intensify this competition during the
next two decades, taxing the conflict-resolu-
tion and problem-solving capabilities of all
levels of government.*

Conflicts over the planning and management of land use in development areas have
surfaced not only on the basis of findings in
this technology assessment, but also on the
basis of analyses conducted by other policy
research groups. For example, a report sponsored by the Rockefeller Brothers Fund found
that a central problem confronting the entire
nation, and the West specifically, was the
unrestricted patchwork of urbanization and
development that sprawls across scenic hills,
valleys, forests, and farms (Reilly, 1973).
Indeed, managing growth and planning land
use have already become major political issues
in some western states (Kirschten, 1977:
783–84); and federal legislation to manage
both state and federal lands has been before
Congress during three of its past four sessions.

Multiple Interests in Land-Use Planning:
Land-use policy in this country has changed
from a trial-and-error approach during early
colonization and development, through periods when management and stewardship were
more important considerations, to the recent
emphasis on zoning and planning (U.S. Cong.,
OTA, 1979:79–99; Stairs, 1977; Jones et al.,
1974; Pierce, 1972:205). In each of these
periods, policymakers have responded to conflicts between divergent interests such as the
needs of homesteaders, mining companies
(Mining Act of 1866; General Mining Law of
1872), conservationists, and those interested
in preserving naturally scenic or historically
significant areas.

Recently, numerous congressional acts and
executive policies have restricted the availability of federal lands for mineral exploration
and development. Since 1968 formal restric-

tions or prohibitions on all federal lands have increased from 20 percent to more than 50 percent. This increase is due primarily to the establishment of wilderness areas and temporary land withdrawals in Alaska (*Oil and Gas Journal,* 1979a:68–69; U.S. DOI, 1978: 88–93; U.S. DOI, 1976:27). In sum, there is currently less land available for the exploration and development of domestic energy resources than has historically been the case (*Oil and Gas Journal,* 1979b:68).

These restrictions underscore a significant problem of land-use planning. In making decisions regarding the development and use of public lands, policymakers must compare the advantages of one type of use (e.g., energy production) against perceived disadvantages to the public. Such determinations are difficult because they involve the weighing of competing and often conflicting values and interests—in particular, divergent federal, state, and local attitudes toward land-use planning.

Initiatives for land-use legislation have come recently from both the executive branch and from Congress; members of Congress from western states have been heavily involved in these actions. For example, former Colorado Rep. Wayne Aspinall was active as chairman of the Public Land Law Review Commission and author of the National Land Policy, Planning and Management Act (HR 7211), proposed in 1972. Although no action was taken on the bill, it reflected a growing recognition of the need to coordinate land-use efforts between federal agencies, other landowners, state and local governments, and private individuals (Congressional Quarterly, 1973:809–10).

A number of conservation-oriented public interest groups support comprehensive land-use legislation, including the national Wildlife Federation, National Audubon Society, and Sierra Club (see box). However, these groups, along with the Isaak Walton League and Friends of the Earth, have opposed some of the land-use bills that have been introduced over the last few years on the grounds that

Interest Groups

The Sierra Club and the National Audubon Society implement objectives through lobbying the courts and technical review. They have been instrumental in delaying coal leasing and have influenced land-use policy. In 1976, for example, the groups petitioned the Fish and Wildlife Service (FWS) to identify critical habitat for the Whooping Crane, and in August, 1978, the FWS added significant new areas, some on the Platte River. This provided a basis for subsequent suits to prevent Basin Electric Cooperative from modifying the flow of the Platte River for use in its 2,000 MWe power plant in Wheatland, Wyoming. Recently, agreement has been reached on water management by Basin Electric to preserve most of this Whooping Crane habitat.—U.S. DOI, FWS. 1978.

the proposed planning, funding, and review procedures were oriented to commercial exploitation of public lands rather than to their preservation and conservation (Congressional Quarterly, 1973:809–10).

Although land-use planning for federal lands is obviously a dominant issue in western energy development, states and localities have traditionally been at the center of land-use control. Historically, wide-ranging authority has been delegated by states to communities (Council of State Governments, Task Force on National Resource and Land Use Information and Technology, 1975:4–6). Yet local efforts are seldom adequate to cope with the regional (and ultimately, statewide or national) land-use consequences and/or planning needs. Furthermore, much of the land in rural areas that will be used for energy development is unincorporated.[9]

In an attempt to respond to these conditions, many western states have established additional planning requirements. Wyoming and North Dakota now require local governments to appoint planning commissions and formulate comprehensive plans. Both Colo-

[9] For further discussion of this point, see Chapter 8.

rado and Wyoming have state land-use commissions. Colorado has a permit system that requires an energy facility to be approved in accordance with a county land-use commission's guidelines or regulations. Wyoming's land-use planning act mandates state and local land-use plans and requires coordination between the state's Industrial Siting Board and the state Land-Use Commission (White, 1975:695–97). And in 1975 Montana passed a law that called for subdivisions to be built in harmony with their environment.

A number of other state and local agencies affect land-use planning and the choices that must be made among different land uses. Attempts to coordinate programs at the state level are typically done through the federal Office of Management and Budget Circular A-95 process, established in the mid 1960s. Substate planning districts in most states coordinate the EPA programs in wastewater and drinking water treatment and conduct area-wide planning for wastewater management. Together with other federal programs, the intent of the substate planning districts is to establish an organized review process, although by one estimate such planning has had limited success and has had only a minimal effect on decisionmaking (Council of State Governments, Task Force on Natural Resources and Land Use Information and Technology, 1975:27).

Diversity of Land Uses: The existence of mineral deposits in forests, on ranges, or in farming areas spotlight concerns about the most appropriate uses of the land. Although multiple uses—that is, the simultaneous use of the surface or subsurface of a tract of land for more than one ongoing productive purpose, including wildlife and recreation, energy development, timber harvesting, agriculture, and grazing—may be encouraged, it is more typical that one use threatens another. These conflicts are usually expressed in terms of individual interests: ranchers and farmers,

for example, worry about the competition for land and water, about the loss of agricultural water to energy development, and about the impact of air pollution on grasslands, crops, and other natural resources (Gisser et al., 1979; Grim, 1976:177–78).

Special concern has centered on the prime farmlands and alluvial valleys in the Northern Great Plains region. Together these lands are extremely productive: they support local cattle ranching and dairy farming and help provide good hay crops. Agriculture in Montana and Wyoming is particularly dependent on the alluvial valley floors (Northern Plains Resources Council, 1979:1–12).

Questions about appropriate uses for the land also include attempts by conservationists to protect archeological sites and the natural beauty of the landscape and environment. Long valued for its pristine condition, the eight-state region could undergo significant transformation if a high percentage of its energy resources were to be developed. Some government administrators (and environmentalists as well) fear that ill-planned and uncoordinated development would threaten the area's historical and aesthetic qualities (U.S. DOI, Heritage Conservation and Recreation Service, 1978).

As discussed above, although comprehensive laws controlling land use have been considered, only land-use planning on public lands has actually been initiated. This has been undertaken principally through the BLM (within DOI) and the Forest Service (in the USDA). The BLM has primary responsibility for administering the resources on public domain and acquired federal lands and is the surface-management agency for most federal lands (except national forests, parks, and Indian reservations). The Federal Land Policy and Management Act (1976) requires BLM programs to provide for land management that promotes "multiple use and sustained yield" while maintaining and enhancing the quality of the environment. The "multiple use and sustained yield" concept stems from the Multiple Use-Sustained Yield Act (1960) which requires the U.S. Forest Service to enhance the availability of wood, water, wildlife, fish, forage, wilderness, and recreation by zoning lands for multiple uses and by analyzing the impact before the development of resources takes place. Similarly, the Federal Land Policy and Management Act requires the development of comprehensive land-use plans for all BLM-administered lands, requires consideration of all potential uses of an area before management decisions are made, and authorizes Congress to veto any BLM decision to exclude tracts of 100,000 acres or more from "principal uses" (defined as domestic livestock grazing, fish and wildlife development, mineral exploration and production, rights-of-way, timber production, and outdoor recreation).

BLM has in the past made coal lands available for development through a joint industry and government nominating system that also weights market interest in new coal production. However, there have been no new leases (except to continue already existing mining operations) since 1971. In that year, a sudden surge in leasing applications and permits due to increased demand for low-sulfur western coal, led to criticism of the entire leasing system and an unofficial self-imposed moratorium by the Department of the Interior. In 1973 the moratorium was made official and extended until a programmatic environmental impact statement (EIS) could be prepared and the leasing system revised. When the programmatic EIS was released in 1974, there was an immediate reaction from environmental groups, especially the Sierra Club, which filed suit to prevent any coal development in the Powder River Basin until a regional EIS could be prepared and approved *(Sierra Club* v. *Morton)*. At about the same time, the Natural Resources Defense Council filed suit *(NRDC* v. *Berklund)* to prevent the Interior Department from automatically giving preference claims to prospectors who have found coal in commercially producible quantities without considering potential environmental damage which would result from

development (Magida, 1975). As a result, the bureau's Energy Minerals Allocation Resource System (EMARS) is currently under review, and the DOI is examining existing leases, applications, and regional environmental studies to determine if new coal production in the West is environmentally acceptable (U.S. DOI, 1978:21; Rattner, 1978; *Old West Regional Commission Bulletin,* 1977:4). The proposals announced in 1979 to renew leasing activities during the 1980s reverse some of the procedures formerly proposed under EMARS. For example, certain tracts can be excluded initially ("disnominated") because they would be difficult to reclaim, because they provide critical habitats for endangered species, or because they are areas of outstanding scenic quality. After this step, coal companies may nominate those tracts that can be most productively developed. Coal exploration and leasing programs for oil, gas, and geothermal energy are proceeding as is the evaluation of uranium mining claims. These programs are, however, under continuing review. Such reviews have already resulted in modifications to the geothermal leasing program to ensure diligent development and more efficient administration (U.S. DOI, BLM, 1979).

Another principal agency involved in the management of public lands is the Forest Service, which oversees surface lands in close cooperation with developers. The National Forest Management Act (1976) established revised standards and guidelines for Forest Service Land-use plans and updated federal policies on managing national forest resources (Weissenborn, 1977). Both the Federal Land Policy and Management Act and the Forest Management Act are intended to develop more comprehensive land-use planning for federal lands. Both address the range of competing interests and conflicts that are the result of managing for multiple uses. Even so, neither of these recent laws seems able to deal with problems raised by the prospecting and claims procedures of the 1872 General Mining Law. For uranium, public domain lands under the jurisdiction of a surface-management agency subsequently move from federal to private ownership, when a claim is patented in the location-patent system.[10] Under these circumstances almost all uranium lands fall under state regulation.[11]

The Wilderness Act (1964), which established a national wilderness preservation system, is also administered by the Forest Service. This act formalized the concepts of wilderness preservation that had been developing since the 1920s. In practice, it eventually will close millions of acres of federal land to energy development (U.S. President, Office of the White House Secretary, 1977; Wagner, 1970:2826-31). On January 1, 1984, lands designated by the Forest Service as wilderness areas will be withdrawn from mining and leasing. Although mineral exploration an production currently is not prohibited on these withdrawn lands, the risk associated with attempting to get development underway in 1984 has already severely restricted energy development ventures in these areas (U.S. DOI, 1976:90).

Evaluation of the wilderness and roadless areas to be withdrawn from development is the focus of the Forest Service's continuing Roadless Area Review and Evaluation (RARE) program. "RARE II," which covers the most recent classification of these lands, proposed the addition of 2,919 roadless areas—some 62 million acres—to the wilderness-roadless system (U.S. DOE, Assistant Secretary for Environment, Office of Technology Impacts,

Coal Strip Mine in Northern Great Plains

1980:5-73). Energy developers reacted to these recommendations, questioning the single-use wilderness designation in view of the multiple-use concept whereby industry and recreationists had historically shared in using the national forests. Following the release of the final EIS, a final review by the Carter administration recommended in 1979 that most of the RARE II lands with potentially significant energy resources be designated as "nonwilderness" or "further planning" areas to allow for mineral exploration and development. These recommendations were pending before Congress as of early 1980.

Finally the National Park Service, the administering agency for the national park system, has an ongoing planning process to develop "general master plans" for the various parks under its control. The Resource Management Plan, one part of the general plan,

---

[10] The location-patent system allows a company or individual who discovers a mineral deposit (iron, copper, uranium, or other "hard rock" minerals) of profitable quantity to file a mining claim which gives the prospector exclusive title to both the mineral and surface rights. This is called a mineral patent. The patent does not require that the deposit be mined, or that the developer pay for any minerals extracted and land used, or that mined land be reclaimed (Choitz, 1928:27).

[11] Also, a number of uranium lands are on Indian reservations under Indian, Bureau of Indian Affairs (BIA), or USGS management.

Strip Mining Operation in Northwest Colorado

deals specifically with mineral development activities. In general, mining is prohibited in national parks; however, the secretary of the interior does have authority to allow use of resources in certain recreation areas. For this to occure, energy developers must demonstrate that their activities will not "conflict or significantly impair recreation and the conservation of the scenic, scientific, or other values of any park in the National Park System" (Moore and Mills, 1977:64).

Objectives and Requirements of Land Reclamation

With the competing land uses and the need for large amounts of land, reclamation objectives and requirements are probably among the most critical problems and issues associated with western energy development. If mined lands can be "satisfactorily" restored, the land-use conflicts presumably would be reduced.

Reclamation: Public interest in the effects on the land of mining, energy conversion, and waste disposal is manifested in a number of general and specific concerns. One is "irreversibility"—the possibility that mining will alter the land so significantly that its value to society will be permanently reduced. As indicated previously, without adequate reclamation, future land-use alternatives will indeed be lost and severe ecological damage can result.

Another concern is the effect of mining on the qaulity and availability of water (Hodder, 1978; Power, Sandoval, and Rise, 1978). Coal seams and oil shale deposits are often major sources of groundwater for a variety of domestic and livestock uses. Mining interests may also come into conflict with environmental interests if reclamation objectives are not carefully prescribed and enforced. For example, improperly restored lands will not satisfy the aesthetic goals of those who want to protect the natural ecology and topography of an area.

These serious concerns are compounded by a tangle of legal and administrative arrangements for land management and reclamation. The requirements for restoring lands affected by energy development differ according to the resource and the jurisdiction. In the case of surface coal mining, the issue of reclamation has been on the public agenda for decades, and more uniform regulations have recently been established (see below). However, reclamation objectives and requirements vary widely for other minerals. Statutory requirements for reclaiming lands disturbed by uranium, oil and gas, and oil shale development do not exist at the federal level or in most states.

Based upon existing state regulations, the revegetation efforts that follow mining have generally involved both native and "introduced" species. Some reclamation policies call for the use of native species because they are well adapted to conditions at the reclamation site. But the limited availability of seeds for native vegetation in arid and semiarid locations may restrict their use.[12] Other reclamation efforts use introduced species because they germinate and grow more quickly. However, it is uncertain whether introduced species will persist and become a stable component in the ecosystem. Problem weeds frequently colonize and dominate other species used in revegetation but are not suitable as food for wildlife or livestock.

From the beginning of mining activities in the nineteenth century, western mining companies have done some reclaiming of mined lands, but with few controls by government agencies. Until the early 1970s government at all levels largely ignored reclamation problems, but during the 1970s a majority of western states enacted legislative reclamation requirements.

The most sweeping changes came with the passage of the federal Surface Mining Control and Reclamation Act (SMCRA) (1977), which called for state implementation of surface mining standards for coal extraction. The specific intent of the act is to provide minimum uniform standards to regulate strip mining. The lobbying for this type of law, by a coalition of environmental groups through several sessions of Congress was strong; opposition by the National Coal Association and other industry groups was equally intense. However, most minerals, including oil, gas, oil shale, and uranium, are still not covered by a uniform reclamation policy.

Strongly supported by Secretary of the Interior Cecil Andrus, SMCRA (1977) substantially changed the federal role in controlling reclamation of surface coal mines.[13] The principal objectives of the act is to provide a uniform system of restoring strip-mined land to its approximate original condition. Table 6-6 outlines selected requirements of the act. Strip mining highly productive farmland was a controversial element of the new law. Environmentalists and some agricultural interests had sought to prohibit all strip mining of farmlands. Coal industry lobbyists argued that the land could be successfully restored and made even more useful after reclamation. As a compromise, the act permits strip mining on "prime farmlands" and

---

[12] While limited availability of seeds is a problem for some species needed for revegetation, not all native seeds are in short supply. Instead, the most significant problem is that wild seeds must be collected in the same general area where they will be planted (U.S. DOI, FWS Staff, 1977).

[13] The legislation covers both the adverse effects of surface mining and, to a lesser extent, the surface impact of underground coal-mining operations.

in alluvial valleys when the developers can show that the proposed mining operations would not preclude later farming and would not adversely affect the quantity of water in the valley floor.[14] Consent of the surface owner is required before mining can take place on the underlying (federal) strippable lands. Revegetation and restoration of the land to its original topography is generally required except in rare circumstances (see box).

SMCRA also establishes uniform reclamation standards on all lands, whether federal or not, and assigns management and enforcement responsibility to the states. Furthermore, the act provides for officials of the OSMRE to periodically inspect all mines and report on the progress of reclamation work. SMCRA also levies a tax on mined coal, regardless of where it is produced, that is used to reclaim abandoned lands, with states sharing 50 percent of the revenues. Finally the administrator of EPA must concur with the regulations published by OSMRE and must also concur in its approval of state programs.[15]

Many other federal agencies are now involved directly in administering and managing reclamation provisions connected with energy development (see Table 6-7). Except for uranium (and other "hard rock" minerals) this involves a surface-management agency that can make stipulations in a lease if the lands are publicly held; the agencies are typically

---

[14] However, the regulations distinguish between pastureland and cropland and do not include grazing pastures or hay meadows in their definition of "prime farmland" (U.S. DOI, OSMRE, 1979:14928-29, 14936, 15318-20).

[15] EPA has both an administrative and a managerial interest in reclamation, particularly in its air, water, and research programs. Mine-dust measurements are a part of EPA's air-monitoring programs, as is measurement of mine runoff. For example, as of July 1, 1977, coal mines that have discharges are required to comply with Final Effluent Limitations Guidelines promulgated by EPA (Heeman, 1977:141).

Table 6-6: *Selected Requirements of the Surface Coal Mining Regulatory Program, 1979*

Scope:	Coal resources that are surface mined (and some surface activities of underground mines); requires studies of reclamation needs of other minerals.
Institutions:	Enforced by the Office of Surface Mining Reclamation and Enforcement (OSMRE) in the DOI; Implemented by states with approved programs.
Abandoned Mine Fund:	Establishes collection procedures for a coal tax of 35¢/ton for surface mined coal, 15¢/ton for underground coal, and 10¢/ton for lignite; 50 percent of the fund goes to states for their reclamation needs.
Procedures:	All surface coal mines require a permit; the permit has reclamation plan requirements; hearing, approval and denial procedures; and provisions for revised plans. A reclamation bond must be posted. Inspection is provided with administrative and judicial review.
Citizen Participation:	Provides for hearings on reclamation plans with participation by interested parties, provides for citizens' suits against agencies and/or the secretary of the interior.
Performance Standards:	Provides detailed performance standards for stabilization, restoration, revegetation, and post-mining land use, and prevention of water pollution or off-site effects.
Designation of Lands Unsuitable for Mining:	Provides for the protection of prime farmlands and alluvial valley floors and for the designation of other lands as unsuitable for mining by interested parties. Public may petition to have certain specific areas protected from mining.
Surface Owner Consent:	Requires written consent from the surface owner and his consultation in developing a comprehensive land-use plan.

Source: U.S. DOI, OSMRE, 1979.

Table 6-7: *Summary of Reclamation-Management Provisions by Resource*

Resource	Statute	Authority	Administering Agency
Coal	Mineral Leasing Act of 1920, Federal Coal Leasing Amendments Act of 1975 (1976)	Provides for controlled development and leasing	BLM Forest Service USGS
	Surface Mining Control and Reclamation Act of 1977	Specifies comprehensive requirements and procedures	OSMRE states USGS
Oil and Gas	Mineral Leasing Act of 1920	Provides limited control of leasing	BLM Forest Service
		Imposes few statutory requirements for reclamation	USGS
Geothermal Energy	Geothermal Steam Act of 1970	Provides for controlled leasing	BLM Forest Service
		Imposes few statutory requirements for reclamation	USGS
Oil Shale	Mineral Leasing Act of 1920	Provides for controlled leasing	BLM Forest Service
		Imposes few statutory requirements for reclamation	
Uranium	General Mining Law of 1872	Provides for patents of public lands largely on private initiative	BLM USGS
	Atomic Energy Act of 1954	Imposes few or no controls on reclamation	DOE

BLM, the Forest Service,[16] or the BIA. In the case of coal, extensive and additional requirements have recently been established.

For oil, gas, oil shale, and geothermal energy resources, most of the reclamation requirements are informal and have been developed on the basis of cooperation between energy developers and the staffs of surface-management agencies. Specific reclamation requirements for uranium mining are largely nonexistent at both the state and federal levels (U.S. Congress, OTA, 1979). Since the early 1970s however, environmental groups (especially the Sierra Club and Wilderness Society) have fought to bring land disturbances from uranium development under closer federal control. Interior Department and Forest Service personnel also favor legis-

---

Modifying Topography

There is an opposite extreme to the original contour concept, the leveling of hills and mountains to tabletop flatness. A recent EPA study indicated that mine operators in rough topography generally prefer this approach, questioning the desirability of reestablishing an original contour. The new Strip Mine Act has provisions for making possible use of exceptionally flat tops for airports or parking lots.—Chironis, 1977:49.

[16] SMCRA (1977:§522) requires the secretary of agriculture to demonstrate that adverse effects from surface coal mining will not occur in national forests, but the act does not contain a similar provision for national grasslands. Energy development is, therefore, easier on grasslands. In addition, fewer revenues from grasslands are distributed locally than are revenues from lands in the public domain, e.g., national forests.

lation and regulations to ensure that mineral development on lands in the public domain are balanced and coordinated with other surface uses. Most of these groups and organizations favor the adoption of a leasing system to replace the current location patent system (Choitz, 1978). They contend that uniform federal controls would increase production from the claims and would minimize abuses by claimants who have no intention of mining the claimed lands (see box). Moreover, a leasing system could contain provisions requiring reclamation and could assure a fair return for the "hard rock" minerals extracted from public lands. The fundamental issues with regard to the extraction of "hard rock" minerals are: How should adequate exploration and development be fostered? How much control should the federal government exercise? What priority should energy development have over other uses of public lands? (U.S. DOI, 1978:69-74).

Reclamation and Waste Disposal: The major concerns about mine wastes have generally centered on the quality of wastes and the expectation that the land used for disposal

is irreversible committed, rather than on the amount of land affected. Many mine wastes, especially uranium tailings, solid wastes from coal, and FGD wastes, can pose real threats to human health and to the environment. And because of its volume, spent shale can also pose environmental and reclamation problems. The seriousness of such threats depends on how the wastes are managed and monitored, including where and how such wastes are stored, treated, transported, and disposed of (ORNL, 1978).

Governmental regulation of mine wastes typically focuses on specific types of wastes or addresses selected consequences. Two of the more important federal laws relevant to managing the impact of energy-related waste disposal are the SMCRA and the RCRA (1976).

The RCRA (1976) provides a framework for dealing with the disposal of all solid wastes. Under this act, land—like air and water—is placed under strict federal regulatory control. In attempting to conserve virgin resources, the law seeks to develop a nationwide system for managing hazardous and solid wastes (via regional solid-waste plans) and for recovering resources from these wastes. Some solid wastes from coal have been classified as hazardous, and this could have a far-reaching effect on the way such wastes are presently handled and on the final regulations of the SMCRA (U.S. DOI, OSMRE, 1979:14902-15485; Engineering-Science, 1979).

Both the regulations dealing with hazardous wastes from most energy extraction and conversion facilities and the regional solid-waste plans are still being developed. EPA is charged with establishing the criteria for regional solid-waste plans, including mine wastes, and with conducting detailed studies of mine wastes.

Control of Recreational and Critical Environmental Areas:

The impact of energy-related population growth on our leisure-time, recreational, environmental, and aesthetic pleasures may be lasting and may prove difficult to deal with successfully. Population encroachments will cause certain species of animals to retreat from their present habitats; and habitat fragmentation and energy development may consume a significant proportion of certain types of habitat. Providing for energy development, for wilderness recreational experiences, and for the more intensive use of national parks is a continuing problem for land managers and policymakers.

The environmental quality of most western states has been stereotyped: New Mexico is the "Land of Enchantment"; Wyoming is a "Sportsman's Paradise"; Montana is "Big Sky Country"; and Arizona is a "Retirement Capital" (Peirce, 1972:290–91). Yet, these areas are also centers for mining and resource exploitation. Congressmen from the western states are active on committees that deal with natural resources and represent frequently conflicting industrial, recreational, and environmental interests (Wagner, 1971:1768–73).

As land-use planning has developed over the past 10 to 15 years, the governmental role in controlling individual uses of natural resources such as wildlife and scenic or recreational areas has also evolved. These policies are part of the government's efforts to planning land use but are concerned more with managing and accommodating the activities of individuals. Many of these controls have been instituted in response to the demands of sportsmen and environmental interest groups to sustain or improve the status of endangered species and wildlife. Interest groups such as Trout Unlimited, Ducks Unlimited, and the National Wildlife Federation are especially active in the West. The National Wildlife Federation, the Sierra Club, and the National Audubon Society historically have been important groups and have lobbied for wilderness areas, wildlife areas, wildlife refuges, and controls for endangered species, especially

since the 1950s. In response, Congress passed the Endangered Species acts of 1969 and 1973 and the Wilderness Act of 1964 and has added significantly to wildlife refuges in the western states.[17]

These same groups have been active in promoting increased recreational and scenic resources for tourists and westerners, in lobbying for legislation (Bureau of Outdoor Recreation Act, 1963) that established the Bureau of Outdoor Recreation (now the Bureau of Heritage Conservation and Recreation) to coordinate recreation planning, and in establishing a fund to acquire lands for recreation (Land and Water Conservation Fund Act, 1965). The bureau is responsible for coordinating wild and scenic rivers, national trails, and state and local recreation programs. It also coordinates government policy restricting recreational activities—an example would be President Carter's Executive Order permitting the designation of areas in which off-road vehicles (ORV) can be used and from which they are prohibited (*Fed. Reg.,* 1977a). In the latter case, the effect has been to antagonize a major sector of the recreational public which frequently uses western lands.

A number of federal agencies deal with wildlife and the quality of the ecosystem indirectly, such as the Soil Conservation Service (SCS) or the EPA. EPA has a direct research mission in understanding the role of plants and animals in the ecosystem, and it regulates air emissions and water quality to minimize adverse effects on wildlife. Russell Train, a past administrator of EPA, has been critical of agencies in the DOI, suggesting that their guardianship of wildlife resources is too often governed by narrow interests and the short-range concerns of income-producing sportsmen. Train cited an estimate by the Council on Environmental Quality that 97 percent of the funds for wildlife management

---

[17] An example of early legislative protection of historical and environmental significance to the West is the Bald Eagle Protection Act (1940).

go to 3 percent of the species—the ones involved in hunting, trapping, or fishing (*Environment News*, 1976).

In the past, both individual energy development companies and their lobbies, such as the American Petroleum Institute and the American Mining Congress, have opposed restrictions on the use of vehicles or that set aside large tracts of land for primitive areas or national parks (see box, "Access to Recreation"). These groups have been especially active in commenting on management proposals that may influence the exploration or development of energy resources (U.S. DOI, Bureau of Outdoor Recreation, 1976).

Control by state governments over recreational activities is diffuse; there are programs in a number of administrative agencies, including surface-management agencies and those involved in managing natural resources. Most states have a parks and recreation department or commission that manages parks, campgrounds, or recreational areas; a sepa-

---

Access to Recreation

There has been considerable opposition to designating lands as national parks. In the case of Canyonlands National Park at the confluence of the Green and Colorado Rivers in Utah, oil, uranium, and other mineral interests as well as stockmen and businessmen opposed creation of the park in the early 1960's. This idea of the park was sold through Park Service films and documents which showed that opening the park to tourists would bring prosperity to Utah residents. In 1977, the National Park Service established the policy that Canyonlands would not be developed for vehicle access, and that it would largely remain a wilderness park. It is now one of the least used national parks, with 79,100 visitors in 1976. —Scher, 1977:10.

---

rate department of fish and game typically controls the access to and catching of game or fish; and a state department of forestry usually controls the use of state forests.

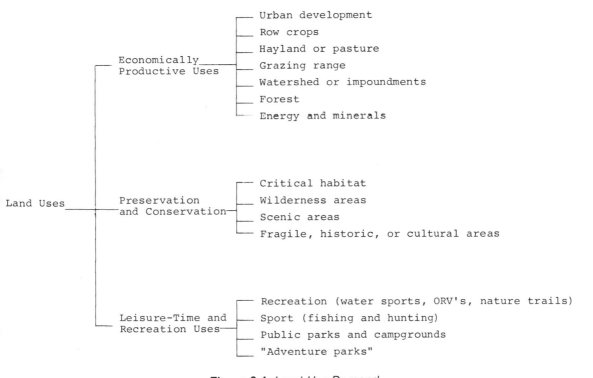

Figure 6-1: *Land-Use Demand*

Summary of Problems and Issues

Land uses can be divided into the three categories shown in Figure 6-1: economic/production, preservation/conservation, and leisure-time/recreation. The development of energy resources must compete for land with other economically productive land uses such as grazing and row crops; and it must compete with preservation and conservation and with leisure-time and recreational uses. Competition already exists within and among these three categories of land use in the West. Large-scale development of energy resources and the associated increase in population that can be expected will intensify this competition.

Resolution of these land-use and reclamation issues will challenge the planning, monitoring, and enforcement capabilities of all levels of government. It is not clear whether existing institutional mechanisms will be adequate to meet the challenge. This situation is complicated by the fact that the federal government and a number of Indian tribes are the major landowners in our eight-state study area. Furthermore, several federal agencies have a management role and responsibility for these lands. Coordinated, cooperative planning and management among these agencies is limited, but the major problem seems to be a lack of coordination and cooperation among the levels of government.

Given both the range of land-use interests and values at stake in the West and the pluralistic nature and federalist structure of our political system, it is unlikely that any single land use or any one level of government will previal over the others. It is much more likely that an effort will be made to accommodate the multiplicity of interests and various agencies and governments involved. Therefore, public policy on land use involves several problems and issues: Which lands will be allocated for which use in what quantity? Can lands disturbed by energy development be returned to their original state or to some economically productive state? How and by whom will allocations be made, access and activities controlled, and reclamation ensured? In the following section, some selected policy alternatives for dealing with these problems and issues are addressed.

Policy Alternatives for Land Use and Reclamation

As the discussion of land use and reclamation in the preceding section showed, the problems and issues that are likely to arise as a consequence of western energy development reflect two kinds of concerns: procedural and substantive. Procedural concerns focus on how policies are made—for example, on whether state and local governments are given adequate opportunities to protect their interests when decisions about the development of federal lands are made. Substantive concerns focus on the content and anticipated consequences of land-use and reclamation policies—for example, on which lands are to be designated for surface mining and what the effects of those designations will be. Thus, it appears that the land-use and reclamation policies related to the development of western energy resources must address two overlapping, general policy objectives:

- To ensure that the institutions and processes for making land-use and reclamation decisions give adequate consideration to all current and future uses; and
- To provide for and protect a diversity of present and future land uses.

To achieve either or both of these objectives, a more comprehensive and better coordinated land-use planning and management system will be required—one which provides participation opportunities for a broad range of interested parties, coordinates cooperative planning and management among the various agencies and levels of governments, and specifies which lands can be used for what purposes. Specific alternatives for creating such a system differ primarily in terms of the extent to which the system would be centralized, its comprehensiveness, and the degree of control over land uses that would be exercised. Specific alternatives range from establishing a centralized, federally coordinated, comprehensive land-use planning and management system to incrementally improving the current policy system and from making case-by-case decisions about certain land uses (such as siting energy facilities) to preparing a comprehensive land-use plan that would designate permissible land uses and require all lands disturbed by energy development either to be reclaimed to their predevelopment condition or redeveloped for some other productive use.

It is not feasible to deal here with the entire range of policy alternatives. The scope of the analysis can be narrowed considerably by focusing on a more specifically western objective, drawn from the two general objectives already described:

- To meet the land-use needs of energy development, including the land requirements of energy-related population increases, while providing for and protecting a diversity of present and future land uses.

As shown in Table 6-8, establishing a better coordinated, more comprehensive land-use management system to control changes in the way land is used is one way to achieve this objective. Alternatively, lands disturbed by the development of energy resources would be either reclaimed or redeveloped. These two alternatives are not mutually exclusive, and a comprehensive land-use planning and management system would probably include both. The terminology of land use and reclamation is not standardized; as used here:

- *rehabilitation* means that disturbed lands will be restored to some productive use;
- *reclamation* means restoring lands to their predevelopment use; and
- *redevelopment* means restoring lands to some other productive use.

Table 6-8: *Alternative Policies for Land-Use and Reclamation*

General Alternative	Specific Alternative
Control changes in the way land is used	Designate permissible land uses
	Control access to and activities on all public and private lands
Require rehabilitation of all lands disturbed by energy resource development	Require "reclamation" (i.e., return to predevelopment use)
	Require "redevelopment (i.e., return to some other productive use)

The first category of alternatives—to control land-use changes—emphasizes a comprehensive planning approach that attempts to ensure an "appropriate" use or "best fit" of land and land-use (see Chapter 11). Based on an understanding of present and future land needs and the compatibility of specific areas and uses, permissible uses for these areas would be identified. This means, of course, that certain land uses may be explicitly prohibited—for example, surface mining on prime agricultural lands. And it may be considered desirable to control access to and activities in designated permissible-use areas. For example, the number of people allowed in a recreation area at any one time may be limited, or snowmobiles may be permitted but not other kinds of ORV's.

This comprehensive planning approach could be centralized, with the federal government formulating land-use goals and the states implementing them. (This is the pattern in water and air quality, as discussed in chapters 3 and 4.) On the other hand, the federal role could be limited to providing financial and technical assistance. However, because of the large amount of federal land in the West, it is unlikely that such a state system could deal effectively with the land-use problems and issues we have discussed unless the federal government were to cooperate and coordinate plans for federal lands with state plans for state and private lands. An intermediate alternative that would address this federal-state problem would be to establish a multi-state land-use planning organization as a forum for cooperative, coordinated land-use management in the region.

The second category of alternatives emphasizes returning all lands disturbed by the development of energy resources to a productive use. Within the current land-use and management system, land would be made available to meet energy needs with the stipulation that it be rehabilitated. The specific alternatives would be to require either reclamation or redevelopment. As with the first category, this approach could give the lead in planning and management to the federal government, to the states, or to a multi-state organization.

In the following section, only reclamation and redevelopment are evaluated and compared. This narrower focus is necessary because of the scope of this study, but it also seems appropriate because of the importance of reclamation for western energy development.

In general, the development of energy resources can be viewed as a temporary or interim land use. Lands used for energy resource development will be available for other uses after development is completed. The basic question is whether these lands will be returned to their predevelopment use or to some other productive use. In general, current state and federal laws and regulations require that lands disturbed by surface coal mining be reclaimed.[18] In the West, this usually means returning them to an agricultural use.

Current regulations permit some variation in meeting these objectives depending on site-specific conditions. However, the clear preference articulated in both state and federal requirements is for reclamation for agriculture

use rather than redevelopment for alternative productive uses. It is by no means certain that reclamation is always the best policy alternative, nor is it clear that reclamation is possible in all parts of the West. What follows is an analysis of reclamation and redevelopment. The information on which to base an evaluation is limited, however, in large part because experience with both reclamation and redevelopment in the West is limited. A final section compares the two alternatives on the basis of the five evaluative criteria defined in Chapter 2.

Reclamation

Almost all energy resource development in our eight-state study area will take place in rural areas where the most common current land uses are agricultural—primarily ranching, grazing, and farming. In this setting, reclamation means returning the lands disturbed by energy development to these predevelopment agricultural uses. Two strategies for implementing this alternative are identified in Table 6-9: to continue current land-use requirements or to make the system more flexible to allow for more site-specific variations.

At the present time, reclamation of surface-mined coal lands goes on simultaneously with mining and, in the West, usually involves backfilling, grading, replacing topsoil (which is typically separated and stockpiled during mining), and establishing a vegetation cover. Revegetating requires soil preparation, seeding and/or planting, mulching, and perhaps fertilizing and irrigating (Hodder, 1978; DeRemer and Bach, 1977). At most western mine sites, seeding can be accomplished by either broadcasting or drilling.[19] If trees and shrubs are planted, they are usually planted by hand.

---

[18] As noted previously, comparable requirements have not been established for other surface-mined resources such as oil shale and uranium.

[19] Advancements in seeding methods include improving hydro-seeding or hydraulic seeding (the application of a seed-and-water slurry to soil) which substantially aids in establishing vegetation.

Table 6-9: *Implementation Strategies for Reclamation*

Specific Alternative	Implementation Strategies
Require all lands disturbed by energy resource development to be reclaimed	Continue to require re-grading to original (or approximately original) contour, replacement of topsoil, and (with some exceptions) revegetation with "native" species.
	Change existing state and federal laws and regulations to set rec-lamation goals and require the states to establish implementation programs which set specific requirements on the basis of site-specific conditions.

A wide variety of seeds, shrubs, and trees are currently being used (Plummer, 1977). Native seeds and seedlings are usually required unless it can be shown that introduced species will have "equal or superior utility for the approved postmining land use, or are necessary to achieve a quick, temporary, and stabilizing cover" (*Fed. Reg.,* 1977b:62691). However, no consensus exists among state reclamation authorities and mining companies as to which varieties are preferable. Sweet clover, crested wheatgrass, smooth broom-grass, and alfalfa are relatively quick and easty to get started, and cattle do well on these varieties. However, there is some question about their long-term survival, and in some instances, state authorities have refused to release the reclamation bonds of coal companies which have used only these kinds of vegetation in their reclamation program (Thoem, n.d.). Most of the states in our study area now require the use of native seeds to produce a vegetation cover even though it usually takes longer to establish.

Success of Reclamation: The effectiveness of reclamation can be evaluated by asking:

- What is the probability of success?
- How long will the disturbed lands be withdrawn from productive agricultural uses?
- How compatible will the agricultural uses of the reclaimed lands be with other land uses and with environmental plans and programs in the area? and,
- To what extent will reclamation requirements obstruct the development of energy resources?

As noted earlier, the current practice is for reclamation to be conducted as an integral part of strip mining. The time required to return surface-mined lands to productive uses depends on such site-specific factors as the climate and soil characteristics. Long-term success in reclaiming land, especially in arid and semiarid areas, is uncertain, but it will depend largely on the management commitment made by the landowners and/or the mining company, including controlling grazing and other potentially damaging activities (Hodder, 1973:85; Green, 1977:46). Current federal and state laws typically require completion of reclamation activities within 12 months of expiration of the mining permit. In areas like the arid Southwest, however, this timetable may not be sufficient to establish successful types of native vegetation (LaFevers, 1977).

In terms of achieving land-use objectives, reclamation ranks high in compatibility with other land uses, particularly if the implementation strategy is designed to meet current requirements with regard to contours and seeding. However, unless proper preventive steps are taken, reclaimed lands can be a "nonpoint" source of water pollutants, particularly if fertilizers are used and if erosion and leaching occur. Sedimentation in local streams can also be a problem. And the reclaimed area is a potential source of blowing dust that can contribute to air-quality problems. If these kinds of problems are not controlled, reclamation may well interfere with other environmental plans and programs

Table 6-10: *Estimates of Reclamation Costs in the West (in 1977 dollars)*

Item	High Desert Area (Sites 1 & 2)[a]	Mountain Area Site[b]	Northern Great Plains		
			Site 1[c]	Site 2[d]	Site 3[e]
Design, engineering and overhead:[f]					
per acre	$480-1,070	$650-750	$350-730	$350-650	$200-400
per ton	0.04-0.06	0.06-0.08	0.01-0.02	0.04-0.07	<0.01-0.01
Bond and permit fees:[g]					
per acre	0-80	25-45	50-70	20-40	60-100
per ton	0.0-<0.01	<0.01-<0.01	<0.01-<0.01	<0.01-<0.01	<0.01-<0.01
Backfilling and grading:[h]					
per acre	1,250[i]-3,420	990-1,970	3,700-6,200	2,200-5,500	1,800-2,900
per ton	0.08[i]-0.18	0.10-0.20	0.12-0.20	0.24-0.61	0.05-0.09
Revegetation:[j]					
per acre	110-470	30-45	100-200	100-150	140-200
per ton	0.01-0.02	<0.01-<0.01	<0.01-<0.01	0.01-0.02	<0.01-<0.01
Total:					
per acre	1,840-5,040	1,670-2,810	4,200-7,200	2,670-6,340	2,200-3,600
per ton	0.12-0.26	0.16-0.28	0.13-0.23	0.29-0.70	0.06-0.10

< = less than.

Source: Adapted from Persse, Lockard, and Lindquist, 1977:7.

[a] Selected from five surface mines operating in the Four Corners area of Arizona and New Mexico.

[b] Representative of four low-sulfur bituminous coal operations in Yampa coal field, Routt County, Colorado.

[c] A single-bed area stripping operation like those in Montana and Wyoming (25-foot seam).

[d] An area strip mine producing lignite (North Dakota).

[e] An area stripping operation like those in Montana and Wyoming (40-foot seam).

[f] Includes development of reclamation plan, preparation of periodic environmental reports, supervision of reclamation work, engineering and surveying for environmental protection, water-quality protection and monitoring, overhead of environmental activities, outside environmental consultants, some surveying and mapping, dust control associated with reclamation, and miscellaneous expenses.

[g] Estimated, using annual charge divided by the number of acres disturbed each year.

[h] Includes removal of vegetative cover where topsoil must be salvaged; removing and stockpiling topsoil; backfilling troughs, ramps, and final cuts; rough and fine grading of spoil; replacing topsoil.

[i] Where available, topsoil is removed and replaced, which more than doubles the cost of backfilling and regrading compared to areas with no topsoil.

[j] Soil preparation, seeding and/or planting, reseeding and/or replanting if required, and irrigation.

in the area. (This facet of reclamation is discussed in more detail below.)

Costs and Benefits: Economic costs and benefits are an important measure of the efficiency of reclamation. As shown in Table 6-10, these costs can vary significantly from site to site. In part, this is because of differences in the accounting procedures used by mining companies; in part it is because of many local factors such as terrain, soil, vegetation, type and thickness of overburden, seam thickness, ground and surface water, climate, type and size of equipment, mining technique, applicable reclamation laws and regulations, and the mining companies' methods of land reclamation (O'Neil, 1977; Persse, Lockard, and Lindquist, 1977:6). As Table 6-10 indicates, the total reclamation costs in the West seem to range from about $2,000 per acre to $7,000 per acre. On a per-ton-of-coal basis, reclamation adds from about 10¢ to 70¢ per ton (which is less than 4¢ per million Btu's).

The 1977 SMCRA is adding to these costs since it requires more extensive and detailed surveys and monitoring as part of the requirement for obtaining a lease or permit. For example, the act requires developers to monitor climate and air quality and to prepare a detailed survey of previous and current land uses and of soil and water characteristics, as well as a plan for improved sedimentation control. However, an analysis of SMCRA's potential impact on energy costs suggests that any increases associated with the new legislation will not be great (ICF, 1977:2). The report does point out that several provisions with regard to alluvial valley floors and other lands declared unsuitable for mining are subject to varying interpretations; however, strict interpretation of these provisions will affect the production of coal more than its cost (ICF, 1977:2).

If an evaluation of the cost of reclamation is to be meaningful, the economics must be weighed against the potential return to the energy developer, landowner, and the public-at-large. In other words, the monetary question is: Can the investment be justified in cost-effective terms? Based on estimates of the amount of land likely to be disturbed and average reclamation costs, reclamation of surface-mined lands in the West could require an investment of from $600 million to $5 billion (1976 dollars). Despite the approximate nature of these figures, it has not been generally shown that reclamation is cost-effective (LaFevers, 1977; LaFevers, Johnson, and Dvorak, 1976:177). However, the existence of current laws and regulations re-

quiring reclamation means that we, as a society, are willing to pay the price—both of administering reclamation programs and of more expensive energy—rather than accept social costs associated with "sacrificed" or abandoned lands.

Reclamation appears to be effective in mitigating many indirect effects of energy development, such as those of an aesthetic nature (King, 1977:174). However, some environmental costs of reclamation are difficult to deal with. For example, regrading and soil compaction may cause higher erosion rates, resulting in increased stream sedimentation and some loss of agricultural productivity. Other direct consequences may stem from the intensive use of fertilizers to stimulate rapid revegetation. Over the long term, the extensive use of fertilizers may have an adverse cumulative effect on nearby or adjacent lands. It does seem, however, that these problems are primarily short-term trade-offs which are unavoidable if energy development sites are to be returned to their previous use (LaFevers, Johnson, and Dvorak, 1976:175–76).

The longer-term trade-offs that affect both the costs and benefits of reclamation are "foregone-opportunity costs," such as the loss of potential benefits if the land were returned to some other use. For example, marginal agricultural lands in some western states may be of such limited economic value that the investment of several thousand dollars per acre needed to return the land to its former use cannot be justified on the basis of economic efficiency alone. One study has pointed out that reclamation costs sometimes are as much as 30 times the value of the land (LaFevers, 1977), thereby underscoring the basic reclamation trade-off.

Distribution of Costs: Overall, reclamation costs will result in increased coal prices. Long-term contracts in 1978 for western coal set an at-the-mine price of between approximately $2.50 and $7.50 per ton (Asbury, Kim, and Kouvalis, 1977:45–61). If reclamation costs are between 10¢ and 70¢ per ton (see Table

6-10), the price of reclamation will amount to from 2 to 28 percent of the price of coal. The distribution of these costs can be an important factor influencing the profit margin of a mining company. In the short term, most of these costs are borne almost exclusively by developers. And if the landowner is someone other than the energy company, the benefits of the investment go largely to the owner. Since reclamation costs may often exceed the economic value of the land after restoration, the issue of equity arises. Mining companies may believe they are being asked to bear an unreasonable share of the financial burden. Of course, in the long term, these costs will be passed on to energy consumers. If reclamation becomes too expensive, monetarily, it could eventually limit production at some locations. In sum, however, the basic question of equity is whether it is fair for future generations to be left with unreclaimed mined lands. As addressed by the SMCRA (1977), the public policy response has been to define this as an unacceptable alternative.

Flexibility and Implementation: Decisions concerning reclamation are generally negotiated between the surface owner, energy company, and the state reclamation authorities (or OSMRE) before development begins. Depending on the discretionary authority granted to the state agency, these decisions may or may not accommodate site-specific factors.[20] As a rule, the more specific the state requirements are in terms of reclamation criteria, the less flexible they are. However, reclamation does keep many land-use options open

[20] In August 1976, Colorado's Mined Land Reclamation Board was confronted with the question of how much discretion should be given to its staff to approve changes in reclamation plans once the board has issued a permit. Colorado's 1976 law appears to require public hearings for any change in mining plans, but the agency seemed to be pushing for authority to approve changes so that mining companies would not be inconvenienced by delays (Wynkoop, 1976).

should the uses of adjacent or nearby land change.

Given current laws, the reclamation alternative should be straightforward and easy to implement. Reclamation appears to be preferred by both developers and environmentalists. The uniformity of its requirements should make administration easy if the rules are closely followed. The costs of administration are likely to be low but will increase if the time period is very long, if success is elusive, and/or if administration becomes more flexible in order to allow site-specific requirements.

Redevelopment

In addition to returning disturbed lands to their predevelopment use (i.e., reclamation), they can be redeveloped for some other productive use such as residential subdivisions, new towns, industrial parks, wildlife refuges, public parks, or waste-disposal facilities. Table 6-11 shows that two general strategies for implementation can be followed: first, to change current state and federal laws and regulations to require approval of a redevelopment plan similar to the current reclamation plan; and second, to include plans for redevelopment when initial land-use plans are prepared, including specifying which redevelopment uses are preferred.

Because of the pressures that energy-related population growth will place on outdoor recreational facilities, we have given special attention here to redevelopment for recreational use. Since experience with redevelopment has been limited in the West, evaluation is based on the experience of several redevelopment efforts in other parts of the country.

Federal agencies (especially the Bureau of Mines), development companies, and mining companies are actively investigating alternative concepts for redeveloping the lands on which energy resources are located (Fowler and Perry, 1973:319; Camin et al., 1972; NAS, 1974; Carter et al., 1974; Matter et al.,

Table 6-11: *Strategies for Implementing Redevelopment*

Specific Alternative	Implementation Strategy
Require all lands disturbed by the development of energy resources to be returned to some productive use other than their predevelopment use.	Change existing state and federal laws and requirements to permit redevelopment for compatible productive uses other than reclamation; require redevelopment plans to be approved by state and federal authorities.
	Include specifications of redevelopment uses in county or area-wide land-use plans, designating where specific redevelopment uses are permitted, encouraged, or prohibited.

1974). Most of the studies are based on actual demonstration projects to redevelop abandoned lands where strip, open-pit, or underground mining occurred sometime in the recent past. These studies provide useful information for policymakers who wish to consider a productive land use other than reclamation. As noted in one of these studies, "possible uses for surface-mined lands have been scarcely 'scratched'" (Andreuzzi, 1976).

One of the redevelopment alternatives which has been discussed in a number of studies is the establishment of public parks with recreational facilities (see *Coal Age,* 1974). In general, redevelopment of mined lands for use as public parks with recreational facilities takes place in several sequential stages: planning, site preparation, and additional project work. The planning phase is critical because it involves the cooperation and assistance of numerous interested parties. For example, in a project conducted in Butler County, Pennsylvania, to develop a state park on 117 acres of abandoned lands, both county officials and private contractors were involved in planning the project (McNay, 1970). In

places where resource development has not yet begun, the planning phase would undoubtedly include participation by citizens and local governments to assist in determining how best to redevelop the disturbed lands. Cooperative agrements must be worked out between federal and state agencies when the land to be mined is owned by the federal government, and similar ownership arrangements must be made between state and/or county governments for lands held privately. Friendship Park, a multiple-use recreational facility in Jefferson County, Ohio, was developed on 1,100 acres of strip-mined land that had been donated to the county by Hanna Coal Company, a division of Consolidated Coal (Zande, 1973:294).

In addition to planning redevelopment and acquiring the land, state and federal reclamation laws and regulations must be followed. For example, regardless of the postmining land use, minimum recontour and revegetation requirements have to be met (where applicable) or variances obtained. As noted earlier, states generally prefer reclamation to redevelopment. Thus, efforts to create a new use may require even closer cooperation among all participants to ensure adherence to the reclamation standards.

Finally, the planning phase includes activities to identify funds available from sources in both the public and private sector. For example, the Appalachian Regional Commission has in the past contributed major financial support to states and counties interested in pursuing redevelopment. Support has also come from the various agencies of the federal government with substantive programs related to the particular type of redevelopment —for example, from the Bureau of Outdoor Recreation to develop picnic areas and from the Corps of Engineers to develop reservoirs.

It is not easy to generalize about site preparation largely because it depends on numerous local factors, including the type of energy resource development which disturbed the land, the intended redevelopment use, the geography of the area, and its climate. However, work accomplished in this phase typically entails backfilling, grading and terracing, compacting, reapplication of topsoils, and the construction of parking lots, reservoirs or ponds, and secondary roads. Except where variances are obtained, redeveloped areas will have to be compacted and graded to conform and blend with the surrounding areas. The soil characteristics of a particular area determine the kind of preparation that will be needed if lakes or ponds are to be built. Topsoil requirements and topsoil amendments will also vary depending on the particular activity (Andreuzzi, 1976:9–14). Land development companies and state and mining-company reclamation officials need to cooperate closely in this phase to ensure that those activities that can be incorporated into the ongoing mining plan are carried out simultaneously.

Additional project work represents the last step in redevelopment. Preparation of the site is completed by revegetating open areas that were not previously stabilized, planting trees where required, and constructing park and recreational facilities. The facilities and attractions to make a public park operational typically have been provided by government agencies. These have included picnic grounds, children's play areas, overnight camping facilities, and even museums. Indeed, a major attraction at Keyser Park, developed on strip-mined land in Lackawanna County, Pennsylvania, is the Anthracite Museum constructed by the Pennsylvania Historical and Museum Commission. This museum presents a comprehensive depiction of anthracite from the mine to the consumer. Besides the 27,000 square feet of exhibit area, there is an auditorium, a library, conference rooms, and staff offices.

The need to consider the redevelopment of mined lands as public parks for some sites in the West appears justified given the anticipated pressure of energy-related population on the region's recreational facilities. This is particularly the case when energy development is taking place, or is projected, close to rural communities where few parks or related amenities presently exist. Of course, the costs and benefits of this alternative must ultimately be weighed against other types of reclamation and redevelopment.

Success of Redevelopment: Redeveloping some lands as public parks with recreational facilities can be evaluated in terms of the probability of success, the amount of time the lands will be out of productive uses, compatability with existing land uses in the area, and the extent to which redevelopment might hinder the future development of western energy resources. A primary reason for considering site-specific redevelopment is that the probability of success is high and the time the land is out of productive use is minimized. For example, at Keyser Park, Pannsylvania, it took three years to redevelop 125 acres of abandoned strip-mined land. This was viewed as an extended period, necessary to allow for seeding, planting, and seasonal construction. The implication is that this schedule could be shortened if redevelopment took place simultaneously with extraction. That is, site preparation could proceed as it would for reclamation so that minimal seeding and planting would be required prior to the actual construction and development of facilities.[21]

Redevelopment alternatives, however, may be limited by recent federal regulations regarding postmining land uses (Murray, 1978: vol. 2; LaFevers, 1977; Reiss, 1977). These regulations require that a postmining use which is not the same as the preexisting use must be approved by the state after consultation with the landowner or the appropriate state or federal land-management agency. In addition, the proposed use must be compatible with adjacent and nearby land uses and must comply with existing land-use policies and plans (*Fed. Reg.,* 1977c). Thus, it appears

[21]Moraine State Park, which included redevelopment of 117 acres of abandoned mine lands, took about two years to revegetate. The schedule for Friendship Park in Ohio was comparable.

that some public uses and facilities could not be deemed compatible with present uses, which are predominantly agricultural. This determination would have to be based on site-specific conditions and needs and on projections of land-use trends in the area. In addition to limiting redevelopment options, such planning requirements could delay the processing of premining reclamation plans, thereby hindering resource development.

Costs and Benefits: Because of a lack of western experience with redevelopment, the costs and benefits are not fully known. In two of the abandoned-land projects cited above, Keyser and Moraine parks, development costs ranged from $1,100 to $6,000 per acre. The higher figure is probably more representative since it includes construction of a pond within the park, something that is likely to be built in many redevelopment plans. Based on our review of studies cited earlier, the only generalizations that can be made here are that costs vary according to the type and variety of planned facilities and that they seem to parallel reclamation costs. Clearly, the financial cost would be higher if redevelopment included a large number of recreational attractions. Also (as in reclamation costs) total costs will depend on how much earth moving is involved (McNay, 1970:23). These expenses vary depending on the contour of the land, the quantity of spoils, and the distance that spoils and topsoil must be moved.

Benefits to states, counties, and individuals in the form of receipts or income will, like costs, vary according to the kind, number, and quality of facilities that involve user fees (e.g., utility hookups, campsites, and cabins). Furthermore, reservoirs or lakes could be stocked with sport fish to produce license-fee income to states and income to private entrepreneurs who sold supplies and equipment to visitors. Other benefits could accrue to the communities through jobs created and gross receipts from the year-round use of park facilities. For example, operation of Friendship Park in Ohio requires a park commis-

sioner and 30 to 50 employees with a yearly payroll of almost $200,000. Annual receipts from the park are estimated to be in the range of $500,000 (Zande, 1973:294).

Viewing these facilities as a source of income for counties and states must be put in a context of risk. Careful planning will be needed to anticipate and provide for recreational needs; the risk involved here is the assessment of the market potential for usage-fee facilities (Camin et al., 1972). And as discussed earlier, the population fluctuations associated with energy development may lead to the overbuilding of facilities, adversely affecting the potential income from the parks.

Distribution of Costs and Benefits: One question about how costs and benefits will be distributed concerns the ratio of public to private funding in redeveloping lands for public benefit. In other words, how much should the energy developer contribute towards developing a park and related amenities? Beyond the basic requirements for regrading and revegetation, the developer, the landowner (if other than the developer), and the various related government agencies will have to determine who pays for what. Aside from the obvious advantages for those who live in the area and for visitors, the energy company could benefit from redevelopment through reduced reclamation costs. For example, a significant part of the cost of reclamation stems from earth moving (backfilling, terracing, and topsoil regrading); certain park facilities, however, such as ponds and motorcycle or ORV trails, could require that some part of the park lands be left unreclaimed to enhance the ruggedness of the terrain. Thus, companies would have to do less regrading and terracing, thereby reducing their total earth-moving costs. The risks of leaving land in an unreclaimed state—especially of environmental degradation and for wildlife habitats—will depend on the careful selection of sites for such activities.

The economic costs of redevelopment for the public sector, although uncertain, will probably be higher than for reclamation due to the costs of acquiring the land from the landowner, of lost real estate taxes, and of administration to maintain the park. But the long-term benefit for future generations is clear. Furthermore, new public parks in energy development areas should relieve part of the pressure that increased populations in the West will place on "national heritage lands." On the other hand, committing lands to recreational uses means foregoing future opportunities because most of these commitments will, for all practical purposes, be irreversible.

Flexibility and Implementation: It appears that redevelopment is best implemented on a site-specific, case-by-case basis and therefore can be applied under a variety of conditions and over a period of time. Redevelopment requires the close cooperation of interest groups in planning for each site and thereby increases local and regional control of land use.

To implement this alternative will require changes in existing state and federal laws and requirements. Consequently, a good deal of innovation will be necessary. In addition, redevelopment increases the need for early planning and public involvement in determining the kind of facilities; it demands closer cooperation among all interested parties, both public and private; and it introduces additional participants such as land developers into the decisionmaking process.

Because of all these requirements, the uncertainties of administering and monitoring redevelopment are high, but this is the case for all rehabilitation activities. Although the administrative costs cannot be estimated from existing studies, it seems that they might be higher for redevelopment than for some other alternatives simply because of the cost of being flexible—it generally takes more people and time to deal with individual cases—and the extensive planning needed for a particular site.

Table 6-12:

Criteria	Measures
Effectiveness:	
How effective will redevelopment for public parks be in returning lands disturbed by energy development to a productive use?	Probability of success
	Duration of withdrawal
	Coverage
	Degree of constraint
Efficiency:	
What are the costs, risks, and benefits associated with redevelopment?	Economic costs
	Cost effectiveness
	Risks
Equity:	
How will these costs, risks, and benefits be distributed?	Costs
	Risks
Flexibility:	
Is the alternative sufficiently flexible to be applicable under a variety of conditions and over a period of time?	Adaptability
Implementability:	
How difficult will it be to implement the alternative?	Acceptability
	Ease of administration
	Cost of administration

Comparison of Alternatives and Conclusion

Table 6-12 summarizes the results of our evaluation of reclamation and redevelopment.

Summary and Comparison of the Reclamation and Redevelopment Alternatives

---

Findings

Overall, redevelopment appears to be more likely to succeed, primarily because of the uncertainty associated with reclamation in arid and semiarid areas.

Depending upon the postmining land use, redevelopment can decrease the time lands are out of productive use, compared to reclamation.

Current reclamation requirements do not apply to oil shale and uranium mining. New state and/or federal legislation would be required for either alternative to apply to lands mined for these resources.

Costs may be a constraint for both in some areas. Planning for redevelopment is likely to take longer than planning for reclamation and may delay development of the new land use.

Economic costs appear to be about the same: the estimates range for reclamation range from $1,000 to $7,000 per acre and from $1,000 to $6,000 for redevelopment.

In both cases, the time required to repay economic costs depends directly on the income that the rehabilitated land generates; if redeveloped to a public recreational use, it may be cost-effective in a social sense even if it isn't in an economic sense. The same may be true for reclamation, but for a smaller segment of society.

Redevelopment may reduce the possibility of irreversible consequences. The economic risk for both depends on the market potential of the rehabilitated land.

Developers and consumers will bear the costs of reclamation. If redevelopment is for public use, public revenues may be lost since the land will be off the tax rolls and will have to be maintained and managed. Revenues may or may not make up for these costs.

The greatest risk with reclamation is that it will not succeed, a risk borne by both the developer and society. Redevelopment seems less likely to fail, but it may not preserve as broad a range of future land uses as does reclamation.

Redevelopment is generally more adaptable, takes into account site-specific differences, and gives officials more discretionary authority in interpreting local needs.

Reclamation is currently required on lands strip mined for coal; redevelopment would require changes in federal and state laws. Both can reduce energy-environmental conflicts. Both can reduce westerners' concerns about bearing the cost of producing energy for other regions.

It is likely to be more difficult to administer redevelopment because of the emphasis on detailed planning to meet local needs. Institutional and administrative needs will probably be greater for redevelopment.

Costs are largely unknown for both; however, based on the more extensive institutional and administrative needs anticipated for redevelopment, its costs are likely to be higher than for reclamation.

It should be emphasized that this analysis is limited, in part because of a lack of adequate information, but also because only one type of redevelopment was considered. However, several underlying points can be made. Neither reclamation nor redevelopment emerges in all cases as the best choice; there are areas where reclamation will be impractical, if not impossible; there are limits on where additional recreational needs exist; and redevelopment for recreation purposes may require the same things as reclamation (for example, revegetation). However, it is not at all obvious that the current policy of most states and the federal government uniformly requiring reclamation is as well informed as it might be. Given the high level of uncertainty about being able to reclaim arid and semiarid lands successfully, the present public policy seems to be an oversimplified response to a complex problem. A blanket policy of returning all disturbed lands to their predevelopment uses largely ignores both the possibility of future changes in land use in the surrounding area due to energy and other developments and the possibility that large per-acre expenditures for reclamation may be ineffective. In particular, the need for expanded recreational opportunities for a growing energy-related population challenges the appropriateness of

a policy that emphasizes only reclamation. A more balanced policy would provide for the determination of local needs and would then choose reclamation or redevelopment (and the specific type of redevelopment) based on those needs.

Clearly we need more information and a review of land-use and reclamation policies on a continuing basis. The experience gained from current developments will provide the basis for this review, as will the numerous research efforts that are currently under way.

Finally, as noted above, current land-use laws and regulations pretty much ignore surface mining for other energy resources such as oil shale and uranium. Given the extent of these resources in the West, this omission cannot continue unless the nation is prepared to have those lands withdrawn from productive uses for an extended time period.

REFERENCES

Andreuzzi, Frank C. 1976. *Reclaiming Strip-Mined Land for Recreational Use in Lackawanna County, Pa.: A Demonstration Project.* Washington, D.C.: U.S. Department of the Interior, Bureau of Mines.

Asbury, J. G., H. T. Kim, and A. Kouvalis. 1977. *Survey of Electric Utility Demand for Western Coal.* Argonne, Ill.: Argonne National Laboratory.

Atomic Energy Act of 1954, Pub. L. 83-703, 68 Stat. 919, as amended by Pub. L. 91-560, 84 Stat. 1472.

Bald Eagle Protection Act of 1940 Pub. L. 76-567, 54 Stat. 250.

Bosselman, Fred, and David Callies. 1972. *The Quiet Revolution in Land Use Control.* Washington, D.C.: Government Printing Office.

Bureau of Outdoor Recreation Act of 1963, Pub. L. 88-29, 77 Stat. 49, 16 U.S.C. 460.

Camin, Kathleen Q., et al. 1972. *Mined-Land Redevelopment.* Lawrence: University of Kansas, State Geological Survey.

Carter, Ralph P., et al. 1974. *Surface Mined*

Land in the Midwest: A Regional Perspective for Reclamation Planning, for the U.S. Department of the Interior. Argonne, Ill.: Argonne National Laboratory.

Chironis, Nicholas P. 1977. "Imaginative Plans Make Mined Land Better Than Ever." *Coal Age* 82 (July):49.

Choitz, Jackie. 1978. "Mining Law Update." *Colorado/Business* 5 (February):27–28.

Clawson, Marion, and Burnell Held. 1957. *The Federal Lands.* Baltimore, Md.: published for Resources for the Future by Johns Hopkins University Press.

Coal Age. 1974. "'Interim Land Use' Keys Amax Coal's Policies." 79 (October):131–38.

Congressional Quarterly, Inc. 1973. *Congress and the Nation, Vol. 3: 1969-1972.* Washington, D.C.: Congressional Quarterly.

Council of State Governments, Task Force on Natural Resources and Land Use Information and Technology. 1975. *Land: State Alternatives for Planning and Management.* Lexington, Ky.: Council of State Governments.

Davis, Grant, Associate Program Manager for

Research, Surface Environment and Mining (SEAM) Program, Forest Service, U.S. Department of Agriculture. November 1976. Personal communication.

Davis, Grant, Associate Program Manager for Research, Surface Environment and Mining (SEAM) Program, Forest Service, U.S. Department of Agriculture. 1977. Personal communication.

DeRemer, D., and D. Bach. 1977. "Irrigation of Disturbed Lands." In *Reclamation and Use of Disturbed Land in the Southwest,* edited by J. L. Thames. Tucson: University of Arizona Press.

Endangered Species Conservation Act of 1969, Pub. L. 91-135, 83 Stat. 275.

Endangered Species Preservation Act of 1973, Pub. L. 93-205, 87 Stat. 884.

Engineering-Science. 1979. *Evaluation of the Impacts of the Proposed Regulations to Implement the Resource Conservation and Recovery Act on Coal-Fired Electric Generating Facilities,* Phase I, Draft Interim Report. McLean, Va.: Engineering-Science.

Environment News. 1976. "Interests Governing Wildlife Too Narrow, Train Says." Boston, Mass.: Environmental Protection Agency, Region I, New England Regional Office, November.

Environmental Protection Agency (EPA), Office of Energy Activities, Region VIII, et al. 1976. *Surface Coal Mining in the Northern Great Plains of the Western United States: An Introduction and Inventory Utilizing Aerial Photography Collected in 1974 and 1975.*
Denver: EPA.

Farmer, E. E., et al. 1974. *Revegetation Research on the Decker Coal Mine in Southeastern Montana,* Research Paper INT-162. Ogden, Utah: U.S. Department of Agriculture, Forest Service, Intermountain Forest and Range Experiment Station.

Federal Coal Leasing Amendments Act of 1975 (1976), Pub. L. 94-377, 90 Stat. 1083.

Federal Land Policy and Management Act of 1976, Pub. L. 94-579, 90 Stat. 2743, 43 U.S.C. 1701 *et seq.*

Federal Register. 1977a. "Off-Road Vehicles on Public Lands." 42 (May 25):26959-60.

Federal Register 42 (December 13, 1977b): 62691-95.

Federal Register 42 (December 13, 1977c): 62681.

Fowler, Dale K., and Charles H. Perry, III. 1973. "Three Years Development of a Public Use Wildlife Area on a Mountain Coal Surface Mine in Southwest Virginia." In *Research and Applied Technology Symposium on Mined-Land Reclamation,* edited by Bituminous Coal Research, Inc. Monroeville, Pa.: Bituminous Coal Research, Inc.

General Mining Law of 1872, 17 Stat. 91.

Geothermal Steam Act of 1970, Pub. L. 91-581, 84 Stat. 1566.

Gisser, Misha, et al. 1979. *Water: Agriculture Versus Energy Development in the Four Corners Area.* Los Alamos, N. Mex.: Los Alamos Scientific Laboratory.

Gits, Victoria. 1978. "Boom Town on the Brink." *Colorado/Business* 5 (February): 30–42.

Green, Becky B. 1977. *Biological Aspects of Surface Coal Mine Reclamation, Black Mesa and San Juan Basin,* Regional Studies Program. Argonne, Ill.: Argonne National Laboratory.

Grim, Elmore C. 1976. "Environmental Assessment of Western Coal Surface Mining." In *Proceedings of National Conference on Health, Environmental Effects, and Control Technology of Energy Use,* pp. 177–78. Washington, D.C.: Environmental Protection Agency.

Grogan, Sterling, Utah International. 1977. Personal communication.

Healy, Robert G. 1976. *Land Use and the States.* Baltimore, Md.: published for Resources for the Future by Johns Hopkins University Press.

Heeman, Michael T. 1977. "EPA Sets Water Pollution Limits." *Coal Age* 82 (July):141.

Hodder, Richard L. 1973. "Surface Mined Land Reclamation Research in Eastern Montana." In *Papers Presented Before Research and Applied Technology Symposium on Mined-Land Reclamation,* edited by Bituminous Coal Research, Inc. Monroeville, Pa.: Bituminous Coal Research, Inc.

Hodder, Richard L. 1978. "Potentials and Predictions Concerning Reclamation of Semiarid Mined Lands." In *The Reclamation of Disturbed Arid Lands,* edited by R. A. Wright. Albuquerque: University of New

Mexico Press.

ICF, Inc. 1977. *Energy and Economic Impacts of H.R. 13950 ("Surface Mining Control and Reclamation Act of 1976," 94th Congress)*, Draft Final Report for the Council on Environmental Quality and Environmental Protection Agency, Executive Summary and 2 vols. Washington, D.C.: ICF.

Jones, Douglas N., et al. 1974. *The Energy Industry: Organization and Public Policy.* Washington, D.C.: Government Printing Office.

King, David A. 1977. "Recreational Opportunity Costs." In *Reclamation and Use of Disturbed Land in the Southwest,* edited by John L. Thames. Tucson: University of Arizona Press.

Kirschten, J. Dicken. 1977. "Converting to Coal—Can It Be Done Cleanly?" *National Journal* 9 (May 21):781–84.

LaFevers, James R. 1977. "Effect of Legislative Change on Reclamation." In *Reclamation and Use of Disturbed Land in the Southwest,* edited by John L. Thames. Tucson: University of Arizona Press.

LaFevers, James R., Donald O. Johnson, and Anthony J. Dvorak. 1976. *Extraction of North Dakota Lignite: Environmental and Reclamation Issues.* Argonne, Ill.: Argonne National Laboratory.

Land and Water Conservation Fund Act of 1965, Pub. L. 88-578, 78 Stat. 897, as amended by the Federal Water Project Recreation Act, Pub. L. 89-72, 79 Stat. 213.

Leydet, François. 1976. "Jackson Hole: Good-Bye to the Old Days?" *National Geographic* 150 (December):768–89.

Magida, Arthur J. 1975. "Environmental Report/Major Revisions Likely in Federal Coal Leasing Program." *National Journal Reports* 7 (August 2):1101–9.

Matter, Fred S., et al. 1974. *A Balanced Approach to Resource Extraction and Creative Land Development Associated with Open-Pit Copper Mining in Southern Arizona.* Tucson: University of Arizona, College of Architecture and College of Mines.

McKell, Cy, Utah State University. 1977. Personal communication.

McNay, Lewis M. 1970. *Surface Mine Reclamation, Moraine State Park, Pennsylvania.* Washington, D.C.: U.S. Department of the Interior, Bureau of Mines.

Mineral Leasing Act of 1920, Pub. L. 66-145, 41 Stat. 457. Lands subject to the Mineral Leasing Act for Acquired Lands of 1947 (30 U.S.C.A. 351-59, 61 Stat. 913), are also covered by the 1920 Act.

Mining Act of 1866, 14 Stat. 251.

Missouri Basin Inter-Agency Committee. 1971. *The Missouri River Basin Comprehensive Framework Study,* 7 vols. Denver: U.S. Department of the Interior, Bureau of Land Management.

Moore, Russell, and Thomas Mills. 1977. *An Environmental Guide to Western Surface Mining, Part One: Federal Leasable and Locatable Mineral Regulations.* Washington, D.C.: Government Printing Office.

Multiple Use-Sustained Yield Act of 1960, Pub. L. 86-517. 74 Stat. 215.

Murray, Francis X., ed. 1978. *Where We Agree, Report of the National Coal Policy Project,* 2 vols. plus summary vol. Boulder, Colo.: Westview Press.

National Academy of Sciences (NAS). 1974. *Rehabilitation Potential of Western Coal Lands,* A Report to the Energy Policy Project of the Ford Foundation. Cambridge, Mass.: Ballinger.

National Forest Management Act of 1976, Pub. L. 94-588, 90 Stat. 2949.

Northern Plains Resources Council. 1979. *The Plains Truth* 8 (February):1–12.

Oak Ridge National Laboratory (ORNL). 1978. *Environmental and Health Aspects of Disposal of Solid Wastes from Coal Conversion: An Information Assessment.* Oak Ridge, Tenn.: ORNL.

Oil and Gas Journal. 1979a. "Congress Resumes Alaska Lockup Push." 77 (February 19): 68–69.

Oil and Gas Journal. 1979b. "U.S. Oil Groups Intensify Blasts at Land Withdrawals." 77 (February 19):68.

Old West Regional Commission Bulletin. 1977. "Interior May End EMARS Leasing." 4 (August 1):4.

O'Neil, T. J. 1977. "Operating Considerations." In *Reclamation and Use of Disturbed Land in The Southwest,* edited by John L. Thames. Tucson: University of Arizona Press.

Packer, Paul E. 1974. *Rehabilitation Potentials and Limitations of Surface-Mined Land in the Northern Great Plains,* General Technical

Report INT-14. Ogden, Utah: U.S. Department of Agriculture, Forest Service, Intermountain Forest and Range Experiment Station.

Pierce, Neal R. 1972. *The Mountain States of America.* New York: Norton.

Persse, Franklin H., David W. Lockard, and Alec E. Lindquist. 1977. *Coal Surface Mining Reclamation Costs in the Western United States,* Bureau of Mines Information Circular 8737. Washington, D.C.: U.S. Department of the Interior, Bureau of Mines.

Plummer, A. P. 1977. "Revegetation of Disturbed Intermountain Area Sites." In *Reclamation and Use of Disturbed Land in the Southwest,* edited by John L. Thames. Tucson: University of Arizona Press.

Power, J. F., F. M. Sandoval, and R. E. Rise. 1978. "Restoration of Productivity to Disturbed Lands in the Great Northern Plains." In *The Reclamation of Disturbed Arid Lands,* edited by R. A. Wright. Albuquerque: University of New Mexico Press.

Rattner, Steven. 1978. "Tough Task in Implementing Crippled Energy Plan." *New York Times National Economic Survey,* January 8, sect. 12, p. 13.

Reilly, William K., ed. 1973. *The Use of Land: A Citizens' Policy Guide to Urban Growth.* New York: Thomas Y. Crowell.

Reiss, I. H. 1977. "Total Utilization of a Land Resource." *Mining Congress Journal* 63 (October):55–59.

Resource Conservation and Recovery Act of 1976, Pub. L. 94-580, 90 Stat. 2795.

Rock Springs, Wyoming, Department of Housing and Community Development, Staff. March 1979. Personal communication.

Ruffner, J. A., and F. E. Blair, eds. 1977. *The Weather Almanac.* Detroit: Gale Research Co.

Scher, Zeke. 1977. "The Great Canyonlands Double Cross." *Denver Post Empire Magazine,* August 21, p. 16.

Stairs, Gerald R. 1977. "Land-Use Planning: An Overview." In *Reclamation and Use of Disturbed Land in the Southwest,* edited by John L. Thames. Tucson: University of Arizona Press.

Surface Mining Control and Reclamation Act of 1977 (SMCRA), Pub. L. 95-87, 91 Stat. 445.

Thoem, Terry, Environmental Protection Agency, Region VIII. N.d. Personal communication.

U.S. Congress, Office of Technology Assessment (OTA). 1979. *Management of Fuel and Nonfuel Minerals in Federal Land.* Washington, D.C.: Government Printing Office.

U.S. Department of Commerce, Bureau of the Census. 1976. "Estimates of the Population of States, By Age: July 1, 1974 and 1975 Advanced Report." *Current Population Reports,* Series P-25, No. 619.

U.S. Department of Energy (DOE), Assistant Secretary for Environment, Office of Technology Impacts. 1980. *Synthetic Fuels and the Environment: An Environmental and Regulatory Impact Analysis.* Washington, D.C.: DOE.

U.S. Department of the Interior (DOI). 1976. *Mining and Minerals Policy Annual Report.* Washington, D.C.: Government Printing Office.

U.S. Department of the Interior (DOI). 1978. *Mining and Minerals Policy Annual Report.* Washington, D.C.: Government Printing Office.

U.S. Department of the Interior (DOI), Bureau of Land Management (BLM). 1975. *Resource and Potential Reclamation Evaluation: Hanna Basin Study Site,* EMRIA Report No. 2. Rawlins, Wyo.: BLM.

U.S. Department of the Interior (DOI), Bureau of Land Management (BLM). 1979. "Geothermal Leasing, General, Competitive Leases, Miscellaneous Amendments." *Federal Register* 44 (March 5):12037–38.

U.S. Department of the Interior (DOI), Bureau of Outdoor Recreation. 1976. *Draft Environmental Impact Statement: Departmental Implementation of Executive Order 11644 Pertaining to the Use of Offroad Vehicles on the Public Lands.* Washington, D.C.: Bureau of Outdoor Recreation.

U.S. Department of the Interior (DOI), Fish and Wildlife Service (FWS). 1978. "Endangered and Threatened Wildlife and Plants: Proposed Critical Habitat for the Whooping Crane." *Federal Register* 43 (August 17): 36588–99.

U.S. Department of the Interior (DOI), Fish and Wildlife Service (FWS), Staff. 1977. Personal communication.

U.S. Department of the Interior (DOI), Geological Survey, Staff. 1977. Personal communication.

U.S. Department of the Interior (DOI), Heritage Conservation and Recreation Service, Staff. August 1978. Personal communication.

U.S. Department of the Interior (DOI), Office of Surface Mining Reclamation and Enforcement (OSMRE). 1979. "Surface Coal Mining and Reclamation Operations, Permanent Regulatory Program." *Federal Register* 44 (March 13):14902-15463.

U.S. President, Office of the White House Press Secretary. May 1977. "Message to the Congress from the President." Washington, D.C., photocopy.

Wagner, James R. 1970. "Agencies Speed Up Review in Push to Expand Wilderness System." *National Journal* 2 (December 26): 2826-31.

Wagner, James R. 1971. "Interior Sub-committees Move Slowly on Legislation to Reform Public Lands Policy." *National Journal* 3 (August 21):1768-73.

Weissenborn, K. 1977. "The Forest Service and Land Use Planning." In *Reclamation and Use of Disturbed Land in the Southwest,* edited by John L. Thames. Tucson: University of Arizona Press.

White, Michael D. 1975. "Constitutional Derivation and Statutory Exercise of Land Use Control Powers." In *Rocky Mountain Mineral Law Institute: Proceedings of the 21st Annual Institute,* pp. 657-719. New York: Matthew Bender.

Wilderness Act of 1964, Pub. L. 88-577, 78 Stat. 890.

Wynkoop, Steve. 1976. "Reclamation Board Ponders Authority." *Denver Post,* August 20.

Zande, Richard D. 1973. "Friendship Park: One Use of Reclaimed Strip Mine Land." In *Research and Applied Technology Symposium on Mined-Land Reclamation,* edited by Bituminous Coal Research, Inc. Monroeville, Pa.: Bituminous Coal Research, Inc.

CHAPTER 7

Housing

During the Sweetwater County, Wyoming, boom's early stages, the housing stock of the county was approximately doubled with 6,000 new units. Fifty-five hundred of these were mobile homes. . . . The housing market demand drove prices and rents up, yet investors and developers were skeptical of putting money into boomtown housing. Even if they had been willing, mortgage money was not available for the typical miner wishing to buy a home. This same pattern is reproduced in most boomtowns. Even when mining or utility companies bring housing into being, little or no provision is made for the local service workers who must also be accommodated in a fast-growing community (Gilmore, 1977).

As this description emphasizes, the population growth associated with the development of energy resources can have a serious impact on housing—both supply and quality—in nearby communities. The majority of these housing problems are the result of a rapid and large increase in population which then quickly declines from the peak level but remains higher than the predevelopment population. The private housing market often does not respond adequately to the new housing needs created by energy development, and government programs generally are not designed to deal with housing shortages in rapidly growing areas. As a result, few new homes are built, available housing is very expensive, and the use of mobile homes usually increases dramatically.

In this chapter, a brief summary of the major factors that affect housing in areas where energy resources are being developed is followed by a discussion of five specific housing problems and issues. The final section of the chapter describes and evaluates some of the policy options that might be used to deal with the problems and issues we have identified. Housing, of course, is but one component of the planning and growth-management problems that occur when energy resources are developed in rural areas. Other local service and facility needs are discussed in Chapter 8.

Factors Affecting Housing Needs

The precise magnitude, or rate, of population growth that begins to seriously affect housing problems is difficult to define, even though recent rapid growth in many western towns has been well documented. A useful rule of thumb seems to be that an annual growth rate of 5 percent (which will double the population in 14 years) tends to produce serious repercussions (Gilmore and Duff, 1975:2; FEA, Region VIII, Socioeconomic Program Data Collection Office, 1977:20); but more extreme cases are not hard to find. Rock Springs, Wyoming, for example, grew at an annual rate of 19 percent from 1970 to 1974, and rapid growth has been a serious problem in many other western towns. Table 7-1 lists 14 towns where the population grew between 5

191

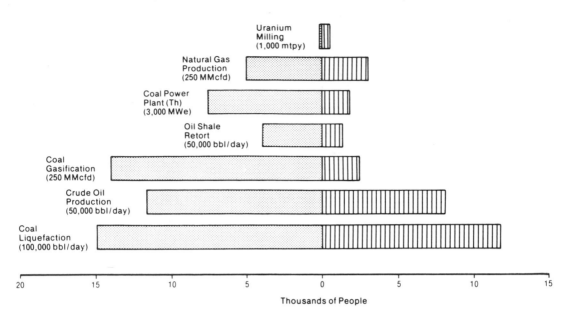

Figure 7-1: *Population Increases from Energy Facilities*

Table 7-1: *Population Growth in Selected Western Communities*

		Population		
Town	Energy Resource[a]	1970	1977	Average Annual Increase (Percent)
Colorado				
Carbondale	Coal	726	1,600	17
Craig	Coal	4,205	6,657	8
Grand Valley	Oil Shale	270	500	12
Rifle	Oil Shale	2,150	3,500	9
Montana				
Hardin	Coal	2,733	3,637	5
North Dakota				
Washburn	Coal	805	1,400	10
Utah				
Cedar City	Coal	8,946	12,000	5
Huntington	Coal	857	1,700	14
Vernal	Oil Shale	3,908	5,200	5
Wyoming				
Douglas	Coal, Uranium, Oil, Gas	2,677	7,200	25
Gillette	Coal, Oil, Gas, Uranium	7,192	10,200[b]	6
Moorcroft	Coal, Gas, Oil, Uranium	981	2,000	15
Rock Springs	Coal, Gas, Oil	10,500	23,250	17
Wheatland	Coal	2,500	4,500	11

Source: FEA, Region VIII, Socioeconomic Program Data Collection Office, 1977.

[a] Resources listed in order of importance to population growth.

[b] Campbell County Chamber of Commerce, 1976.

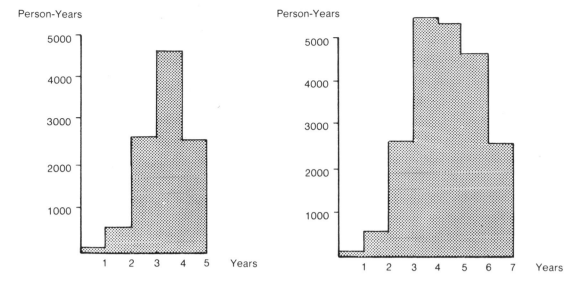

7-2.1: One Facility - Normal five-year construction schedule

7-2.2: Two Facilities - Second facility, started two years after first, compounds impact

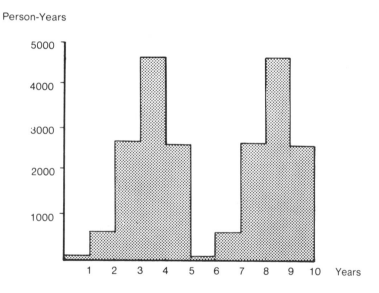

7-2.3: Two Facilities - Second facility started when first is completed creates severe peaks and valleys in impact

Figure 7-2: *Personnel Requirements for Coal Gasification Plant Construction and Impact of Various Schedules (Populations associated with each year would be about 3.0 times greater than employment).* (Source: Carasso et al., 1975:6-30-31.)

and 25 percent annually from 1970 to 1977. In general, rapid growth affects small communities more than large towns, because the same energy facility (and, therefore, the same population increase in absolute numbers) has a proportionally greater impact.

In addition to the rate of population growth, another major factor affecting the severity of housing problems is the relationship between the size of the construction work force and the operations work force. Figure 7-1 shows the peak employment levels of both for a range of energy technologies. The construction/operation employment ratio gives an indication of the boom-bust cycle associated with large construction projects. More capital-intensive facilities create larger booms in local employment, but they also result in greater employment declines when the construction work is completed.

Scheduling can also play a major role in determining the actual magnitude of population growth and, therefore, of housing problems. For example, construction of a coal gasification facility normally takes about five years. The increase in population during the construction phase is typical of large-scale energy projects: thousands of people are needed for only a short time (Figure 7-2.1). Simultaneous construction of two facilities, of course, brings even larger numbers of people into an area, thus increasing the needs for all services, including housing. Accordingly, if construction of a second gasification facility were begun two years after the first, a longer period of temporary impact would occur (Figure 7-2.2). In this situation, there is a risk that the expanded needs created during a seven-year construction period would be expected to continue over the long term. This occurred in several places along the route of the Alaskan pipeline (see Gilmore and Duff, 1975). Conversely, consecutive construction projects cause a smaller absolute impact, but can result in more severe peaks and drops in population and its impact (Figure 7-2.3). If construction of the second facility is delayed, planning for increased housing and services on the basis

of an uncertain schedule may be more difficult. Smaller facilities would somewhat reduce the labor needed for construction, but because of economies of scale, the impact from a facility one-half the typical size tends to be greater than half the full-scale impact.

Problems and Issues

Housing in the U.S. has traditionally been provided by the private sector. In most circumstances, the private market operates satisfactorily, although some government programs have been established to help meet the needs of low- and moderate-income families. In virtually all cases of government action, however, programs sponsored in the public sector have worked through the private housing market, allowing private firms to construct housing and private lenders to deal with financing. However, in situations involving rapid or large population changes, such as occur in western communities located near areas of energy resource development, the private housing market does not respond adequately. A lack of programs to deal with these housing shortages has meant that mobile homes are the principal type of housing for both temporary residents, who are employed during the construction phase, and permanent residents, who will operate the new facility.

Although rising home prices have been felt in most parts of the country, incomes in western towns are well above the national average (Mountain West Research, 1976a:46–51, 103), suggesting that low income is not the reason for poor housing conditions in the region's boomtowns. Overall, residents of these towns have less choice in housing than do people elsewhere in the nation.

In the next section, we will focus on five of the housing issues prevalent in western towns:

• Risks for private investors;
• Inadequate government housing programs;

- A predominance of mobile homes;
- A lack of alternative types of housing;
- Severely inadequate housing on Indian reservations.

Risks for Private Investors

Risks of housing construction in boom-towns are apparently too great for lenders, housing developers, and energy firms to invest in these towns. Housing is traditionally provided by the private sector, and without direct financing of home construction and mortgages by local financial institutions, few new homes can be built.

The market for new housing in energy development communities is almost completely dependent on the employment prospects. Energy development projects involving advanced technologies may be cancelled, mines may be shut down (before mortgages would be paid off), or projects may be delayed for as much as several years, all of which makes the risk of being left with unsold homes too great for builders (University of Denver, Research Institute, Industrial Economics Division, 1977). The uncertainty involved in western boom-town housing pervades all aspects of financing housing, beginning with the estimation of actual housing demand (Figure 7-3). Lenders

and housing developers, like builders, cannot afford to rely on announced schedules for the development of advanced energy technologies, which can leave them holding unsold homes when projects are delayed. The private, secondary mortgage market (Figure 7-3, Step 9), made up principally of linkages between savings and loan associations, on the one hand, and government agencies, insurance companies, and large investors, on the other hand, may also be unwilling to back investments outside the low-risk categories. However, even when permanent financing is available, construction financing is often difficult to obtain in small towns (Figure 7-3, Step 5). For example, a mine projected to be active for 20 to 25 years provides the town with an employment base that is shorter than the term of a typical (30-year) mortgage on a new home.

The potential demand for new homes is somewhat more certain in the larger towns of the region, such as Gillette, Wyoming, and Farmington, New Mexico, where some executives or administrative personnel connected with the energy facility tend to live. Construction personnel, however, are temporary residents in the community, and the uncertainty about future energy development leads to a fear of overbuilding for the long term. This is particularly true of towns that have previously had booms followed by busts; these events have made some local financial insti-

Step	Factors Considered
1. Market Research	Demand for housing
2. Land Acquisition	Location, area, drainage, zoning, utilities, and other considerations
3. Subdivision and Development	Streets and utilities
4. Building Permit	Local laws and regulations
5. Construction Loan	Local loans watched closely by local lenders
6. Construction	Surety bond to guarantee completion
7. City Certification of Occupancy	Local building standards
8. Sale of Home and Permanent Financing	Local long-term loan
9. Sale of Mortgage to Government or Private Investors	Secondary mortgage market

Figure 7-3: *Typical Sequence of Housing Development*
(Source: Adapted from Nelson and Whitman, 1976.)

Table 7-2: *Average Price and Total Monthly Costs of Housing Types in Impacted Western Communities, 1977*

State	Average Total Monthly Cost			Typical Down Payment: Three-Bedroom Single-Family Home	Average Price	
	Three-Bedroom Single-Family Home	Two-Bedroom Mobile Home	Unfurnished Two-Three Bedroom Apartment		Three-Bedroom Single Family Home	Two-Bedroom Mobile Home
Arizona	$302	$210	$125-250	20%	$39,000	$10,000-15,000
Colorado	404	239	$200-250	20%	46,000	10,000-15,000
Montana	337	213	150	20%	40,000	10,000-15,000
New Mexico	425	176	250	10%	50,000	10,000-15,000
North Dakota	395	195	200-240	20%	50,000	10,000-15,000
Utah	385	191	250-300	20%	50,000	10,000-15,000
Wyoming	410	197	285-385	10%-20%	50,000	10,000-15,000

Source: Carlisle, 1977:492; U.S. Department of Commerce, Bureau of the Census, 1977a; U.S. Department of Commerce, Bureau of the Census, 1977b: 48.

tutions extremely cautious about the current development of energy resources. If homes might not be needed in the future, it is difficult to obtain financing for construction even though real housing shortages may exist at present. Jack Presnell of Phillips Petroleum summarizes the view of lenders in northwestern New Mexico: "There is a natural reluctance on the part of the traditional private lending agencies to lend mortgage money in situations like this until they can see more [clearly] what the future holds" (Presnell, 1977).

The reluctance of local lenders and the resultant low level of home-building activity have had a dramatic effect on prices (Table 7-2). The average price of new homes in 1977 in our eight-state study area (around $50,000) was about 30 percent higher than the national average for nonmetropolitan areas, and apartment rents in the region were nearly double the national figure of $120 per month.

The reluctance of local lenders is attributable in part to their isolation from investors who operate at the national level. Local financial institutions in the West are unable to accommodate the demand for new mortgages and construction loans in rapidly growing small communities. And external financing, a common practice for urban housing, is constrained in the West both by the absence of financial relationships with out-of-state institutions and by the inability of local financial institutions to accumulate sufficiently large blocks of mortgages to sell in the secondary mortgage markets (Carlisle, 1977:492; Wyoming, Department of Economic Planning and Development, Office of the Chief of State Planning, 1977). These obstacles are intensifying: from 1970 to 1976 the secondary mortgage market tripled in volume nationally, making small-town institutions even less competitive in the national system (Brockschmidt, 1977).

Factors other than financing also contribute to housing shortages in the West. For several reasons, housing developers and builders have not been adequately meeting the needs of energy boomtowns. Local builders frequently lack experience with large-scale developments and are not organized to manage them. Outside developers often reject smaller cities and towns as unsafe risks or as inadequate markets for large-scale projects (see box).

In addition, housing construction is difficult to maintain in the face of inflated wages, driven up by competition from energy firms which offer higher pay. Thus, recruiting and keeping a skilled construction labor pool is difficult and costly. The difficulty of hiring workers and their high payrolls restrain local firms from expanding. The development of energy resources can also cause out-of-town firms to stay away from the communities involved because there, costs are higher than in major urban areas (University of Denver, Research Institute, Industrial Economics Division, 1977).

The homebuilding industry consists largely of small firms that build homes primarily in their local area. A few large homebuilders account for about one-third of all new homes, but these are located almost exclusively in large metropolitan areas. Outside major urban areas, builders tend to operate within a fairly small area—frequently, part of a state. Local building codes, zoning regulations, labor markets, and union work rules vary from area to area. More importantly, construction financing is usually obtained in the local area, and builders often have trouble obtaining loans outside their usual locus of operation (Henderson and Kohlhepp, n.d.).

Large-scale housing developments are still uncommon in western energy areas, remaining instead in the more profitable suburban areas of major cities. Large developers also tend to have better access to sources of financing in major cities. These large developers and their investors have not been operating in small towns, and small developers and builders have been unable or unwilling to take risks on speculative (as opposed to custom-built) homes.

Several energy developers, however, have taken actions to provide housing (primarily mobile homes) for construction workers and their families because of evidence of lower productivity, worker (or spouse) dissatisfaction, and increased labor turnover without adequate housing (Metz, 1977; Jacobsen, 1976). Some firms still tend to resist housing commitments, generally on the grounds of cost and the fact that housing has not been the firms' responsibility historically (Arthur Young and Company, 1976); but a trend toward housing activity by the industry is clear.

Much of the housing provided by energy companies is in the form of planned mobile-home developments, either on a temporary basis in a permanent subdivision or in sites removed from other towns (Atlantic Richfield's development of a mobile home park at Wright, Wyoming, 40 miles from Gillette, is an example [see box]). Some commercial, social, and recreational amenities are included in such developments, but often not while construction workers' families are the only residents.[1]

Largely because of the overwhelming preference of families for single-family housing (Mountain West Research, 1976b), permanent housing projects financed and planned by energy development companies are also becoming more common in the West. The best example is the new town of Colstrip, a company town owned by Montana Power Company. There, older homes were modernized, and both single-family and multifamily units were built, in addition to permanent mobile home spaces. Urban amenities such as recreation areas and commercial development are also part of the modern company town (Myhra, 1975; White, 1977). In some ways the company town approach is an advantageous solution to housing shortages as well as to the local impact of energy development. The cost of building and operating such a town, however, is prohibitive for most firms, and there is little sentiment among companies for repeating the Colstrip development in other areas (Arthur Young and Company, 1976). Other firms have become indirectly involved in providing housing for their employees, especially in isolated, small towns in Utah and Wyoming. The common approach is to provide a guarantee to private developers for a certain number of homes. However, these developments typically give rental or purchasing preference to company employees (Metz, 1977). As a result, local workers usually employed by a construction contractor and service workers in nearby towns are not aided by these activities.

Finally, some states are working with private industry to alleviate the impact of energy development. The Wyoming Industrial Siting Administration (WISA), which has taken the lead in this area, will issue a siting permit

for major facilities only after studies and hearings with state agencies and the local community provide information on the local impact (Industrial Development Information and Siting Act of 1975). In the only permit processed by late 1977, Basin Electric Power Cooperative agreed to provide approximately 1,900 housing units and other services in connection with the construction and operation of its power plants near Wheatland (WISA, n.d.; Valeu, 1977).

Inadequate Government Housing Programs

Public sector programs have not dealt effectively with the housing problem, and no federal programs are directed toward the specific housing shortages in rapidly growing areas of energy development. Although several western states have established housing finance programs designed to overcome the lack of mortgage money and federal assistance, these actions have had little effect on local housing markets.

The federal government's major role in the housing market is through federally chartered agencies that purchase mortgages initiated by local institutions. This secondary mortgage market thus makes more funds available to local mortgage lenders and is intended to balance out the supply and demand of mortgage funds. The Federal National Mortgage Association (FNMA, called Fannie Mae), the Government National Mortgage Association (GNMA or Ginnie Mae), and the Federal Home Loan Mortgage Corporation (FHLMC or Freddie Mac) all purchase mortgages, often for resale. Although the three institutions are able to purchase conventional mortgages, they have tended to concentrate on those loans insured by the Federal Housing Administration (FHA) or guaranteed by the Veterans Administration. FHLMC, however, is specifically intended to operate in the conventional mortgage sector and is becoming increasingly ac-

[1]Mobile homes are discussed at greater length in the next section.

tive in that direction. In addition, the Federal Home Loan Bank Board extends credit in the form of advances to its member institutions, most of which are savings and loan associations (Nelson and Whitman, 1976:483–87; for additional information on the activities of each agency, see Brockschmidt, 1977).

These federal institutions reduce the risk to lenders, but only indirectly affect the overall availability of home mortgage money. As home prices and interest rates rise, federally insured loans at reduced rates become potentially more important in alleviating housing shortages.

Nationwide, FHA loans have accounted for 15 to 30 percent of the home mortgage market, but that percentage has been declining in the 1970s (Nelson and Whitman, 1976:483–87). The use of FHA loans is falling off for several reasons: periodic construction inspections, slow processing times, interest rates lower than conventional rates which discourages lenders, heavy documentation requirements, and "red tape." Builders must pay closing costs (called points) to bring the FHA interest rate up to market rates. In western rural areas where energy resources are being developed, the scarcity of mortgage bankers and of savings and loan associations (which tend to be linked to the secondary mortgage market) further decreases the use of FHA programs (Carlisle, 1977:492; Schafer, 1977).

Retirees and other people on fixed incomes are especially hard-hit by the increase in housing costs and rents during energy development. Low-income housing programs sponsored by the Department of Housing and Urban Development (HUD), such as the Section 235 Program, limit loans to $32,000 for a three-bedroom home ($38,000 in high-cost areas)[2] for families earning less than the area's median income. Since average housing prices are well above these limits in the West, (see Table 7-2), the program is of little benefit in

---

[2] These are the 1978 limits; they are raised periodically to meet rising housing costs.

> ### The Housing Committees System
>
> *Housing in rural areas and small towns is not a major focus of existing congressional committees and subcommittees. In the Senate, housing is dealt with primarily by the Urban Affairs Committee and the Subcommittee on Housing and Urban Affairs, which has no representatives from the eight western states. In the House of Representatives, the Committee on Banking, Currency, and Housing looks primarily at large-scale financial matters and the Subcommittee on Housing and Urban Development is oriented toward urban problems. Housing in small towns tends to be dealt with by committees responsible for agriculture or Indian affairs.—Congressional Quarterly, 1977*

areas experiencing the impact of developing energy.

Moreover, HUD is oriented in general toward large cities and metropolitan areas rather than small towns. Programs such as the Community Development Block Grant Program place little emphasis on the needs of rapidly growing areas (Carlisle, 1977:492). HUD also is under considerable political pressure to distribute an even greater proportion of its funds to large, older cities, especially in the Northeast (*Denver Post,* 1977). In addition, due to the committee structure of Congress, legislative attention to housing has been focused in the Senate on large cities and in the House on national financial markets (see box).

This big-city focus is borne out by the relatively low usage of HUD programs in energy impacted areas in the West. Only one HUD program was used in the western areas of energy development during Fiscal Year 1976: a loan program for developing low- and moderate-income housing projects. And this program was used in only 7 of 75 affected counties in the West. On a state-by-state basis, the West's share of HUD funding was well below the regional share of aid for all federal programs (3.02 percent). Only Colorado received more than this percentage, with the money

Mobile Home Park in Energy Development Area Near Farmington, New Mexico

going almost entirely to the Denver area (U.S. DOE, Socioeconomic Program and Data Collection Office, Region VIII, 1977).

On the other hand, the Farmers' Home Administration (FmHA), USDA, is intended to deal with small towns and rural areas. Even though severe income limits are imposed on loan qualifications, FmHA programs are used in 71 of the 75 western counties where energy resources are being developed (U.S. DOE, Socioeconomic Program and Data Collection Office, Region VIII, 1977). Families with annual incomes less than $15,600 can obtain a maximum loan of $33,000 (Carlisle, 1977: 489–90). A large proportion of families in these rapidly growing communities have incomes above the lending limit, but the programs do assist the area's families who are on fixed incomes.

In recent years, the western states have attempted to provide conventional housing, largely because of housing pressures generated by the growth in population. State financing of housing is restricted in all states to low- or moderate-income groups (Carlisle, 1977: 494–505). Colorado, Montana, and Wyoming have actually generated money through the sale of bonds to be passed through lending institutions in rural areas. Montana's Board of Housing has the broadest financing authority, because it takes into account "the ability of persons and families to compete successfully in the normal housing market," as well as the availability and cost of housing in particular areas (Montana Revised Codes Annotated, 1975; Daley, 1976:119–20).

Too Many Trailers

Because of the failure of lenders, federal and state governments, and the energy industry to successfully address housing problems, mobile homes have become the most common type of housing found in towns near the sites of energy development. Mobile home parks are usually not very satisfactory for the people living in them because they are crowded, not well maintained, and lack the typical advantages of property ownership. Mobile homes are also not attractive to local government, since they do not pay their own way in property taxes.

Mobile homes are a common solution to the inadequate supply of housing in the West. They fill a useful need for temporary residents during the construction of energy facilities. Construction always requires a short-term workforce, but the fact that residents who are not involved in construction also have no alternative to mobile homes attests to the area's lack of housing choices (Zelenski and Cummings, 1977; FEA, Region VIII, Socioeconomic Program Data Collection Office, 1977).

The reliance on mobile homes has been unsatisfactory to both residents and local governments. In many cases, the homes in a development are placed very close together; yards are small or nonexistent; and paved streets are rare. Those living in mobile homes often find them small, poorly constructed, depreciating in value, or simply possessing an "aura of shabbiness" (McKown, 1976:76–77, Luxenberg, 1977). In the words of a Gillette, Wyoming, resident, "Mobile home parks can be nice places to live, unlike the slums ours have turned out to be" (Pernula, 1977:34). In company-provided, employee-only mobile home developments, the social segregation from other townspeople, while solving some problems of maintenance, also creates dissatisfaction (Fradkin, 1977; University of Montana, Institute for Social Science Research, 1974).

Mobile home development is similar to the single-family home development process outlined in Figure 7-1. A developer still must go through the required development steps, but usually the only construction required is a concrete pad and utility hookups for each mobile home. The developers of mobile home parks frequently retain ownership of the park and lease individual lots. It is uncommon for mobile home residents in the West to own the lot they occupy, although some mobile home owners buy small rural lots on which to put their mobile homes.

The method of financing mobile homes also contributes to residents' dissatisfaction with this type of housing. Except in Wyoming, the customary practice is to finance mobile homes as personal property (like automobiles), rather than as real property. Interest rates, therefore, are substantially higher (by 2 to 3 percent) for mobile home loans than for home mortgages. And since down payments range from 10 to 40 percent, a large initial sum is still required on an average $15,000 unit.

Designating mobile homes as personal property and separating them from the land has two other important effects. First, mobile homes tend to depreciate rather than increase in value over time as conventional home/lot combinations do. Second, mobile homes are not taxed as real property, and thus local governments can tax only the value of the land. The tax revenue generated by mobile homes, therefore, are well below those for conventional real estate. Furthermore, the property owner is the legal landowner, and mobile home residents pay no property taxes directly (Luxenberg, 1977).[3]

These legal and financial aspects of mobile homes lower their desirability in the eyes of local governments. Little revenue is added to the tax rolls, and the high density of mobile home parks also conflicts with aesthetic values. As a result, mobile home parks are usually relegated by zoning to the less desirable areas of towns or are banned entirely (Luxenberg, 1977). Nevertheless, the demand for housing still tends to be met by mobile homes. Although municipal zoning often restricts mobile home parks, in some states there is little or no land-use control and zoning in

---

[3] In some retirement type mobile home parks, this legal separation of home from land does not take place. However, in the West and much of the rest of the country, the situation is as described here.

unincorporated county areas.[4] Mobile home parks clustered near a town's edge and along highways have become a common sight throughout much of the West since the large-scale development of energy resources began in the early 1970s (Federal Interagency Team, 1976).

Government programs concerning mobile homes have little positive effect on the housing situation in western towns. Federal mobile home construction and safety standards have upgraded the quality of the units in the manufacturing stage (Fed. Reg. 1975a; 1975b; 1977); the quality issues that remain are those intrinsic to the design of mobile home parks. Although spacing, distance from street, and utility hookups are specified criteria for FHA-insured loans (U.S. HUD, Office of Community Planning and Development, 1976), conventional loans are usually used to develop mobile home parks in the West; therefore, the federal standards rarely come into play. State standards could supplement the lack of other regulations, but only Montana and North Dakota have guidelines for mobile home park design.

Little Choice in Housing

Few other housing choices, such as apartments, are available in western towns because of the fear of an unstable short-term market. The reasons are similar to those for the lack of investment in single-family housing, and emphasize the overwhelming problems for adequate housing in energy development areas.

The ability to choose among a variety of housing alternatives is one measure of housing quality that is, all too often, absent in western towns. The lack of choices stems in part from the financing conditions that con-

strain single-family housing and the resultant heavy reliance on mobile homes. Choices other than these, such as apartments, townhouses, condominiums, and various modular designs are discussed briefly below.

There are two principal reasons why apartment developments are not yet common in communities where energy resources are being developed. First, apartment construction involves a sizable investment (at least $30,000 per two- to three-bedroom unit) which is difficult to finance in a small town. In towns such as Farmington and Gillette, the large population does attract some investment in apartments. For example, some apartment development is being financed or guaranteed by energy companies, but, as with single-family housing, rental priority is given to company employees (Metz, 1977). Second, the typical monthly rents for apartments, where they are available, are rather high, especially in comparison with mobile homes (Table 7-2). In smaller towns (with populations under 15,000), multifamily housing is scarce—if not impossible to find.

Modular or factory-built wood-frame housing has become a major source of single- and multifamily housing in the West, particularly in larger towns. Relatively little labor is required to set modular homes on their foundations, reducing the need for skilled workers to finish them on-site. In most towns, the cost of modular housing is about the same as the cost of houses constructed on-site (about $55,000 in 1976 for a 1,100 square-foot house in Gillette). Land costs, local wage levels, subdivision and development costs (such as sidewalks, street lights, and utility connections), and financing are the same for both types (Federal Interagency Team, 1976: 4). Some of the advantages of modular units are lost in transportation costs, which average $4.50 to $5.00 per mile for each unit (Kaiser Engineers, 1976:Sec. 10). Cities and rural areas not on a rail line or near an interstate highway (such as Farmington or southern Utah) face even higher delivery costs and delivery delays due to inadequate roads.

---

[4]The question of zoning and land-use control is discussed in Chapter 8.

Employee barracks also have been used in the West, principally for construction workers who do not have their families with them. For example, the Wyodak power plant project being built by Pacific Power and Light near Gillette uses prefab quarters for unmarried construction workers. Many men work a five-day week and commute hundreds of miles to their homes each weekend (*Civil Engineering —ASCE,* 1977). Living conditions in such quarters can be quite unsatisfactory to some workers (*Denver Post,* 1976b).

Indian Housing: Opportunities for Improvement

The development of energy resources in the West provides a unique opportunity to improve the often wretched condition of housing for Indians. New housing and re-habilitation are likely to be among the improvements, funded largely by the federal government.

Housing on Indian reservations raises several problems that are somewhat different from those discussed above. In addition to a housing shortage, the principal problem for Indians is quality: the lack of plumbing and the crowding of large families in small quarters (Navajo Tribe, Office of Program Development, 1974: 48–49; U.S. DOI, BIA, Billings Area Office, 1976:II-131–34). Table 7-3 presents some Indian housing conditions in the region, including the Navajo Reservation, the largest reservation in the West, where housing needs are most severe. As energy development proceeds on and near Indian lands, the energy companies will be under pressure both to employ Indians and to contribute to improvements in the Indian housing situation.

Mobile homes are also becoming common on Indian reservations, even in areas where water is not available, and this trend is likely to continue. Few houses in the Four Corners area, particularly on the Navajo Reservation, have running water (U.S. DOI, BIA, Planning

Table 7-3: *Housing Conditions of Indians in the West, 1973*

Condition	Housing Units Affected
Total Indian Families[a]	38,221
Standard Units	9,641
Substandard Units	24,352
Families Doubled Up	4,228
New Units Needed	12,517
Rehabilitations Needed	16,063
Total Need/Percent of Families	75.0%

Source: U.S. Congress, Senate, 1975:3

[a] In the Billings, Navajo, and Albuquerque Bureau of Indian Affairs areas.

Support Group, 1976:II-89–93), and few towns have sewer systems. Tribal housing developments with running water will probably be able to provide only a small fraction of the new homes needed for Indians.

A further problem connected with the development of energy resources concerns employing both Indians and non-Indians. Individual land ownership is commonly prohibited on reservations, including the Navajo Reservation. Therefore, conventional residential and commercial developments are not possible (Commission on Civil Rights, 1975:19). Thus, non-Indians tend to settle in nearby towns, while Indians tend to live in mobile homes and substandard housing scattered throughout the reservation.

Summary of Problems and Issues

The housing problem in western communities with new energy development is one of both quantity and quality. Housing is in short supply because of rapid population growth and the financing constraints in the private housing market. Energy development involves uncertainties about the long-term housing demand, and these uncertainties have resulted in a very low level of local lending for construction. Lending institutions in the West

Table 7-4: *Alternative Policies for Housing*

General Alternative	Specific Alternative
Decrease the number of workers living on-site	Adjust project schedules
	Encourage long-distance commuting
Increase housing construction	Make state siting permits conditional on housing provisions
	Stimulate industry investment in housing
Increase the availability of construction and home-mortgage financing	Seek government guarantees in the secondary mortgage market
Improve the mobile home situation and diversify housing choices	Changing financing and tax structure for mobile homes
	Develop quality mobile home parks
	Institute local land-use controls for all areas of a county
	Provide financial assistance for lenders and developers of apartments, condominiums, and townhouses
	Offer tax incentives for rental housing

generally are not linked effectively to the secondary mortgage market, the national system for financing housing. Energy developers have provided some housing in larger towns for their employees, but a shortage of home construction prevails throughout the areas affected by the development of energy resources.

There are no federal programs specifically designed to deal with housing shortages in rapidly growing areas. Because of an orientation toward low-income families in urban areas, state actions, like federal programs, have little effect in the West. Federal actions have been more effective at improving the quality of housing on Indian reservations; however, a lack of coordination of the various Indian housing programs is blamed for slow progress in improving Indians' general housing situation.

In the absence of adequate public or private response to the housing demand, mobile homes have filled the gap for the majority of newcomers to western areas of energy development. Mobile homes, however, provide little property tax benefit to local government because they are bought and owned as personal property. Partly as a result of this, mobile homes are often restricted by local zoning regulations to the less desirable parts of town. In most areas, mobile home developers can avoid zoning restrictions completely by using unincorporated land outside municipal boundaries; mobile home parks that cluster on the edges of a town are generally of low quality. But few small towns can offer any other housing choices, such as apartments or townhouses.

Policies for Improved Housing

As we have seen, the shortage and quality of housing, particularly in small, isolated western communities, is a major problem resulting from the development of energy resources in the West. Policies for dealing with this problem can attempt to achieve one or more of the following objectives:

- Reduce the peak demand for temporary housing;
- Utilize the existing private housing market to increase the availability of financing and the numbers of houses being constructed; and
- Improve the quality of all housing.

These three objectives are not necessarily mutually exclusive; for example, decreasing demand for temporary housing and increasing the amount of new housing available would presumably contribute to an overall improvement in housing quality in these communities.

In this section we will identify and evaluate

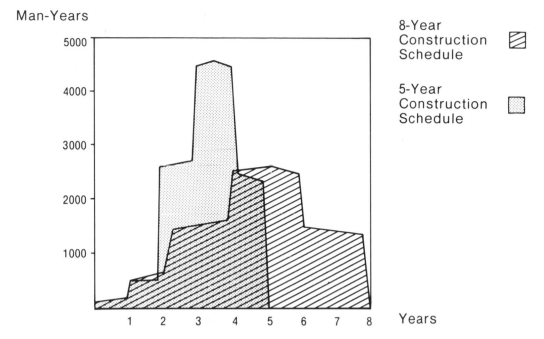

Figure 7-4: *Personnel Requirements for Coal Gasification Plant Construction in Two Alternative Schedules.* (Source: Carasso et al., 1975:6-30-31.)

several alternative policies for achieving these three objectives. The policy categories to be considered are shown in Table 7-4. For example, the first objective, to reduce the peak demand for temporary housing, is addressed by policies that would decrease the number of workers living at or near energy development sites in the West. The second objective, to increase the availability of financing and the amount of construction available through the private market suggests two policy categories: increase housing construction; and improve the mobile home situation. Specific alternatives within each of these categories are also listed in Table 7-4. The four categories of policies and the specific alternatives within each category are described and evaluated below.

Policies to Decrease the Number of Workers Living On-Site

This category of policies is intended to reduce the number of temporary workers who would otherwise move into the local community from outside the area and require housing. Two specific policy alternatives address this objective: Adjusting project schedules and encouraging long-distance commuting.

Adjusting Project Schedules: The construction schedule can be a major factor in reducing the population increase in the vicinity of an energy project and, thus, in reducing the demand for housing. For example, construction of a coal gasification facility usually takes about five years and requires more than 4,600 workers during peak construction activity. And the impact would be compounded if two or more facilities were constructed simultaneously. Peak labor requirements, however, could be reduced to about 2,600 workers if the construction schedule were lengthened to eight years (Carasso et al., 1975:6–30 to 6-31). This would reduce the severity of the impact on local housing (Figure 7-4).

The responsibility for implementing this

policy would largely rest with private industry, but it could be stimulated by requirements such as state siting laws that put an upper limit on the levels of local impact. The effect of such a policy would be to reduce peak employment during construction by about one-third and to reduce housing shortages by a similar amount.

Both labor and construction costs, however, would increase considerably if the construction of an energy facility were extended from five to eight years. The increased interest costs on a $50 million coal mine built over eight years instead of five years would amount to $2.5 million per year, or $7.5 million for the three-year extension. The interest costs on a coal-fired power plant with a $1 billion price tag would increase by $100 million if construction were lengthened from eight to ten years.[5] Thus, the increased costs associated with reducing peak construction employment would be reflected in the increased price for energy paid by consumers and in the inflationary increases of construction costs.

The adjustment of project schedules for constructing energy facilities also raises questions about the differences in the impact of developing western energy on the West and on other regions of the country. For example, as noted above, the costs of energy facilities will increase when the construction period is extended; this will, in turn, increase costs for both utilities and consumers, many of whom are outside the West. Consequently, policymakers have to consider regional trade-offs between the cost of energy to consumers nationwide and the costs of mitigating the housing impact in development areas of the West. If construction schedules are lengthened to reduce the peak demand for temporary housing, energy consumers in other parts of the country will likely pay higher prices for energy from the West. The scheduling changes

may also mean delays in converting some utilities from gas and oil to coal. In this case, the nationwide effort to convert to coal would conflict with the attempts of coal-producing states to reduce the impact of energy development on local housing.

Project schedules for a single facility are generally rather flexible. However, if two or more construction projects are proposed for the same area, coordination is considerably more difficult. The first firm to begin construction of an energy facility may be able to a large extent, to follow its preferred project schedule. Other firms that enter the area later would receive their siting permits contingent upon schedule adjustments (in most cases, delays) to reduce the local housing squeeze caused by simultaneous construction of two or more facilities.

Primary responsibility for implementing these scheduling adjustments rests with energy developers in conjunction with state and local government officials. Although the goal of reducing the temporary housing demand is somewhat general, conditions specific to individual states and local areas would make federal involvement difficult.[6] Individual projects must be assessed to determine their impact on communities, and the impact of alternative schedules also need to be compared. State industrial-siting agencies generally already have the authority by issuing conditional siting permits to acquire information on scheduling and to require some adjustments to schedules (Industrial Development Information and Siting Act, 1975). By tying compliance on schedules to a state-issued construction permit, the construction schedule becomes part of the definition of the project from the very beginning.

Long-Distance Commuting: Regardless of the construction schedule established for an

[5] This is calculated as one-half of the number of years added to construction multiplied by the nominal project cost ($1 billion) multiplied by the interest rate (10 percent).

[6] Without national energy-facility siting legislation, federal control over project schedules would be unlikely. See Chapter 11, "Energy Facility Siting," on this point.

energy project, the peak demand for temporary housing in the local area can be reduced if the number of workers who live near the construction site is kept to a minimum. The second specific policy (Table 7-4) suggests that workers be encouraged to live in larger population centers and commute to construction sites, either individually or in carpools and buses. Offering incentives to energy companies, such as tax breaks or faster permit approval, would be one way to implement this housing solution for construction laborers.

Encouraging commuting would reduce the peak demand for temporary housing in the local community in direct proportion to the number of workers who lived elsewhere. Cities large enough potentially to provide adequate housing include Denver, Grand Junction, Billings, Albuquerque, Salt Lake City, and Provo. Long-distance commuting could be accomplished with minimal costs either to government or to energy consumers because the costs of commuting would be assumed either by the individual workers, if they continue to drive their own vehicles, or by the company, if developers provide transportation. Even if industries establish fleets of mass transit vehicles or aircraft to transport up to 50 or 75 percent of their labor force from distant population centers, the cost would be lower than that required to meet the temporary peak demand for housing in the development area. Some form of mass transportation would be preferred, of course, in the light of the national goal to reduce oil consumption.

Compared to a policy of adjusting project schedules, the risks of delaying the start of energy production would be minimal under a policy of long-distance commuting, since new plant construction could proceed on schedule. The effects on worker productivity also appear to be positive because construction employees and their families could live in the area's larger cities where they would have access to a wider choice of housing, as well as to a broader range of services and amenities.

Larger cities also are better equipped to deal with some population influx.

The commuter policy is very flexible and could be adapted for individual projects, since no new mechanisms are required. The voluntary nature of this option allows companies to experiment with various transportation modes to meet the workers' needs. Plant shift schedules could be accommodated by this proposal to enable workers to set up carpools and make other transportation arrangements.

The initiative for implementing long-distance commuting would rest primarily with private industry, although state and local governments may need to offer some assistance in the form of tax incentives and informal cooperation. The primary stumbling block is likely to be the negative attitude of some workers toward any form of organized transportation; many will prefer to live in the immediate vicinity of the development and to drive their own vehicles. Industry might make long-distance commuting more appealing to workers by organizing company buses to and from the major population centers.

Policies to Increase Housing Construction

In Table 7-4, two general policies are listed that respond to the general objective of increasing the number of houses being constructed—first, to increase building and, second, to make financing more available; this section considers the first. The specific options that will be discussed involve actions by industry or industry and government to increase the level of homebuilding by the private sector and assume that the energy industries should bear part of the cost of providing housing for their workers. Providing ample quality housing may also contribute to workers' satisfaction and help stabilize the work force by reducing turnover (Richards, 1978). Three specific policies for preserving and strengthening the role of the private housing market in the West are:

- Make state siting permits for energy facilities contingent upon industry agreement to provide housing for the workers it expects to employ;
- Provide incentives and inducements to stimulate investment by the energy industry in housing or to guarantee the financing of housing construction; and
- Provide guarantees of home loans in the secondary mortgage market by the federal or state government.

Housing as a Condition for Siting Approval: Siting legislation for energy facilities in some states stipulates that permits may be made conditional, based on an industry's willingness to mitigate the anticipated impact of construction and operation. However, in addition to dealing with a proposed facility's socioeconomic impact or its effect on population, state siting laws have other objectives, such as balancing environmental concerns with the desire to develop energy resources. The legislation in some states often addresses the impact on the public sector, such as local government problems, but usually there are no provisions for avoiding the impact on the private sector in such areas as housing. In contrast, the policy we will consider here is intended to deal with the impact on housing in the manner of the WISA. A prominent example is the agreement by Basin Electric Power Cooperative to furnish 1.900 housing units and other services to help mitigate the impact of its proposed facility in Wheatland (WISA, n.d.; Valeu, 1977). Similar requirements could be included in the siting legislation of other states or in federal siting legislation.

It is difficult to generalize about the amount of new housing that would be made available under such a policy. Although many states have siting laws, few states have had any experience with this kind of conditional siting. Generally, if policymakers choose to implement siting legislation that requires

industry to agree to provide housing for its employees before receiving a permit, the affected area can ensure that the housing will be adequate to lessen at least the primary impact of the proposed facility.

The cost of housing measures can escalate quickly, depending on the policy alternative chosen. For example, costs can range from about $5,000 per unit for housing site development, to more than $30,000 per unit for house construction. Lot preparation and development for 1,900 homes in Wheatland cost about $10 million, or over $5,200 per unit (Valeu, 1977). Thus, even 100 lots without homes can cost the industry as much as $500,000 to develop. Beyond the cost of construction, firms also face the risk of holding unsold homes if energy development plans are delayed or cancelled. However, if development does occur, most if not all the housing in western towns can be sold as long as financing is available.

The cost of mitigating the impact on local housing could utlimately be passed on to consumers, but it would add to total construction costs before it could be recovered. Thus, western towns would benefit from these housing measures, but the cost would be borne by consumers of western energy in other areas.

Conditional siting permits, industry investment in housing, and state and federal housing-assistance programs can generally be quite flexible in dealing with the housing needs specific to a given locality and within a state as a whole. Siting permits contingent upon industry providing housing are especially suited to meet the specific type and degree of impact of energy development in particular communities. In addition, state-level supervision may be preferred to federal control throughout most of the West.

Finally, industry acceptance of housing construction requirements depends to some extent on the procedure for a siting permit. If the state contingency permit supersedes many or all other permits (called one-stop siting), then the savings in time and money can be used to provide housing or other community facilities. (If federal permits could also be consolidated, the savings would be even greater.) On the other hand, an additional permit requirement would find little industry support, even without contingencies.

Incentives for Housing Investment: In contrast to requiring adequate housing through state siting permits, there are many cases in which investment in western housing by private industry could take place without a mandatory policy. This approach would also be appropriate for firms or projects which are below the size covered by siting laws. The most promising incentive is a tax credit for the amount of housing built in boomtown areas. If implemented as a guarantee by industry that is, in turn, guaranteed by the federal government, adequate housing for energy booms could be provided in many locations.

The amount of new housing provided by the industry guarantees will also vary on a case-by-case basis. Energy industries in the West have been increasingly willing to provide guarantees to private developers for a specified number of homes. Housing built with the backing of industry has included both single- and multifamily new units; industry has also provided assistance ranging from the purchase of permanent mobile home spaces to financial help for modernizing older homes. The amount of housing provided is likely to be adequate to meet the needs of workers employed by the energy developer but usually will not be available to contract or service workers and their families. Federal guarantees or tax credits would address that set of housing needs in the West.

The potential costs of a federal assistance and guarantee program are exemplified by Colorado's sale of $50 million in bonds to finance mortgages and $34 million of bonds to finance construction loans (Carlisle, 1977: 497). A full-scale initial program in all western states could require at least $600 million over a ten-year period; that would build about 10,000 homes. Most of these would be needed in Wyoming, which would supplement the larger program at the state level with the Wyoming Community Development Authority.

When state and federal programs are used to increase the availability of financing and the construction of homes, taxpayers bear both the risks and costs. In the case of state programs financed by severance taxes, utilities pay the increased costs when they buy the fuel and pass them on to consumers by increasing the rate base. If industry guarantees to provide housing, only those homes left unsold would be part of the tax bill. Tax credits are an intermediate policy, providing partial support by taxpayers nationally for additional housing in the western states. Politically, this may be hard to institute until western energy becomes indisputably prominent in the U.S. energy picture.

Stimulation of the Secondary Mortgage Market: The secondary mortgage market is the ultimate source of financing for all housing in the U.S. (Brockschmidt, 1977). Large investors who are able to purchase mortgages indirectly and in large blocks are most frequently drawn to lenders who deal in large volumes of homes. Consequently, small lenders—a group that includes virtually all nonmetropolitan institutions—are unable to compete with lenders from large urban areas where hundreds of homes are built each month. Federal housing agencies could take steps to direct some funds to areas where energy resources are being developed, in effect making more money available.

Based on the assistance currently provided by state and federal housing programs, it is estimated that federal and state support for the secondary mortgage market potentially could be doubled to provide loan guarantees for housing in western communities (U.S. DOE, Energy Impact Assistance Steering Group, 1977). The overall cost of such a policy would not be noticeable and would probably not affect the financing of housing significantly in other parts of the country.

The federal housing agencies described earlier are independent agencies, however, and respond relatively slowly to directives from Congress or HUD. Federal legislation may be necessary to channel government funds in the secondary mortgage market to western areas suffering the impact of energy development.

Policies to Improve the Quality of Housing: The set of alternative policies to meet this objective will be discussed only briefly since we did not evaluate them in detail for this study. Diversifying the housing mix would certainly open up the choice of housing in western towns. Both single-family homes and multifamily units are in short supply in most smaller western towns, and there are a disproportionately large number of mobile homes. One specific policy would be to encourage the construction of alternative housing types, such as apartments, condominiums, and townhouses, by providing both mortgage assistance to lenders and developers and tax incentives to both developers and renters.

Even if policymakers are successful in diversifying the kinds of housing available in western boomtowns, mobile homes will still remain the market's predominant feature in the short- to midterm future, especially for construction workers who live in the area only temporarily. Therefore, this category of policies is basically aimed at improving the quality of mobile home housing.

Thus, the first specific policy calls for improving the quality of mobile home housing by changing their tax status and, thereby, the way in which they are financed. Since they are generally classified as personal property in the western states, this change would also require mobile homes to be reclassified as real property, making them eligible for lower-interest loans. It could also lower the depreciation rate of mobile homes (they might even appreciate in value) and enable local governments to derive property tax revenues

that could be used to finance local services and facilities.

The generally objectionable nature of mobile homes has restricted their location to high-density developments in less desirable parts of town or in fringe locations immediately outside municipal jurisdictions. The last two specific policies listed in Table 7-4 are designed to address these problems by controlling the way in which mobile home parks are developed and by instituting local land-use controls for all areas of the country, ensuring that the parks cannot escape this control. Mobile home parks could certainly be developed with less congestion and many of the amenities usually associated with residential lots, such as paved streets and neighborhood parks. The same standards can be enforced in the surrounding county by adopting a local land-use law that applies countywide.

Summary of Alternatives and Conclusion

Population growth associated with the development of energy resources in the West can seriously affect both the quantity and quality of housing available to energy and service workers, their families, and long-time residents in nearby communities. In this chapter, we have considered a variety of alternatives to avoid or mitigate those problems, and the major categories of alternative policies are summarized in Table 7-5.

At the outset, it should be noted that the housing policies we evaluated respond to two different housing situations perceived in the western United States. First, policy alternatives were judged in light of their ability to mitigate those impacts anticipated for western communities near areas of potential, future energy development. Second, the same alternatives were also evaluated in light of the fact that energy development, and accompanying housing problems, are already occurring in numerous areas of the West. Thus,

Town and Power Plants at Colstrip, Montana

the best alternatives will be those that can be applied to both existing and future housing impacts.

It would seem that those alternatives most effective for the immediate short-term could serve as temporary checks to an existing, deteriorating housing situation. Certainly, policies such as decreasing the number of on-site workers or improving the mobile-home situation would provide some relief to areas already experiencing energy development. Of course, long-term solutions to housing problems should be sought for these areas, so efforts toward increasing construction or financing availability should be made. In the area of increasing construction, the most promising and flexible alternative would be the provision of industry incentives. In conjunction with industry incentives, the federal agencies could greatly assist by shifting loan guarantee and financing monies into these hard-pressed communities.

On the other hand, communities not yet experiencing the growth problems of energy

development would most likely want to prepare long-term plans. In those communities, a mix of policies to deal with the boom-bust cycle would be appropriate. Before development begins, communities could explore adjustments in project scheduling, mobile-home usage and commuting schemes in order to alleviate the temporary influx of construction personnel. Then, to deal with the long-term influx of permanent operations personnel and local service workers, communities could negotiate conditions for siting, set incentives for traditional housing development and attempt to obtain a greater commitment of federal home loan funds and loan guarantees.

In conclusion, some of the peak demand for housing could be avoided if policymakers located fewer workers in small western communities near the site where energy resources are being developed. However, such options would increase the cost of development in order to reduce the impact on housing. Policies to increase the supply of housing, by providing finance and construction funds, are often less costly and can be used in communities where the impact is already being felt. The housing supply alternatives we have evaluated are also more flexible and can be tailored to the housing situation in particular states and towns.

The role of energy developers in reducing the boom-bust cycle will need to be increased. Particularly in housing, the private sector has ample opportunity to alleviate the problems and, indeed, to make a sound investment. Federal programs are much more difficult to direct at the specific needs of individual communities, although the cost can be spread more broadly among all taxpayers. Overall, no single alternative is the best choice; rather, several policies will have to be implemented together to improve the housing situation in western towns.

REFERENCES

Arthur Young and Company. 1976. *Problems of Financing Services and Facilities in Communities Impacted by Energy Resource Development.* Washington, D.C.: Arthur Young.

Brockschmidt, P. 1977. "The Secondary Market for Home Mortgages." *Monthly Review,* Federal Reserve Bank of Kansas City (September–October):11–20.

Campbell County Chamber of Commerce. 1976. *Economic Impact of Anticipated Growth: City of Gillette and Campbell County, Wyoming.* Gillette, Wyo.: Campbell County Chamber of Commerce.

Carasso, M., et al. 1975. *The Energy Supply Planning Model,* vol. 1. San Francisco: Bechtel Corporation.

Carlisle, N. 1977. "Boomtown Housing: The Problems and Possible Solution." In *Financial Strategies for Alleviation of Socioeconomic Impacts in Seven Western States,* edited by L. Bronder, N. Carlisle, and M. Savage. Denver: Western Governors' Regional Energy Policy Office.

Table 7-5: Summmary Evaluation of Housing Alternatives

Criteria	Measures	Decrease Workers Living On-Site	Increase Housing Construction	Increase Financing Availability	Improve Mobile Home Situation
Effectiveness	Amount of Housing Added	*Adjusting Project Schedule* can reduce peak employment. *Commuting* can disseminate large percentage of workers to off-site housing.	*"Condition of Permit"* alternative could lessen most of primary impact, but not secondary. *Incentives* could induce industry investment which might meet all primary and secondary housing needs.	Federal housing authorities could put up enough guarantees to meet all housing needs.	Provide inexpensive but decent housing for most temporary workers.
	Short- or Long-Term Policy (How long alternative could be implemented)	Short-Term	Long-Term	Long-Term	Short-Term Only

Efficiency	Dollar Expense of Alternative	Adjusting Project Schedule raises capital, interest and labor costs significantly. Commuting costs are minimal, whether paid by company (mass-trans) or employee (carpooling).	Expensive, from $5,000 to $30,000 per unit. Some estimates of $600 million for 10,000 homes.	No real increase in cost, just shift of funds by federal agencies toward energy development areas.	Lenders could lose higher interest payments. Local governments could benefit by collecting property taxes on mobile homes.
Flexibility	Adaptability to local growth patterns	Adjusting Project Schedule only good for areas without multiple project development. Commuting only good in areas with larger population centers nearby, but various schemes can be experimented with.	Somewhat flexible. Company built or guaranteed housing can be tailored to local conditions, but long-term commitments of financing detracts from flexibility.	Very fluid mortgage market created with high degree of adaptability to local conditions.	Very flexible. Local control assures local outlook.
	Reversibility	Adjusting Project Schedule is generally irreversible unless high level of local dislocation will be tolerated. Commuting schemes easily changed.	Long-term financial commitment makes reversal almost impossible.	Federal housing authorities can easily reverse policies.	Local mobile home ordinances and tax regulations can be easily changed.
Equity	Who Will Pay Costs?	Employees pay cost of commuting, unless provided commuter transportation by company. Adjusting project schedule will ultimately be paid for by energy consumers.	In the case of tax credits or federal guarantees, taxpayers bear costs. Otherwise energy consumers pay costs.	Federal and state taxpayers	Mobile home owners and park developers
Implementability	Institutional Constraints	Energy developer may not want to increase project time.	Condition of Siting only good where certainty exists that institutional action will not block completion of project. Most needed legislation already exists.	Intra-agency adjustments required but machinery already exists.	Some opposition from local park developers. Most machinery already exists at local level.

Civil Engineering—ASCE. 1977. "Wyoming Grassland Transformed to Coal Mining Center." 47 (September):50–56.

Commission on Civil Rights. 1975. *The Navajo Nation: An American Colony.* Washington, D.C.: Commission on Civil Rights.

Congressional Quarterly, Inc. 1977. *Washington Information Directory 1977-78.* Washington, D.C.: Congressional Quarterly.

Daley, J. B. 1976. "Financing Housing and Public Facilities in Energy Boom Towns." In *Rocky Mountain Mineral Law Institute: Proceedings of the 22nd Annual Institute,* July 22–24, 1976, pp. 47–144. New York: Matthew Bender.

Denver Post. 1976a. "Town Revived in Coal Boom." July 13.

Denver Post. 1976b. "Public Utility's Work Barracks Criticized." November 1.

Denver Post. 1977. "House Seeks $14.5 Billion in Urban Aid." May 13, p. 5.

Enzi, Michael. N.d. Personal communication.

Federal Energy Administration (FEA), Region VIII, Socioeconomic Program Data Collection Office. 1977. *Regional Profile: Energy Impacted Communities.* Lakewood, Colo.: FEA.

Federal Interagency Team. 1976. *The Energy Boom in Southwest Wyoming.* Washington, D.C.: U.S. Department of Housing and Urban Development.

Federal Register 40 (September 1, 1975a): 40260–303.

Federal Register 41 (December 18, 1975b): 48752–92.

Federal Register 42 (January 4, 1977):960–66.

Fradkin, P. L. 1977. "Craig, Colorado: Population Unknown, Elevation 6,185." *Audubon* 79 (July):118–27.

Gilmore, John S. 1977. Prepared statement in U.S. Congress, Senate, Committee on Environment and Public Works. *Inland Energy Development Impact Assistance Act of 1977. Hearings* before the Subcommittee on Regional and Community Development, 95th Cong., 1st sess., August 2 and 27, pp. 266–67.

Gilmore, John S., and Mary K. Duff. 1975. *Boom Town Growth Management.* Boulder, Colo.: Westview Press.

Henderson, Arn, and Daniel B. Kohlhepp. N.d. "Analysis of the Housing Mix Decision on the Future Housing Production in the Kaiparowits Region." Unpublished report prepared for the University of Oklahoma, Science and Public Policy Program, Technology Assessment of Western Energy Resource Development.

Industrial Development Information and Siting Act of 1975. Wyoming Statutes §§35-502.75 *et seq.* (Cumulative Supplement 1975).

Jacobsen, J. 1976. "Coping with Growth in the Modern Boomtown." *Personnel Journal* 55 (June):288–89, 303.

Kaiser Engineers. 1975. *Kaiser Coal Project, Interim Report June 30, 1975,* vol. 6: *Community and Public Relations (Area D).* Oakland, Calif.: Kaiser Engineers.

Luxenberg, S. 1977. "Mobile Homes Growing Up." *New York Times,* May 22, p. 1.

McKown, C. 1976. "Personal and Social Acceptance of Manufactured Homes and Other Innovations." In *Quality Housing Environment for Rural, Low-Income Families,* Bulletin Y-102. Muscle Shoals, Ala.: Tennessee Valley Authority, National Fertilizer Development Center.

Metz, W. C. 1977. "Residential Aspects of Coal Development." Paper presented at the Conference of the American Institute of Planners, Kansas City, Missouri, October.

Montana Revised Codes Annotated. 1975. §35-503(8) (Cumulative Supplement 1975).

Mountain West Research. 1976a. *Construction Worker Profile,* Final Report. Washington, D.C.: Old West Regional Commission.

Mountain West Research. 1976b. *Construction Worker Profile: Community Reports.* Washington, D.C.: Old West Regional Commission.

Myhra, David. 1975. "Colstrip, Montana—The Modern Company Town." *Coal Age* 80 (May): 54–57.

Navajo Tribe, Office of Program Development. 1974. *The Navajo National Overall Economic Development Plan.* Window Rock, Ariz.: The Navajo Tribe.

Nelson, G. S., and D. A. Whitman. 1976. *Cases and Materials on Real Estate Finance and Development.* St. Paul, Minn.: West.

Pernula, Dale. 1977. *City of Gillete/Campbell County: 1977 Citizen Policy Survey.* Gillette, Wyo.: Gillette/Campbell County Department of Planning and Development.

Presnell, Jack. 1977. Testimony in U.S.

Congress, Senate, Committee on Environment and Public Works. *Inland Energy Development Impact Assistance Act of 1977. Hearings* before the Subcommittee on Regional and Community Development, 95th Cong., 1st sess., August 2 and 27.

Richards, W. Robert. 1978. "Mining Industry Housing and Community Relations." Paper presented at the American Mining Congress Coal Convention, St. Louis, Missouri, April.

Schafer, R. 1977. "Housing in America." *Technology Review 80 (October/November):* 10.

U.S. Congress, Senate, Committee on Interior and Insular Affairs. 1975. *Indian Housing in the United States,* A Staff Report on the Indian Housing Effort in the United States with Selected Appendices. Washington, D.C.: Government Printing Office.

U.S. Department of Commerce, Bureau of the Census. 1977a. *Annual Housing Survey: 1975, United States and Regional.* Washington, D.C.: Government Printing Office.

U.S. Department of Commerce, Bureau of the Census, 1977b. *Characteristics of New Housing: 1976,* Construction Report Series C-25-76-13. Washington, D.C.: Department of Commerce.

U.S. Department of Energy (DOE), Energy Impact Assistance Steering Group. 1977. *Report to the President on Energy Impact Assistance,* Draft Report. Washington, D.C.: DOE.

U.S. Department of Energy (DOE), Socioeconomic Program and Data Collection Office, Region VIII. 1977. *Federal Aid Programs in Region VIII for the Fiscal Year 1976: Programs Available in Areas of Concern and the Usage of Such Programs by Energy Impacted Counties in Region VIII.* Denver: DOE

U.S. Department of Housing and Urban Development (HUD), Office of Planning and Development. 1976. *Rapid Growth from Energy Projects: Ideas for State and Local Action, A Program Guide.* Washington, D.C.: Government Printing Office.

U.S. Department of the Interior (DOI), Bureau of Indian Affairs (BIA), Billings Area Office. 1976. *Final Environmental Statement: Crow Ceded Area Coal Lease, Tracts II and III, Westmoreland Resources.* Billings, Mont.: BIA.

U.S. Department of the Interior (DOI), Bureau of Indian Affairs (BIA), Planning Support Group. 1976. *Draft Environmental Impact Statement: Navajo-Exxon Uranium Development.* Billings, Mont.: BIA.

University of Denver, Research Institute, Industrial Economics Division. 1977. *Methodology Papers: Housing.* Denver: University of Denver, Research Institute.

University of Montana, Institute for Social Science Research. 1974. *A Comparative Case Study of the Impact of Coal Development on the Way of Life of People in the Coal Areas of Eastern Montana and Northeastern Wyoming.* Missoula: University of Montana, Institute for Social Science Research.

Valeu, R. L. 1977. "Financial and Fiscal Aspects of Monitoring and Mitigation." Paper presented at the Symposium on State-of-the-Art Survey of Socioeconomic Impacts Associated with Construction/Operation of Energy Facilities, St. Louis, Missouri, January.

White, M. 1977. "Colstrip Power Project—Its Monitoring and Mitigation Programs." Paper presented at the Symposium on State-of-the-Art Survey of Socioeconomic Impacts Associated with Construction/Operation of Energy Facilities, St. Louis, Missouri, January.

Wyoming, Department of Economic Planning and Development, Office of the Chief of State Planning. 1977. *Housing Finance: Implications for the State of Wyoming.* Cheyenne: Wyoming, Department of Economic Planning and Development.

Wyoming Industrial Siting Administration (WISA). N.d. Docket Number WISA-75-3.

Zelenski, C., and J. Cummings. 1977. *Low to Moderate Income Housing Need in the State of Wyoming.* Cheyenne: Wyoming, Department of Economic Planning and Development.

Growth Management

Wheatland, Wyoming—Ask the people here what it's like to live in the midst of a boom, and you get a variety of answers.

To City Clerk Jim Dunham, it's meetings— lots of meetings, a bigger office, a bigger budget and phones that never stop ringing.

Greed. That's how Tom, a construction worker, typifies every boom town he's been to, and Wheatland is no exception. It's 90 cents now for a beer that was 65 cents when he arrived a year ago.

Schuyler Lucas, recreation director, is excited about having enough players for 62 softball and baseball teams.

Acres and acres of productive wheat land that went under concrete for Black Mountain village where the construction workers live. That's what Tyler Dodge, a local farmer, sees when he thinks about the boom.

To Police Chief Buck Evans, it means recovering a stolen mini-bike and then discovering the owner left town a week ago without leaving a forwarding address [Ambler, 1979: 1].

The development of energy resources in the West can seriously affect the quality of life, particularly in small, isolated communities and especially for energy workers and their families. In addition to inadequate housing (both in quantity and in quality), which was discussed in the previous chapter, public services and facilities in most of these communities will almost certainly be inadequate to meet the needs of the greatly increased population which the development of energy

resources will bring. The problems that people and communities can expect to face are compounded by a number of factors: fluctuations in population due to changes in the size of the work force; a lack of adequate funds, especially for capital expenditures; a general lack of professional management and planning capability; and a characteristic absence of assistance programs.

Depending on the characteristics of the energy development technology and the location at which it is constructed and operated, the impact on growth management can vary significantly. The two significant technological factors or characteristics that influence growth management are labor requirements (both the size and stability of the work force) and scheduling, both for the energy facility itself and in relation to other facilities. Significant geographical factors are the size and location of the community that will be affected and the distribution of revenues among the various jurisdictions—for example, counties, school districts, and municipalities.

The most important factor influencing the impact of an increased population is the size of the community before energy development begins. Larger cities (with populations above 10,000) generally serve as major market centers for a wide area and, therefore, have more diversity and capacity for growth in both the public and private sectors. A professional planning staff is also more likely to be employed by larger towns. A variety of services and facilities are more likely to be available, and thus are easier to expand, than in smaller communities.

In addition to a town's size, the number of communities in the vicinity of an energy development site will help determine the magnitude of impact that any one town will experience. An isolated town where nearly all workers and their families live will feel the impact much more than will any single town in an area where the new population is distributed among several towns. For example, Gillette, a small isolated city in northeastern Wyoming, has absorbed virtually all of the impact of large-scale coal development in Campbell County. On the other hand, the impact of energy development in western Colorado have been distributed among Rifle, Grand Valley, Meeker, and Rangely.

In much of the eight state area we have studied, the benefit of increased revenues that comes with energy development accrues to a jurisdiction other than the one that experiences the most serious impact on its growth. Property tax revenues usually go to counties and school districts, not to the municipalities where most of the workers and their families live. Increased sales tax revenues most often go not to these communities but to the small cities which are regional trade centers. In nearly all energy-development communities, therefore, local government has been unable to keep up with demands for services, and planning for future rapid growth has been even more difficult.

Energy development can also adversely affect the everyday lives of residents of these communities. Inadequate services, for example, translate directly into perceptions by the residents of an unsatisfactory quality of life. However, shortages of public services are common in small towns everywhere, including those in the West, and the actual effect of energy development is often difficult to determine.

Growth management is the major concern we address in this chapter. We begin by discussing the background of problems and issues that surround local services and facilities and then present and evaluate alternative policies for these problems and issues.

Problems and Issues

The rapid growth in population associated with energy development greatly affects the ability of nearby local governments to respond to the demands for services generated by that growth.[1] Municipal services and facilities quickly become inadequate both for long-time residents and for people who move into the area because of the development of energy resources. Many public services, such as water supply, sewage treatment, and medical care, can accommodate an increased demand only if new facilities are constructed. Building these facilities takes time as well as money and results in a lag between the time when services are needed and the time when revenues are available. This problem of "lead time" is particularly serious in the case of energy-related construction near small towns, where most of the development of western energy resources will take place (Colorado, Governor's Committee, 1974; Bolt, Luna, and Watkins, 1976: Vol. 1).

In addition to the "lead-time" problem, disparities also arise between local and state jurisdictions in the distribution of the costs and the revenues associated with the development of energy resources. New property taxes and other revenues from an energy project often go to the county, school district, and/or state, while the major need for services and facilities occurs in the towns where the workers and their families actually live.[2] Thus,

---

[1] See "Factors Affecting Housing Need" in Chapter 7 for documentation of this growth.

[2] We assume that workers will live in the vicinity of their employment. This is certainly the current situation and is likely to be the dominant one for the medium-term future. (For information on the origin and location of energy workers in the West, see Mountain West Research, 1976.)

these needs can arise in a county or even a state other than the one in which the development of energy takes place. For example, the Decker mine in southeastern Montana, one of the nation's largest producing coal mines, has a work force of about three hundred, but nearly all the workers and their families live just south of Decker—in and around Sheridan, Wyoming. Thus, Montana receives a severance tax on the coal produced, whereas Sheridan and the state of Wyoming receive no coal-related revenues from the mining operation. And there are numerous examples of similar jurisdictional conflicts—between towns and counties and between two or more counties—throughout the West.

The public services and facilities that are needed in a community vary somewhat according to population. For example, courts, detention facilities, and juvenile treatment facilities tend to be found in all larger towns (of 10,000 population and above), such as county seats, but may not exist in smaller towns. Other services and facilities, however, are needed in all towns. One recent study of boomtown financing listed the following needs in order of priority: (1) fire protection; (2) law enforcement; (3) water; (4) sewage treatment; (5) solid waste collection and disposal; (6) hospital and medical facilities; (7) detention facilities; (8) juvenile treatment and custody facilities; (9) county and municipal courts; (10) recreation facilities; and (11) administrative space (Bolt, Luna, and Watkins, 1976: Vol. 1, p. 14).[3]

The cost for the set of facilities listed above ranges from $1,500 to $2,300 (in 1975 dollars) per new resident. A population increase of 500 people, for example, will cost a town

at least $750,000. Water- and sewage-treatment facilities typically account for about 75 percent of the total funds needed for new public facilities. Annual operating expenditures also increase with additional population and average about $130 per capita, or $65,000 for each 500 additional people (Lindauer, 1975:63–68; THK Associates, 1974:30–41). Operating costs can also involve sudden jumps, for example, to change from a volunteer fire department to a full-time department.[4]

This section presents the issues and current responses regarding local services and facilities in western towns. The three issues to be covered here are:

- Financial shortages for local governments;
- Health care and social services; and
- Land use and financial planning.

Financial Problems and Assistance

In the West, energy facilities, such as mines and conversion plants, are usually located oustide of any nearby towns. These locations incur an ad valorem *or property tax obligation to the county and school district in which the facility is located. The workers at the facilities nearly always live in towns, which do not receive a corresponding increase in tax base. Federal and state actions have done relatively little to alleviate the impact of local growth. While state severance taxes have provided major assistance to local governments, most federal programs have failed to address directly the issues of growth management in the West.*

In the absence of property tax receipts, municipal governments usually rely on sales

[3] In the original list, primary and secondary classrooms and other educational facilities were designated as the tenth item. This was omitted here because school financing typically is managed exclusively by school districts rather than by municipal, county, or multicounty governments (see Mountain West Research, 1976). Streets and roads are usually maintained, but not built, by local governments.

[4] James Donoho, mayor of Hartsville, Tennessee, testified that changing from a volunteer force to a ten-man department will cost his community $100,000 per year (1977:182). The higher wage levels in western towns would probably require even greater costs, in the range of $150,000 per year.

taxes, property taxes on assessed valuation within town boundaries, and other charges and fees that tend to increase with the population. These revenues may be adequate to cover rising operating costs, but generally they are not sufficient to fund major capital improvements such as water and sewer facilities. As Mike Enzi, mayor of Gillette, Wyoming, puts it: "We simply need money (and) reliable information" (Strabala, 1977).

Town residents do benefit when their local school district receives property tax revenues from a new energy facility.[5] County governments also provide town residents with hospitals, recreational facilities, county roads, and libraries, but little sharing of the county's new wealth occurs to pay for municipal services and facilities, such as water and sewage systems, city fire and police protection, and sanitation. The city of Gillette, for example, is facing an annual deficit of $3–$4 million at least through 1985. Campbell County, on the other hand, expects an increasing surplus beginning in 1979 (at the present property tax rate of 5.269 mills) (Campbell County Chamber of Commerce, 1976); and the situation in Hayden, Colorado, is similar (Gits, 1978).

The construction of energy facilities exacerbates these local financial problems. Construction is a short-term, labor-intensive activity that makes it difficult for local governments to anticipate what long-term needs will be as opposed to the temporary needs of a construction related population. Furthermore, assistance measures based on energy production, such as severance taxes, cannot be collected until energy is produced; they provide no relief during construction (Guenthner, 1977:179–81). Finally, as discussed in Chapter 7, when construction workers (or others) live in mobile homes, their contribution to

the municipal tax base is generally not equal to their demands on local governments.

Local governments have little control over the distribution of revenues needed for public services and facilities. The control of property tax revenues perpetuates disparities between towns, on the one hand, and counties and school districts, on the other. Wyoming, for example, has recognized this problem by creating a Joint Powers Act that permits cities and counties to take steps jointly in providing public services and facilities. This law essentially extends urban powers to counties when they act in conjunction with a city (Daley, 1976:123–24). However, the initiative lies with the counties, and they are generally reluctant to become responsible for municipal finances and unwilling to accept liability on debts incurred as a result of the increase in a town's population (Bronder, Carlisle, and Savage, 1977).

Responses to the local financial impact of an energy facility frequently rely on the existing property tax base, rather than on alternatives to it.[6] In Utah and Montana the prepayment of taxes gives local governments a head start in providing for increased needs while maintaining the distinction between municipalities and the jurisdictions within which energy facilities are located (Daley, 1976: 117–19, 123–24). As a result, the principal tax benefit accrues to school districts and to county services such as roads; little, if any, revenue is available for municipal services. Experience with prepayment in the two states is still too limited to provide much insight into how effective this solution will be. A major disadvantage of tax prepayment is that the U.S. Internal Revenue Service (IRS) has refused to allow local taxes to be deducted from federal returns until the year in which the local payment is due. This significantly decreases the likelihood that voluntary pre-

[5] This occurs most frequently where there are county-wide school districts. Even in this situation, however, residents do not always live in the county (or state) in which the energy facility is located.

[6] Alternatives to the current methods of local property tax collection are discussed below in this chapter.

Table 8-1: State Energy-Related Tax Revenues and Redistribution to Local Governments in Seven Western States

State	Energy Mineral Tax Revenues					Distribution to Local Governments
	Oil and Gas	Coal	Mining	Oil Shale	Other	
Arizona			3% of gross value			44% overall to local governments, ⅝ of that to school districts; no energy-related revenue is earmarked for local governments
Colordao	2%-5% depending on production	30¢/ton underground 60¢/ton surface	$50 plus $15/acre on surface mines annually	Up to 4% of gross proceeds	Oil shale lease royalties	45% of severance tax revenues to local governments; 60% of oil shale lease royalties to local governments thus far
Montana	2.1%-2.65%-oil 2.65%-gas	12¢/ton or 4% underground 40¢/ton or 30¢ surface (whichever is higher)			1.688% of sales of electricity	17% of severance tax revenues to local governments; most goes to trust fund and general fund
New Mexico	3.75%	.05% on new contracts; 17% on old			.04¢/kWh of electricity generated	No requirement on local distribution, but 43% of all state revenues went to local government (primarily schools) in 1975-76
North Dakota	5%	50¢/ton +1¢ escalator clause geared to wholesale price index			0.25¢/kWh of electricity produced	40% of coal severance tax to local governments; small amount of coal conversion tax, based on complicated framework
Utah						State aid largely to school districts; overall, 32% of state money sent to local government in 1974-75
Wyoming	4%	8.5% + privilege tax of up to 2% until $160 million is raised for capital facililities			5.5% of gross produce on uranium	27% of all state revenues went to local government in 1974-75; little energy-related revenue is earmarked for local government

Source: Bronder, Carlisle, and Savage, 1977.

payments will be made (Bronder, Carlisle, and Savage, 1977:557–58).

States with severance taxes on coal and other energy resources can alleviate the financial imbalance at the local level by establishing a central collection and disbursement system. Table 8-1 shows the range of severance taxes in the West, excluding any state receipts of federal mineral royalties. The seven states listed in the table vary considerably in the extent to which each taxes energy production. For example, given a coal production level of 100 million tons and an average selling price of $12 per ton, Montana would collect $360 million whereas North Dakota would collect only $50 million. At the other extreme, Utah has no severance tax and has a constitutional prohibition on revenue collection by the state for local governments.

From the point of view of local impact, however, the more important variation is in the distribution of severance tax revenues to local governments. Two states in our study area, New Mexico and Wyoming, have no specific allocation of mineral revenues to local governments; consequently, most of their severance tax transfer goes to school districts. Colorado, Montana, and North Dakota all earmark a percentage of their severance tax receipts for local government. It is more common among the western states for the state government to retain the bulk of mineral tax revenues for state purposes, such as the state's general fund, capital improvements, and trust funds (Bronder, Carlisle, and Savage, 1977:557–58).[7] The allocation formulas in mineral tax legislation may prove inflexible in some cases for dealing with local impacts that have yet to be felt. In addition, some towns have difficulty persuading the state disbursing authority of the magnitude of the population impact they have experienced. For example, Miles City, Montana, was unable to

[7] Although the authorized capital improvements have included municipal water and sewage treatment facilities for improved areas, the emphasis is on roads, highways, schools, and universities.

Table 8-2: *Limitations on Bonded Debt in Six Western States (percent of assessed valuation)*

State	Counties	Munici-palities	School Districts
Colorado	1.5	3	20
Montana	2.5	5[a]	...[b]
New Mexico	4	4	6
N. Dakota	5	5[c]	5[c]
Utah	2	4	...
Wyoming	2	4[d]	10

Source: Compiled from Bronder, Carlisle, and Savage, 1977.

[a] May be increased by 10 percent for water and sewer systems.

[b] State makes up shortfalls through statewide levies.

[c] May be increased an additional 3 percent by two-thirds approval of state legislature.

[d] Limit of 8 percent for sewer projects; no limit on water projects.

afford the special census necessary to document its population growth during the 1970s and thus was unable to qualify for assistance.

State limits on bonding and the amount of debt that local governments may incur can hinder local attempts to finance new facilities with bonds which would be repaid using future tax revenues. Table 8-2 shows the limitations on bonded debt in six western states. While in some circumstances these limitations are not overly restrictive, they can be too inflexible to meet the needs of energy development areas, given the problem of lead time. There seems to be little popular support for increasing local debt limitations —at least in Wyoming, where a constitutional amendment to that effect failed to pass in 1976 (Savage, 1977:559–60).

Some western states attempt to deal with this problem by having either a state agency or a quasi-governmental entity sell tax-free bonds to finance the construction of local public facilities. Montana's Board of Housing is able to finance community services and facilities, as well as housing, because Montana statutes broadly define housing development to include "streets, sewers, utilities, parks . . . and other nonhousing facilities as the Board determines to be necessary, convenient, or desirable" (Montana Revised Codes, 1975).[8]

Federal programs have been less responsive than state programs to the conditions of energy development areas. Two well-intended programs based on federal land or mineral ownership are not nearly as effective as they might be. For example, the Public Lands–Local Government Funds Act (1976) or "In Lieu of Tax Payment" Act provides payments to local government units in which federal entitlement lands are located. These are almost exclusively county and school district lands, and the revenues generally do not go to municipalities. Under the Bureau of Land Management Organic Act (Federal Land Policy and Management Act, 1976) one-fourth of the royalties from federal minerals are earmarked for local governments, and loans may be made to provide for the social and economic impact of energy development in anticipation of future mineral revenues (Federal Land Policy and Management Act, 1976; Savage, 1977:568–71).[9] Loans to municipalities tend to be rather small, because the future revenues on which loans can be based are limited. The result, therefore, is that the loans do little to help fund the greatly increased services and facilities required.

A new program to assist areas affected by the production, processing, or transportation of coal or uranium may prove helpful. Section 601 of the Powerplant and Industrial Fuel Use Act (1978), called the Energy Impacted Area Development Assistance Program, gives the FmHa (in the USDA) funds to grant to local or state governments in areas where employment related to the development of energy resources has or is projected to increase by 8 percent or more annually (Fed. Reg., 1979). Such grants may help governments develop growth-management and housing plans, as well as assist in the areas of housing and public facilities and services by providing money for planning and for site development and acquisition. The nationwide first-year funding level is set at $60 million and will rise to $120 million annually thereafter. Thus, this program will be able to address the pre-impact planning needs and initial development costs, but still leaves the major expenses for construction of streets, water and sewer facilities, hospitals, and other public facilities in the hands of local governments.

The other federal programs relevant for general funding in energy development areas are primarily loan programs. Some specific programs, such as for water and sewage treatment and planning, include grant funds, but general-purpose funding tends to be based exclusively on loans. The reliance on laons, rather than grants assumes that the local government will be able to repay in the future (U.S. DOE, 1977). This assumption is valid for many county governments, school districts, and other jurisdictions that have large property tax receipts; but it is not true for many towns in the West, where the service needs are far greater than in rural areas.

While officials of state and local governments agree on the need for grants, they do not agree on how the grants should be distributed. In most cases, local governments greatly prefer to receive federal money directly instead of having the funds channeled through the state government. Mayor Larry Kozisek of Grand Junction, Colorado, attributes this to a lack of concern on the part of state officials and to the minimal political clout that rural areas and small towns have at the state level (1977:178). He cites as evidence of this the delays in state distribution of federal funds

[8] For Wyoming's response, see Daley, 1976:58–62. In Wyoming, the state supreme court ruled on February 13, 1978, that the Wyoming Community Development Authority was unconstitutional because it would allow bonds to be issued prior to the collection of mineral tax revenues to repay them (see *Witzenburger* v. *State of Wyoming,* 1978).

[9] In practice, however, loans have not been made against future revenues because of low interest rates and constitutional prohibitions in states such as Wyoming.

Table 8-3: *Water and Sewage Systems in Communities of Six Western States*

	Colorado	Montana	North Dakota	South Dakota	Utah	Wyoming	Total
Water Treatment							
No system[a]	0	11	5	3	47	8	74
At or over 100% of capacity	12	2	4	0	--	5	23
Incomplete data	1	1	2	0	9	8	21
							118
Sewage Treatment							
No system[b]	5	6	2	0	32	4	49
At or over 100% of capacity	11	5	1	0	10	15	42
Incomplete data	0	2	9	1	1	7	20
							111
Number of communities considered	38	23	29	5	56	37	188

Source: FEA, 1977.

[a] A few communities have only private wells.

[b] A few communities have private septic tanks or lagoon systems only.

and the allocation of a large portion of federal assistance to counties rather than to municipalities (Kozisek, 1977:220–33). The success of the new Section 601 program still remains to be proven in western towns.

Water- and sewage-treatment facilities are the most expensive items among municipal expenditures. Sewage treatment is almost exclusively a municipal function, since in the West few areas outside of towns have public sewage service. In addition to needing an expanding water supply and sewage-treatment facilities as a consequence of population growth, recent water pollution legislation requires—after several delays from earlier deadlines—that existing public works install secondary waste treatment by 1983 (FWPCA, 1972; Clean Water Act, 1977).

As shown in Table 8-3, over one-half of the towns affected by the development of western energy either lack a water or sewer system or have reached the capacity of their existing systems. Upgrading treatment works

to meet the new standards adds further to the financial burdens of local governments. This upgrading is aided by the Wastewater Construction Grants Program, administered by the EPA. Various delays and complicated administrative procedures in the program have been blamed for the delays in meeting the recently changed 1977 clean-water deadline (CEQ, 1975:71–72; Kirschten, 1977).

The Construction Grants Program, the largest recent public works program in the U.S., is intended to help communities meet the costs of new water- and sewage-treatment works. Through fiscal year 1976, the eight western states received only 1.55 percent of the total national funds for this program, even though the eight states account for about 4 percent of the nation's population (Kirschten, 1977). The Construction Grant Program appears to be aimed more at large cities than at small towns, since large cities are somewhat favored by procedures that allow metropolitan communities to apply directly to EPA, whereas nonmetropolitan communities are required to go through a state agency for funds. This can cause some unusual results for western communities requesting federal assistance (see box). Most other federal economic assistance is intended for urban areas and high-unemployment areas, not for areas of rapid growth and low unemployment, such as Gillette (Enzi, 1977).

If EPA grant funds are not available to meet continuing needs, towns may sometimes obtain assistance from other agencies. The Economic Development Administration (EDA) and the Title V Regional Commissions have frequently provided grants to communities for water- and sewage-treatment plants. The EDA usually does this under the Special Economic Development and Adjustment Assistance Program—called the Economic Adjustment Program (Public Works and Economic Development Act, 1965, 42 U.S.C. §§ 3241, 3243, 3245). Beulah, North Dakota, and Carbondale, Utah, are among the western towns to receive construction grants from EDA (FEA, 1976).[10] And unlike EPA, EDA applications from small towns need not go through a state agency; this often means funds are available quickly.

State funding has also helped some towns meet their needs for water and sewer systems. For example, over $3 million of the $11.6 million allocated through March 1977 by the Montana Coal Board to Big Horn, Rosebud, and Treasure counties was for water and sewer projects (Bronder, Carlisle, and Savage, 1977:129–34). The state of Wyoming also has allocated coal tax grants to Wyoming communities—several for water and sewer facilities (Bronder, Carlisle, and Savage, 1977: 376–92).

Out of Step

Mayor Mike Enzi of Gillette, Wyoming has a string of stories to tell about being turned down for federal money. EPA rejected a request for sewage treatment facilities because Gillette already had a treatment plant and must wait until 27 other towns in Wyoming get their first ones. Most other economic assistance is intended for urban areas and high-unemployment areas, not for areas with growth and low unemployment such as Gillette.— Enzi, 1977.

---

rapidly growing communities that accompany energy development. A shortage of doctors and nurses is particularly difficult for rural areas to overcome. Boomtown growth also creates social stresses in areas that do not have services available to deal with such problems as alcoholism, child abuse, and juvenile delinquency.

Medical needs typically expand at least as rapidly as the population and even faster when a large proportion of the new residents are children, as is commonly the case in communities affected by energy development (Mountain West Research, 1976). Limited health services can result in a smaller proportion of children being immunized against preventable diseases, fewer screening tests for tuberculosis and other diseases, and less followup on active cases of tuberculosis and venereal disease. The inability to obtain early treatment can permit other diseases to become more serious (Copley International, 1977).

The shortage of doctors is a particularly seriuos problem. Many small towns have no doctor; many more communities have no hospital (see Table 8-4). Frequently the only hospital is in the county seat, and long distances separate towns. The doctor/popula-

Health Care and Social Services

Adequate health and social services are among the most difficult to provide in the

---

[10] However, both EPA and EDA rejected a request by Gillette, Wyoming, for funds for treatment facility expansion (see Enzi, 1977).

Table 8-4: *Doctors and Hospitals in Energy Development Communities in Six Western States*

State	Affected Communities	Towns With No Doctor	Towns With No Hospital
Colorado	38	14	22
Montana	23	12	14
North Dakota	28	18	21
South Dakota	4	0	0
Utah	56	38	45
Wyoming	36	17	23
Total	185	99	125

Source: FEA, 1977.

Note: Communities with a population over 40,000 were excluded from this table.

tion ratios are much lower in western towns than in the country generally. For example, Gillette has less than one doctor per 900 people, compared to the national average of one doctor per 660 people. Public health clinics with staff nurses often supplement the medical system, but cannot fully substitute for resident doctors.

The shortage of doctors in small towns is not unique to the West, of course, but it is exacerbated in this region by the increasing medical needs that are caused by the development of energy resources. Furthermore, in the U.S. medical care is largely a private enterprise. Hospitals and clinics may be available, but doctors must be attracted to small-town practices instead of large cities with full-service hospitals (Coleman, 1976; Lankford, 1974:244-45). One of the serious impacts on the quality of life in energy boomtowns is the added difficulty of attracting doctors (U.S. DOE, 1977). Getting nurses to work in rural areas appears to be less of a problem, but nevertheless, it is more severe in energy development communities (Sloan, 1975:141-55; Rapp, 1976:24-27). State loan programs for medical training are beginning to include forgiveness provisions to induce doctors to practice in rural areas. This approach has been cited as the most effective public policy means of attracting doctors away from large cities (Coleman, 1976).

Although medical care is a service of the private sector, all levels of government have taken part in programs to provide health care to the population. County governments generally provide some funds for emergency medical services (EMS) in rural areas, supplementing EMS grants from the U.S. Public Health Service (EMS Systems Act, 1973). In some states, state funds support EMS programs, which are very extensive in most of the West (Rapp, 1976:23-27). In many cases, priorities imposed at the federal and state levels for health planning preclude adequate attention to substate areas. Health planning tends to be done either very locally or statewide, and neither scope provides effective planning for energy development areas (Copley International, 1977:25).

Among the services most needed in boomtowns are social services, such as treatment and counseling for alcoholism, child abuse, and juvenile delinquency. These services are often not available in communities where the population has been stable for many years, but become necessary as the disruption of rapid growth and an influx of newcomers occurs (Kneese, 1975:74-76; Richards, 1976). Growth alone spurs demands for social services, because a larger number of people need such services. These services generally require little in the way of capital improvements, but they do require a continuing operating budget for professional staff members. The benefits of a social service program have been well documented from the experience of the Wyoming Human Services project, which operated social service teams in Gillette for several years and, more recently, in Wheatland (Uhlmann, 1976).

The problem of inadequate social services has received less attention than more costly municipal facilities. However, the problem has been recognized by New Mexico's governor, Jerry Apodoca: "Impact assistance should not be limited to traditional infrastructure needs such as water and sewer, but should extend to health, housing, social services, and other areas" (1977:78). In the region we studied, all states except Wyoming have state-funded mental health, alcohol, and drug-abuse programs, but generally they have been inadequate due to insufficient funding (Rapp, 1976:23).

In some western communities, energy companies have contributed to social services. The Missouri Basin Power Cooperative and Atlantic Richfield have provided support for the Wyoming Human Services Project in Wheatland and in Gillette, respectively. A general reluctance on the part of energy companies to assist communities, evident a short time ago, is now beginning to change (Arthur Young, 1976).

Planning Responses

The most basic need in planning for the impact of energy development is adequate and timely information. Local government officials often do not have access to critical planning information, such as construction schedules, accurate employment estimates, and projected capital investment. The absence of information is compounded by a lack of adequate land-use planning that is capable of responding to large-scale growth from energy development.

It is admittedly difficult for energy firms to provide local government with accurate information, since they themselves must deal with

uncertainties involving markets, suppliers, and government regulations.[11] As a result, the interaction between the private and public sectors frequently is not sufficient to solve the growth-related problems of energy development.[12]

The extent of the planning problem is indicated by the designation of 188 communities in Federal Region VIII as affected by the development of energy resources (FEA, 1977).[13] Very few of these towns have a full-time municipal staff, especially planners, who are needed to deal with rapid growth.

Planning generally includes two principal activities: land-use planning and management, or program planning. In energy areas, both activities require information concerning the scheduling, magnitude, and related impacts of energy development. Land-use planning is the common concern of the planning profession, especially for urban regions. In small western towns, program (or financial) planning is at least as important; it sets budget priorities, establishes the sequence of projects, and identifies revenue sources to meet needs. These two types of planning are discussed below.

The lack of land-use controls at the local level is a pervasive condition in the West. Most incorporated municipalities have zoning ordinances, building codes, and subdivision regulations, but land-use control outside of municipalities is rare.[14] Towns can zone only

[11] See the discussion of this point in Chapter 9, Capital Availability, and Chapter 11, Energy Facility Siting.

[12] There are notable exceptions, however: Energy Fuels Company has contributed 30 percent of the salary for the first city manager in Hayden, Colorado (see Gits, 1978).

[13] The other two states in our study area, Arizona and New Mexico, are in Federal Region X and VI, respectively. These regions have not done similar studies of their states.

[14] Building codes and zoning regulations are more common in larger towns, but comprehensive, or master-plan, documents are less prevalent than individual regulations. Many of the existing plans are more than ten years old, making them of questionable usefulness in a time of rapid community change.

within their boundaries, and county-wide plans (if they exist) frequently remain unenforced and unenforceable. Rural residents, especially landowners, in the West strongly believe in personal sovereignty over their land and resist planning that suggests or embodies government control of private property (Lamont and Jones, 1976; Christiansen and Clack, 1976; Siegan, 1976).[15]

Zoning and other land-use controls, where they exist, also have built-in problems (Woolfe, 1976). Zoning is a mechanism that depends on prior decisions regarding the appropriate use of land. However, rezoning, variances, and exceptions tend to prevail as better or more opportune uses of land arise. Zoning also is not able to undo past mistakes since it must recognize the prior rights of existing uses. New or modified zoning may not be retroactive, and nonconforming uses usually are allowed to remain. Furthermore, zoning requires an intricate administration that can absorb planning staff time, thus neglecting general or comprehensive planning which reflects community values.

Planning at the state level is fairly common in the West. In particular, Colorado, Montana, and North Dakota have statewide control structures for land-use planning and regulation (Lamont and Jones, 1976). In general, the other states in the study area allow local landowners (who influence county commissions) to choose whether to adopt or avoid land-use controls. (As discussed above, county governments usually avoid controls or adopt unenforceable planning mechanisms.) However, each of the eight states have a department responsible for some degree of state planning, for technical assistance to local governments in obtaining state and federal aid, for planning, and for community development. The same department frequently serves as the state administering agency for

federal grants. Furthermore, all of the states except Wyoming are divided into multi-county planning districts, or regions, although in some states, such as New Mexico, regional planning organizations are not active.

State planning actions can conflict with local desires, as illustrated by the long-time conflicts between Western Slope communities in Colorado and state agencies in Denver (Kozisek, 1977:220–33). In addition, state agencies are generally not sufficiently coordinated to deal with a wide range of local problems such as those resulting from energy development (Lamont and Jones, 1976; Rapp, 1976:40–46).

The federal government has a large role in local land-use planning, primarily through financial support for planning and through legislated planning requirements, such as EPA's Water Quality Management Program. The HUD Comprehensive Planning Grants (701) Program has been the principal source of federal support for local land-use planning (National Housing Act, 1954; Wachs, 1976). The objective of the 701 program is to strengthen the planning and decisionmaking capabilities of governors, local governments, and area-wide planning organizations and thereby to promote the more effective use of the nation's physical, economic, and human resources. All local governments—except metropolitan cities, counties, and councils of government (COGs)—must apply through their state planning agency, which allocates funds received from HUD. Thus, all western communities affected by energy development must receive these funds through their state agencies or COGs. Of about $60 million funded nationally for the 701 planning program in 1975, about $630,000 (approximately 1 percent) went to affected communities in six of the eight states in our study area (FEA, 1976).[16] However, the plans produced with 701 funds need not be enforced by any level of government and may tend merely to provide

---

[15]The view of zoning as an infringement of landowners' property rights is not unique to the rural West; the city of Houston, for example, has no zoning regulations (see Siegan, 1976).

[16]More recent data have not been compiled for the West.

reference documents for a community's professional planners. The same may happen to plans developed under the 601 program, the Energy Impacted Area Development Assistance Program passed in 1978.

EPA's 208 Water Quality Management Planning Program provides funding for comprehensive planning on a statewide or area-wide basis with an emphasis on improving water quality and controlling pollution sources (FWPCA, 1972; EPA, 1976). Under this program, area-wide and state planning agencies must develop water-quality management plans that include both regulatory programs to control the location, construction, and modification of facilities that can cause pollution and programs to control runoff and other pollution from agriculture, mining, forestry, construction, and urban stormwater (Donley and Hall, 1976). In effect, the 208 program requires land-use planning indirectly as one means of achieving the program's goals (Supalla, 1976).

In addition to HUD and EPA, EDA in the Department of Commerce has a number of programs that provide planning assistance to state and local governments and to Indian nations.[17] These generally tend to supplement other federal programs, but they have become very important to certain governments such as the Navajo Nation (Navajo Tribe, 1976). Technical Assistance Grants that provide skilled professionals have been widely used throughout the West, particularly in Colorado, North Dakota, and by the Navajo Nation. Indian tribes are a unit of local government eligible for the federal programs described above, as well as for several special loan and grant programs, many of which are administered by the Interior Department's Bureau of Indian Affairs (BIA). For example, road construction and maintenance, school construction, and housing development grants all are available through the BIA (OMB, n.d.).

The demands on a community's attention and its revenues change rapidly in areas where energy resources are being developed. As one need is taken care of, another equally important "brushfire" usually awaits attention. The unstable pattern of population growth which communities experience because of large construction projects temporarily inflates the need for facilities and services beyond what is projected for the long term. These short-term needs make planning difficult, because it is possible to overbuild if the long-term population level is uncertain.

In many cases, it is more practical to provide only temporary facilities to meet the short-term needs (Howard, Needles, Tammen, and Bergendoff, 1976). In general, planning for such needs, without greatly overestimating or underestimating, requires a substantial amount of accurate information from the energy developers. However, this information, which should provide a fairly accurate schedule for each of several developments, is not uniformly complete from all developers. The result is a substantial information gap for local planners that greatly hinders the effectiveness of planning.

A final hurdle lies in the problematic nature of state and federal funding. The time when promised funds will actually be received is rarely predictable and even eligibility for some funds is often uncertain or difficult to assess (Enzi, 1977). As a result, financial planning currently is done largely on an ad hoc basis as possibilities and conditions change. Finally, the choice between state and federal funds may be made on the basis of such factors as the ability to hire local people for planning, rather than having to accept state personnel who are not committed to the specific area.[18]

---

[17]Among these are two programs established by the Public Works and Economic Development Act (1965): Economic Development-Support for Planning Organizations, for redevelopment areas (42 U.S.C. §§3151, 3152), and the State and Local Planning Program, for state and local governments (42 U.S.C. §§3121 et seq.).

[18]This point has been mentioned by officials in western Colorado and in Montana.

Energy Development Offers Job Opportunities

One mechanism for dealing with information gaps and the local impact of energy development on communities is comprehensive facility-siting legislation. Private industry can be induced—by state legislation, for example —to take an active role in providing assistance to affected communities. The WISA concentrates on the mitigation of socioeconomic impacts as a condition for granting site permits for large energy and industrial facilities. After studies and hearings by state agencies and the local community, a siting permit can be issued with conditions specifying the mitigating steps required (Industrial Development Information and Siting Act, 1975). In the only application that WISA processed through mid 1977, Basin Electric Power Cooperative agreed to provide 37 specific types of measures, amounting to several million dollars, for capital and operating costs as well as for

housing in connection with alleviating the impact of its power plant near Wheatland, Wyoming (WISA, 1975; Valeu, 1977). This approach allows industry to propose assistance in its own way to the point where a permit is granted.

Summary of Problems and Issues

Local services and facilities are in short supply and are difficult to plan for in areas of energy development because of several factors, including: (1) rapid and unstable population growth; (2) a paucity of information about the timing, magnitude, and location of impacts; (3) a dearth of local funds in the towns, where they are most needed; (4) a lack of adequate state and federal programs to provide financial assistance to communities; and

Energy Development Will Impact Small Communities

(5) a shortage of staff expertise to deal with requirements for federal and state programs.

In this regard, a number of issues have arisen in the West as a result of the development of energy resources. There are persisting revenue imbalances between the units of local governments, such as counties and school districts, that receive revenues from energy facilities and the municipalities which experience most of the population growth, but receive little additional revenue. There is inadequate cooperative planning in the public and private sectors generally; and developers tend to deal only with the problems of their employees rather than with the overall impact of energy development. Finally, there are conflicts between state and local governments over direct access to federal funds. Most federal guidelines require small towns to apply through state agencies, which many town officials believe are unresponsive to local needs.

Policies for Growth Management Problems and Issues

Our discussion in the previous two sections leads to the conclusion that policymakers confront basically two kinds of growth-management problems as a consequence of the development of energy resources in the West: inadequate funding to meet the needs of a greatly increased population for community services and facilities; and a lack of planning expertise and information. In dealing with these problems, policymakers can attempt, first, to ensure that adequate funds are available to provide services and facilities at a level which will minimize the adverse impact on

Table 8-5: *Alternative Policies for Growth Management*

Policy Objectives	General Alternative	Specific Alternative
To provide adequate financing for local services and facilities to assist western towns in accommodating energy-related population growth	Change state and local tax structure	Municipal annexation of county land
		Prepayment of property and sales taxes
		State collection and distribution of property taxes on energy facilities
		State severance tax
		Tax credits for private-sector assistance
	Increase assistance	Special programs for water and sewer facilities in nonmetropolitan rapid-growth areas
		Distribution of federal land and mineral payments directly to affected towns
		Federal and regional assistance for municipalities
To assure that affected communities receive adequate and timely information	Improve local planning capabilities	Federal growth-management planning assistance
		State siting legislation requiring information on possible impacts

the quality of life of energy development. And, second, policymakers can attempt to ensure that the affected communities receive planning assistance and adequate and timely planning information. In the following analysis, we will consider two categories of policy alternatives to meet the first objective and one policy to meet the second objective; these are:

- Changes in the state and local tax structure;
- Increased impact assistance; and
- Improved local planning capabilities.

Specific policies for each category of alternatives are outlined in Table 8-5. If policymakers elect to change state and local tax structures, they can annex county land to increase the tax base of municipalities, encourage the prepayment of property and sales taxes, provide for state collection and distribution of property taxes on energy facilities, impose severance taxes on the extraction and processing of natural resources, and provide tax incentives that would encourage private

developers to take the initiative in mitigating the local impact of energy development. Three alternative policies to increase assistance to affected communities are also discussed: special programs for water and sewer facilities in rapidly growing areas; distribution of federal land and mineral payments directly to towns; and federal and regional assistance for municipalities.

Policymakers can respond to the planning needs of western communities by assuring local officials of accurate and timely information prior to the development of energy resources. The options facilitating the sharing of information that is necessary for planning include planning assistance and reporting requirements contained in state siting legislation.

Changes in State and Local Tax Structure

The policy alternatives we will consider in connection with the first objective outlined above assume that the financial impact of energy development can and should be redistributed, either by changing the state and local tax structure or by altering federal assistance programs. As mentioned earlier, energy facilities are often located in rural jurisdictions, whereas the workers who operate these plants usually live with their families in nearby towns, which are unable to meet the increased demands of their new residents. Specific policy alternatives that address this issue are:

- Municipal annexation of county land;
- Prepayment of property taxes;
- State collection and distribution of taxes;
- Severance taxes (where not yet in existence); and
- Tax credits for assistance in mitigating adverse impacts.

Annexation: Municipal annexation of county land where energy facilities are located would allow local communities to increase their tax base and to extend their zoning and planning powers to energy development sites. The subsequent increase in municipal revenues could then be used to finance higher levels of services. Many western states have constitutional provisions, however, that allow municipalities to annex county property only with the concurrence of the residents involved and of the county government.

One version of this alternative provides for cooperative, power-sharing agreements—between a county with a large tax base and cities and towns—to participate in financing necessary public facilities. For example, the Wyoming Joint Powers Act (n.d.) permits counties, municipalities, school districts, or special districts to plan, create, finance, and operate jointly in a number of functional areas (Bronder, Carlisle, and Savage, 1977:340). Utah has enacted a similar measure, the Interlocal Cooperation Act (1953), which allows two government units to exercise jointly any powers either of them can exercise separately.

Proposals to change state or local tax structures are often difficult to implement, because they reallocate existing tax revenues. The municipal annexation of county land is designed to enable communities to include nearby energy facilities within their jurisdiction and thus to tax energy developers at the same level as municipal residents. However, this option has generally proved difficult to implement, since most states require the consent of the county and of the residents involved. The county government is likely to oppose such annexation measures because it stands to lose considerable tax revenues; and citizens are likely to be opposed, fearing that they will pay higher taxes under municipal jurisdiction without benefiting from municipal services. Therefore, annexation appears to be a workable strategy only if policymakers can modify existing annexation procedures or can offer substantial inducements to county citizens.

The principal effect of an annexation policy is to redistribute the benefits of energy development to cover the costs imposed on the towns. Local conditions and needs can be allowed to influence the actual mechanism and timing of annexation and tax distribution. A portion of the property tax on an energy facility could still be allocated for county purposes.

In theory, municipal annexation of county lands could be an effective option for providing revenues to cities and towns. But it is unclear just how much assistance annexation would actually provide to financially hard-pressed communities. For example, the additional taxes that can be collected on new property annexed by a municipality may or may not be adequate to finance the increased levels of services and the new capital facilities needed (Bronder, Carlisle, and Savage, 1977: 23); taxes are more likely to meet increased demands for services and facilities when the new energy plants are capital-intensive. However, to date, county governments have not been willing to relinquish their control over taxing authority on county property; therefore, it is difficult to determine the tax revenues that could be made available if this alternative were implemented.

Prepayment of Taxes: Even if municipalities and county governments have a sufficiently large tax base from which to draw revenues for financing the increased levels of public services and facilities, new demands often arise before energy development facilities become operational and begin paying taxes. To solve this problem, the second specific alternative would require energy firms to prepay property taxes, giving local governments a head start in financing higher levels of services, as well as capital improvements. The state could collect the estimated tax obligation that a facility is expected to incur during the construction or operation period and spend the money on local needs. Then when actual tax assessments are made, energy companies would receive a tax credit for the portion they had prepaid.

The prepayment of sales and property taxes would require that state legislatures adopt a change in collection procedures similar to Utah's 1975 Resource Development Act, under which energy developers may prepay the state sales tax obligations they would expect to incur during construction or operation (U.S. HUD, 1976:33–34). The money is then spent on traditional state-related improvements such as roads, in the affected areas. This option might also allow direct disbursement to local governments for development-related problems that have a high priority. However, there are several potential problems which make this proposal unattractive unless incentives or inducements are adopted. First, tax prepayment is costly to companies because they pay the taxes before they are owed; and second, the IRS prohibits the deduction of the full amount of the taxes in the year in which they are prepaid (U.S. DOE, 1977:99).

Prepayment of sales and property taxes does not increase the total amount of money available for assistance; however, it does make the funds available earlier and thus can overcome or mitigate the problems of lead time. If prepaid taxes are distributed to affected cities and towns rather than to counties, the prepayment alternative can also be a fairly effective strategy for increasing the amount of revenue available for municipal assistance.

Of all the changes in tax structure discussed here, tax prepayment is the most expensive. This is because increased costs in the form of interest payments are incurred by industry when tax assessments are paid before they are normally due.

The social benefits derived from this policy option, however, could be quite high. The increased revenues can be used to improve the social infrastructure in affected communities, thereby improving the quality of life. If a community is able to develop a sufficient tax base early enough, it will be able to provide the services and facilities demanded by both new and long-time residents, thus reducing the costs of dealing with boom-related social problems such as crime, alcoholism, drug abuse, delinquency, and mental illness. The net effect of tax prepayment is to shift the costs for the initial impact of energy development from communities to energy firms. The need for this will vary with specific local conditions. However, prepayment of sales and property taxes is difficult to implement at the state level without concomitant revisions in the federal tax codes. As we have noted, private energy developers have little incentive at the moment to pay taxes before they are due because the IRS does not allow companies to take the full amount of the deduction for the year in which taxes were prepaid.

State Collection and Distribution of Taxes: The property tax has traditionally been a revenue source primarily for local governments, and no state government in the West derives a major proportion of its revenue from this mechanism (Bronder, Carlisle, and Savage, 1977; Stinson and Voelker, 1978). However, the third specific alternative which calls for changing state and local tax structures would allow the states to collect and distribute the property taxes on energy facilities as a means of equalizing tax revenues among jurisdictions. Since one of the problems we have identified is not an overall shortage of tax revenues within the state, but rather an inequitable distribution among jurisdictions, policymakers seeking to achieve a better use of available revenues might choose this method. This policy alternative is patterned on a Wisconsin plan that provides for the statewide sharing of all property tax revenues from energy plants (Wisconsin Statutes, 1971; U.S. HUD, 1976:33). A tax rate equal to the statewide average of all state, county, and local taxes could be applied to the value of energy companies' taxable property, and the revenues could then be distributed to affected communities. This alternative is likely to spark considerable opposition for two reasons: local governments fear loss of control over the taxing power; and jurisdictions with higher revenue collections do not want to lose tax monies to other areas of the state.

Nevertheless, the state collection and distribution of property taxes ranks as a highly effective alternative because it allows policymakers to draw on the entire state as a revenue base, with the power to reallocate tax monies statewide, and to distribute these monies on the basis of need. This strategy is also particularly effective in alleviating the problem of jurisdictional mismatches—that is, by ensuring that the jurisdiction bearing the impacts receives some of the benefits. Thus, this policy ensures that the costs and benefits of energy development are spread across the entire state, rather than being localized in a single community or jurisdiction. Because of this, the units of local government which would receive tax revenues under existing structures tend to oppose the state collection and distribution of "their" revenues. The overall equity of this plan may be difficult to implement in some areas, unless municipal needs are given a higher priority than traditional rights to taxation.

Table 8-6: *State Severance Tax Collections (millions of dollars)*

State	1974	1975	1976
Colorado	$ 1	$ 2	$ 4
Montana	10	15	31
New Mexico	44	72	87
N. Dakota	4	7	13
Utah	5	6	12
Wyoming	5	18	41

Source: U.S. DOE, 1977:107.

Severance Taxes: A fourth specific alternative for financing needed services and facilities is for policymakers to adopt a state severance tax on the extraction and processing of natural resources. This tax can be based either on specific units of production (cents per ton produced) or on the gross value of the production (percent of gross value or proceeds). All of the states examined in this study except Arizona, South Dakota, and Utah have imposed a severance tax on coal, oil, and gas; and various states also tax electricity generation, mining, and oil shale activities (Bronder, Carlisle, and Savage, 1977). Thus, this policy is addressed primarily at those states without such taxes as well as states with low rates of taxation.

The effectiveness of state severance taxes in meeting the increased needs in energy development areas will depend upon both the rate of taxation and the level of mining and processing within a particular state. Table 8-6 compares the current severance tax collections of the six states in the region that have such taxes; this appears to be a very effective mechanism for raising funds, especially in New Mexico, Wyoming, and Montana. However, the effectiveness of the severance tax as a way to counteract or mitigate the impact of energy development also depends upon the manner in which revenues are distributed to the affected areas.

Severance taxes on energy resources tend to be passed along to customers; therefore, residents outside the West will share in the cost of lessening the adverse effects of energy development when they purchase power generated from western coal, oil, gas, or uranium upon which the tax has been imposed.

Of all the alternatives discussed in this section, the severance tax raises the most far-reaching questions about equity. Current tax rates in our eight-state study area range from about 2 percent for oil and gas production in some states to 30 percent on coal production in Montana (Bronder, Carlisle, and Savage, 1977). The wide variation in the tax rate suggests that there are very different opinions about what is a "fair" or allowable rate. One report argues that states in the Northern Great Plains use the severance tax as an opportunity "to shift a substantial part of their state tax burden onto coal consumers in other states and [onto] coal-producing companies" (Nehring and Zycher, 1976). Montana's severance tax on coal, which is the highest in the U.S., is currently being challenged by Wisconsin Power and Light on the basis that it is excessive and a burden on commerce (*Electrical World,* 1978). Officials in Montana, however, describe another side of the problem. Jim Nybo, a former official of the Montana Energy Advisory Council, defends the 30 percent coal-severance tax on two grounds:

> It does two main things. It sets up a structure where we can let coal development pay for the adverse social and economic effects that are imposed on our state because of the development. Second, it sets aside some of the revenues from the one-shot exploitation of coal and invests it in the future [Schneider and Gilmore, n.d.] .

Severance taxes allow a high degree of flexibility in meeting the increased need for services and facilities because the taxes incorporate a statewide collection and distribution system for mineral production taxation. However, like the property tax, severance taxes do not generate revenue until the operation stage of an energy facility. Thus, funding shortages during the construction of a facility are still likely unless considerable energy production elsewhere in the state is already under way. In this situation, funds from operating mines are redistributed to communities near new facilities; Montana, North Dakota, and Wyoming all use this method of statewide allocation.

Tax Credits: The fifth specific alternative for changing state and local tax structures is to allow energy companies to claim a tax deduction for the value of any community assistance they provide. This option is designed to increase the amount of financial help from the private sector to combat the adverse effects of development. Although energy developers have provided some assistance and in-kind services to municipalities, they have been reluctant to increase their role without some recognition in the form of tax incentives. Companies have also been hesitant to provide assistance in connection with their projects because state regulations often place a limitation on the allowable costs of business that may be recovered through increases in the rate base. Thus, one version of this policy alternative might be to broaden the definition of costs that electric utilities can pass on to their customers.

This alternative can also be effective, especially on a site-specific basis. Energy companies already have played a role in setting up and maintaining numerous services and facilities in energy development communities. At a minimum, these usually include streets, water and sewer distribution or collection networks, and parks. For example, Basin Electric Power Cooperative has provided about 1,900 units of housing, together with other services, in connection with its power plant near Wheatland, Wyoming (WISA, 1975; Valeu, 1977); and similar assistance could be expected in other areas.

However, tax credits for private assistance are likely to encounter difficulties in implementation since they are viewed by some as special favors or "tax breaks" for business. It might also be difficult to arrive at an acceptabel value for some services, such as in-kind assistance, furnished by energy developers. Tax credits also run counter to the philosophy that the full costs of energy development should be internalized—that is borne by the developers as a development cost.

Assistance Alternatives

The second category of alternatives for providing financial assistance to communities in energy development areas focuses on improvements in and additions to the existing array of federal aid programs. Although there are currently a large number of federal programs that could provide assistance, they are usually subject to competing claims from communities with needs unrelated to the development of energy resources. Furthermore, the eligibility criteria for existing federal programs favor rapidly growing metropolitan areas. The specific policies analyzed here include:

- Special water and sewer grant programs for areas of rapid growth;
- Distribution of land and mineral payments directly to affected cities and towns; and
- Increases in federal assistance directly to municipalities or through the Title V Regional commissions.

Special Water and Sewer Programs: One of the major local problems caused by the development of energy resources can be overcome by instituting special water and sewer programs. Furthermore, by earmarking funds for nonmetropolitan areas policymakers could target the money more directly to energy development areas, rather than to rapidly growing suburban areas. This plan could be implemented by changing the eligibility criteria for EPA grants for water and sewer facilities.

For example, special water and sewer programs could meet the needs of rapidly growing communities by adopting criteria that allo-

cated funds specifically on the basis of population increases. Over one-half of the towns affected by the development of energy resources in the West lack either a water or a sewer system or are at the capacities of their existing systems. A program of special water and sewer grants for these communities could meet their needs if the construction allotments to western states were increased significantly above the amount they now receive. Such special water and sewer programs would also yield substantial environmental benefits—cleaner lakes and streams and purer water supplies—for energy development communities.

The equity of special water and sewer programs will depend on the particular implementation strategy adopted. If water and sewer grants are provided to all rapidly growing areas, the existing trend of grants going primarily to metropolitan suburbs would continue. On the other hand, if growth criteria are defined to direct the assistance to energy development communities outside the metropolitan areas, the problem of inadequate facilities could be mitigated.

Special programs to provide water and sewer grants to areas of rapid growth is likely to be one of the most difficult policy alternatives to implement. This will depend, however, on the particular strategy that is employed. If eligibility criteria restrict special grants to nonmetropolitan areas, those groups with a stake in the present setup, such as governments of rapid-growth metropolitan suburbs, who now receive the majority of such grants, are likely to make strenuous objections. This constraint could be reduced or eliminated by making grants available in metropolitan areas as well, but that would defeat the purpose of assisting areas that suffer the impact of energy development.

Federal Land and Mineral Payments: The distribution of federal land and mineral payments directly to towns on the basis of population growth involves the reallocation of existing funds rather than any increase in

---

Regional Commission Impact Assistance

At about the same time as the introduction of the Inland Energy Impact Assistance Bill, Gov. Thomas L. Judge of Montana proposed a somewhat different solution. In a meeting with President Carter, Judge proposed increasing the funds available to the Old West and Four Corners Regional Commissions to $20 million annually. In addition, he proposed that the Commissions be authorized to provide 100 percent funding for energy impact projects, 20 percent more than is presently authorized. The legislative requirements would be simpler than in the case of the Hart Bill, which would set up a new bureaucracy modeled on the Office of Coastal Zone Management. Judge's proposal gives strong support for the Title V Regional Commissions as regional institutions in which the states and the federal government are both active.—Old West Regional Commission Bulletin, 1977.

---

the total amount available. A portion of these payments currently is distributed to counties and school districts (Public Lands-Local Government Funds Act, 1976), thus exacerbating the inequality in revenues between the various jurisdictions of local government. By paying these funds directly to local communities, the jurisdictions would be freed from a dependence on loans which are currently made in anticipation of revenues (Federal Land Policy and Management Act, 1976; Savage, 1977:568–71). The funds could be used to finance a wide range of local needs associated with the development of energy resources. One proposed variation on this option would allow states to borrow against future mineral royalties to meet initial or construction-phase costs.[19]

Just as urban communities will likely voice moderate to strong opposition if special water and sewer programs are established for non-metropolitan areas, state governments will likely be reluctant to give up their control of federal land and mineral revenues. Since these funds currently are handled by the

states, the communities affected by energy development must compete with other, unaffected towns for a share of the revenues; and when the needs of affected communities are compared with other state priorities, the available assistance is often insufficient.

Federal Assistance: The third policy option for changing federal assistance programs would be to increase the funds available through EDA and the Title V Regional Commissions and to disburse them directly to local communities affected by the development of energy resources. These two agencies have frequently provided grants to communities for water- and sewage-treatment plants, and some state officials think the agencies' role could be enlarged. Although the emphasis here is on providing financial assistance to communities suffering the impact of energy development, the Title V Commissions could also help in providing planning assistance and information to local communities (see box). The role of EDA in economic development programs is highly regarded in the West, probably higher than the USDA, which has been given the major funding role in the Energy Impacted Area Development Assistance Program of the Power Plant and Industrial Fuel Use Act (1978).

The increase in federal assistance specifically for energy-related problems that can be made available out of existing and new federal sources could exceed $150 million annually if all three of the policies outlined above are implemented.[20] This cost would

---

[19] Section 317(c) of the Federal Land Policy and Management Act of 1976 already allows states to borrow against future mineral royalties. However, since the Secretary of the Interior has refused to lend money at the 3 percent interest rate mandated in the act, measures have been introduced in both the House of Representatives and the Senate to raise the rate to that at which states now issue tax-exempt municipal bonds: 4½ to 5½ percent (see U.S. Congress, House of Representatives, 1978).

be offset by social benefits deriving from the improved level of services and facilities, and other social and cultural amenities would increase residents' satisfaction with the quality of life. This, in turn, could increase worker productivity by an estimated 25–40 percent and reduce labor turnover and construction delays (U.S. DOE, 1977:34). Because labor turnover, construction delays, and lowered productivity significantly increase the total costs of an energy facility (U.S. DOE, 1977: 32), the increases in federal assistance programs would yield economic as well as social benefits.

Federal and regional assistance to combat the impact of energy development is a fairly flexible alternative since the regional commissions and the EDA are able to accommodate a wide variety of administrative options for assistance and deal with a broad range of needs. The amount of assistance can be increased or decreased among various states, depending on specific community needs.

Information Alternatives

We have identified two specific policy proposals for policymakers who wish to ensure that local communities have accurate and timely information about energy development. We did not evaluate these proposals in detail, but they are presented here briefly to show the range of policy options that are available.

Federal Planning Assistance: The first specific alternative calls for comprehensive information, planning, and assistance at the federal level. One attempt to create such a

program is the proposed Energy Impact Assistance Act of 1978, introduced by Sen. Gary Hart of Colorado and cosponsored by Sen. Jennings Randolph of West Virginia. Some features of this bill are incorporated in the Powerplant and Industrial Fuel Use Act (1978), but the bill's emphasis appears to be jointly on planning and other assistance to mitigate the impact of energy development. Also, the Hart-Randolph bill would make the Commerce Department responsible for the program; the enacted bill channels funds through the USDA.

Federal funds for local planning also require a concomitant effort on the part of local officials to take planning seriously, as a means of solving future difficulties. Federal planning funds have, in the past, produced (and still do produce) many documents that remain only as evidence of expenditures of federal dollars.

Siting Information: The second specific policy alternative for providing accurate and timely information on energy development relies on the states to adopt facility-siting legislation with requirements for reporting the impacts of development. Local areas that have not previously experienced energy development or for which new types of energy facilities are planned have a particular need to obtain a description of the various impacts that can be expected. At a minimum, the siting legislation would require that information be reported on the size of the construction and operating work forces, on water use, on land use, on the amount of land needed for mining and waste disposal, and on air and water residuals. Although the kind of comprehensive facility-siting legislation that this option proposes is not yet common in the West, the Wyoming Industrial Development Information and Siting Act (1975) is an example of the mechanisms policymakers could adopt (WISA, 1975; Valeu, 1977). In terms of implementation, the initiative for including such information requirements in siting legislation would rest with the indi-

[20] The minimum figure of $150 million annually reflects the Carter Administration's proposal for assistance to energy development communities. The addition of loan guarantee programs and other assistance allocated through general revenue-sharing or the Economic Development Administration would increase this total substantially (Office of the White House Press Secretary, 1978).

Table 8-7: *Summary and Comparison of Growth Management Alternatives*

Criteria	Factors	Changes in Taxation System	Federal Assistance	Information Alternatives
Effectiveness	Additional revenue applied to mitigation efforts	Annexation and severance tax add additional monies to mitigation effort. State collection, tax credits, and prepayment change only time or method of receiving revenue.	Increased federal funding will add revenue to mitigation efforts. Federal land and mineral payments reallocate rather than increase mitigation funds. Special water and sewer programs target some increases in funds.	Under federal planning assistance alternative, some fund increases available for planning. Little, if no money, directly added to mitigation as part of a siting information program.
	Short- or long-term alternative	Prepaid property taxes and tax credit program is short-term. Annexation, severance tax system, and state collection of property taxes would take longer to implement.	Reallocation of federal land and mineral payments is short-term. Creation of special water/sewer programs or increases in federal funding are longer-term.	Siting information program could be implemented quickly. Federal planning assistance would be longer-term.
Efficiency	Cost of utilizing alternative	Annexation costs higher than all other alternatives at local level. Severance tax and state collector require large bureaucracy. Prepayment and tax credits utilize existing machinery.	Cost varies depending on size of program desired. Generally, increase in federal funding or special water/sewer programs denotes increase in bureaucracy and expense. More expertise than information alternatives.	Bureaucratic and financial costs of siting information program would prove negligible. New program in planning assistance would cost considerably more, but both should cost less than other categories of alternatives.
Equity	Control over mitigation resources and programs	All alternatives provide for state, county, or local control over mitigation resources. Prepayment leaves revenue with counties. State collection and severance taxes leave revenue with states. Annexation and tax credits directly fund municipal programs.	Federal land and mineral payments go to affected towns automatically. Special water/sewer programs and increased federal funding under control of agencies and Congress.	Funding comes from federal sources who control it, but local decisionmakers responsible for program. Subject to budget priorities of Congress.
	Who pays?	Under severance tax, energy consumer pays. All other taxation alternatives based on same payment of property taxes by energy facility.	Energy company pays federal land and minerals assessments. Taxpayer pays for special programs or increases in federal funding.	Taxpayer pays cost.
Flexibility	Responsiveness to change in growth patterns	Annexation irreversible. Tax credits and prepayment can be revised as situation warrants. Severance taxes and state tax collection have limited flexibility.	Federal land and mineral payments are constant. Special water/sewer programs and increased federal funding are very flexible.	Siting information and planning assistance are both flexible.

Implementability			
Institutional constraints	State collection and severance taxes require increased state bureaucracy and expense. Annexation would be severely opposed by counties. Prepayment would require sophisticated assessment and estimation techniques. Tax credits require revision of tax laws.	Federal land and mineral payments will be transferred from control of state and into control of cities and towns. Programs need continuous approval by Congress. Agency in-fighting over who will handle special water/sewer programs.	Siting legislation would be required. Federal planning assistance requires commitment of local officials to serious planning effort.
Major opposition	State collection and annexation would be opposed by governments. Severance taxes opposed by energy purchases outside state of export. Tax credits would be opposed by consumer groups.	Except for increases in federal funds, other alternatives will be opposed by non-impacted areas. Increase in federal funding might be opposed by Congress. State governments will oppose loss of federal land and mineral payments revenue.	Siting information legislation, depending on scope, might be opposed by energy developers. Federal assistance in comprehensive planning might be opposed by some prodevelopment resource areas.

vidual states. Private energy developers also have a role in this plan because they would be required to show that efforts had been made to mitigate the impact of development before they received a siting permit. The opportunities for interaction and cooperation between government and industry make this option particularly attractive.

Summary of Alternatives and Conclusion

The key factors in managing the growth problems experienced in areas near energy development are money and planning information, both of which can be directly used by local municipalities to improve governmental services to its residents. Table 8-7 summarizes and compares alternatives for managing growth problems through the direct raising of revenue by governmental units, the establishment of requirements which will produce such revenue in the private sector, or the provision of planning information and assistance.

How well an alternative can contribute to mitigation efforts, then, is a function of when and where it can deliver the needed revenue. Specifically, three major evaluative inquiries should be applied to each alternative examined. First, does the alternative have a reasonable chance of being implemented? Second, does the alternative overcome "lag-time" and provide revenue when it is needed most? Finally, does the alternative provide funds directly to the affected municipality?

Those alternatives with the goal of increasing information would be the easiest to implement. Both the siting information strategy and federal planning assistance represent the least expensive and least intrusive of all the alternatives considered. Even so, the siting program would not directly add any revenue to municipal mitigation efforts and any assistance to local planners by the federal government would be earmarked for that purpose and little else. Thus, although those alternatives add to the knowledge of managing growth problems, their effectiveness at mitigating those problems is limited.

Making changes in state and local taxation policies would prove more difficult than establishing information-gathering programs, but the amount of revenue brought to bear on mitigation efforts would be much higher. In addition, programs could ultimately be controlled by local officials. With the exception of municipal annexation of county land, most taxation strategies could be implemented in a straightforward, rapid manner. State collection of property taxes on energy facilities, severance taxes, and tax credits could be mandated by state legislatures. Prepayment and annexation are strategies that could be established by county and municipal government action, respectively.

One difficulty is that the "lag-time" problem still exists with some of these alternatives, most notably the severance tax, annexation, and the state collection of energy facilities' property taxes. Only the tax credit or prepayment program directly transfers revenue to municipalities when it's needed. Of course, the goal of the other taxation strategies is to eventually get more funds to affected areas, but such arrangements can easily become subject to economic and political restrictions.

Federal assistance programs have the most potential in terms of the amount of money which could be designated for mitigation efforts. However, besides relying on the uncertainties of the congressional appropriations process, each strategy faces some important implementation constraints. Special water/sewer programs can alleviate what is usually the most serious (and expensive) growth problem. In addition, water/sewer programs result in little or no bureaucratic expansion and offer opportunities to simultaneously improve environmental problems. Implementation could only be carried out after overcoming intraagency resistance and restructuring programs, however, and money would have to be taken from generalized programs.

Nonenergy municipalities would most likely oppose the alternative vigorously due to the expanded use of available money.

Similarly, disbursing federal land and mineral payments directly to municipalities would certainly assist those areas significantly. Yet, such a program would result in the loss of those funds to the states which have traditionally received them. More than likely, the states would work against such a change. The advantage of special programs and land/mineral payments is that both minimize "lag-time."

Overall, the best strategy would seem to be the selection of a mix of these alternatives; many of these alternatives are mutually supportive. For example, maximization of revenue directly available to municipalities could be obtained through a mix incorporating municipal receipt of severance taxes and federal land/mineral payments, special water/sewer programs and corporate payments in lieu of taxes (tax credit program).

In summary, we have discussed some of the growth-management problems and issues associated with accelerated energy development in the West. Traditionally, however, energy development research has concentrated on environmental concerns, while the impacts of energy development on people have frequently been neglected or minimized. Partially, this is because important information is lacking. Consequently, the investigation of the socioeconomic impacts of the development of western energy resources is still an evolving area of study, and there is considerable room for improvement in both the descriptive and analytical fields. In some cases, we simply have too little experience to judge what effects a specific policy will have on the magnitude of energy related problems. Thus, our knowledge of the social dynamics in energy development communities is still quite limited, and this, in turn, affects which alternatives are evaluated, as well as how they are evaluated.

The policy alternatives discussed in this chapter attempt to achieve one of two policy objectives. The first is to ensure that funds available to affected communities are adequate to provide services and facilities at a level which will minimize the adverse impact of energy development on the quality of life. The second is to ensure that affected communities receive planning assistance and adequate and timely planning information. It is unlikely that any single policy will have a major effect on the overall impact of energy development. There are trade-offs among the alternatives, and policymakers will have to consider all of these when dealing with specific locations in the West. Yet, at a time when government action is needed on a number of levels—federal, state, and local—in conjunction with efforts by the energy industry, antagonism between those governmental units over the receipt and distribution of revenue has led to conflict over most aspects of growth management. Thus, until the institutional mechanisms for state and local planning are improved, the problems of housing and growth management that are related to energy development will continue to be dealt with on a piecemeal and uneven basis in different states and local communities. Conversely, growth management at the local level could prove to be one of the most manageable problem areas, providing that each party to energy development shares in the responsibility.

REFERENCES

Ambler, Marjane. 1979. "Wheatland Strives for Boom Town Perfection." *High Country News,* June 1, pp. 1, 4.

Apodoca, Jerry, Governor of New Mexico. 1977. Statement in U.S. Congress, Senate, Committee on Environment and Public Works. *Inland Energy Development Impact Assistance Act of 1977. Hearings* before the Subcommittee on Regional and Community Development, 95th Cong., 1st sess., August 2 and 27, pp. 73–79.

Arthur Young and Company. 1976. *Problems of Financing Services and Facilities in Communities Impacted by Energy Resource Development.* Washington, D.C.: Arthur Young.

Bolt, Ross M., Dan Luna, and Lynda A. Watkins. 1976. *Boom Town Financing Study,* 2 vols. Denver: Colorado, Department of Local Affairs.

Bronder, L. D., N. Carlisle, and M. D. Savage. 1977. *Financial Strategies for Alleviation of Socioeconomic Impacts in Seven Western States.* Denver: Western Governors' Regional Energy Policy Office.

Campbell County Chamber of Commerce. 1976. *Economic Impact of Anticipated Growth: City of Gillette and Campbell County, Wyoming.* Gillete, Wyo.: Campbell County Chamber of Commerce.

Christiansen, Bill, and Theodore H. Clack, Jr. 1976. "A Western Perspective on Energy: A Plea for Rational Energy Planning." *Science* 194 (November 5):578–84.

Clean Water Act of 1977, Pub. L. 95-217, 91 Stat. 1566, 33 U.S.C. §§1251 *et seq.*

Coleman, Sinclair. 1976. *Physician Distribution and Rural Access to Medical Services,* R-1887-HEW. Santa Monica, Calif.:Rand Corporation.

Colorado, Governor's Committee on Oil Shale Environmental Problems, Subcommittee on Regional Development and Land Use Planning. 1974. *Tax Lead Time Study for the Oil Shale Region: Fiscal Alternatives for Rapidly Growing Communities in Colorado.* Denver: Colorado Geological Survey.

Copley International Corporation. 1977. *Health Impacts of Environmental Pollution in Energy-Development Impacted Communities,* Executive Summary for the Environmental Protection Agency. La Jolla, Calif.: Copley International.

Council on Environmental Quality (CEQ). 1975. *Environmental Quality,* Sixth Annual Report. Washington, D.C.: Government Printing Office.

Daley, J. B. 1976. "Financing Housing and Public Facilities in Energy Boom Towns." In *Rocky Mountain Mineral Law Institute:*

Proceedings of the 22nd Annual Institute (July 22-24, 1976), pp. 47-144. New York: Matthew Bender.

Donley, D. L., and K. L. Hall. 1976. "Section 208 and 303 Water Quality Planning and Management: Where Is It Now?" *Environmental Law Reporter* 6 (October):165-70.

Donoho, James, Mayor of Hartsville, Tennessee. 1977. Statement in U.S. Congress, Senate, Committee on Environment and Public Works. *Inland Energy Development Impact Assistance Act of 1977. Hearings* before the Subcommittee on Regional and Community Development, 95th Cong., 1st sess., August 2 and 27, pp. 181-90, 234-42.

Electrical World. 1978. "The Fuels Outlook." 189 (March 1):49.

Emergency Medical Services (EMS) Systems Act of 1973, Pub. L. 93-154, 87 Stat. 594 as amended by Emergency Medical Services Amendments of 1976, Pub. L. 94-573, 90 Stat. 2709.

Energy Impact Assistance Act of 1978 (Hart-Randolph Bill), Senate Bill 1493.

Environmental Protection Agency (EPA). 1976. *EPA Programs in Support of State, Regional and Local Planning.* Washington, D.C.: EPA.

Enzi, Michael B. 1977. "Energy Boom: Wyoming's Coal Veins Just Bring Troubles." *Los Angeles Times,* May 18 , p. V-3.

Federal Energy Administration (FEA). 1976. *Socioeconomic Impacts and Federal Assistance in Energy Development Impacted Communities, Federal Region VIII.* Denver: FEA.

Federal Energy Administration (FEA), Region VIII, Socioeconomic Program Data Collection Office. 1977. *Regional Profile: Energy Impacted Communities.* Lakewood, Colo.: FEA.

Federal Land Policy and Management Act of 1976, Pub. L. 94-579, 90 Stat. 2743, 43 U.S.C. §§1701 *et seq.*

Federal Register 44 (January 31, 1979):6197.

Federal Water Pollution Control Act (FWPCA) Amendments of 1972, Pub. L. 92-500, 86 Stat. 816, 33 U.S.C. §§1251 *et seq.*

Gits, Victoria. 1978. "Boom Town on the Brink." *Colorado/Business* 5 (February): 30-42.

Guenthner, Max, Mayor of Underwood, North Dakota. 1977. Statement in U.S. Congress,

Senate, Committee on Environment and Public Works. *Inland Energy Development Impact Assistance Act of 1977. Hearings* before the Subcommittee on Regional and Community Development, 95th Cong., 1st sess., August 2 and 27, pp. 179-81.

Howard, Needles, Tammen, and Bergendoff. 1976. *Temporary/Mobile Facilities for Impacted Communities.* Washington, D.C.: Old West Regional Commission.

Industrial Development Information and Siting Act of 1975, Wyoming Statutes §§35-502.75 *et seq.* (Cumulative Supplement 1975).

Interlocal Cooperation Act (1953), Utah Code Annotated §11-13.

Kirschten, J. Dicken. 1977. "Plunging the Problems from the Sewage Treatment Grant System." *National Journal* 9 (February 5): 196-202.

Kneese, Allen V. 1975. "Mitigating the Undesirable Aspects of Boom Town Development." In *Energy Development in the Rocky Mountain Region: Goals and Concerns,* pp. 74-76. Denver: Federation of Rocky Mountain States.

Kozisek, Larry, Mayor of Grand Junction, Colorado. 1977. Statement in U.S. Congress, Senate, Committee on Environment and Public Works. *Inland Energy Development Impact Assistance Act of 1977. Hearings* before the Subcommittee on Regional and Community Development, 95th Cong., 1st sess., August 2 and 27, pp. 177-79, 220-33.

Lamont, W., and M. B. Jones. 1976. *Land-Use Planning in the Rocky Mountain Region.* Denver: Federation of Rocky Mountain States.

Lankford, Phillip I. 1974. "Physician Location Factors and Public Policy." *Economic Geography* 50 (July):244-55.

Lindauer, R. L. 1975. "Solutions to the Economic Impacts of Large Mineral Developments on Local Governments." In *Energy Development in the Rocky Mountain Region: Goals and Concerns,* pp. 63-74. Denver: Federation of Rocky Mountain States.

Montana Revised Codes Annotated §35-503(8) (Cumulative Supplement 1975).

Mountain West Research. 1976. *Construction Worker Profile,* Final Report. Washington, D.C.: Old West Regional Commission.

National Housing Act of 1954, Pub. L. 83-438, 68 Stat. 320, 12 U.S.C. 24, §§1430 *et seq.,* as amended, 40 U.S.C. §461, 42 U.S.C. §§1407 *et seq.*

Navajo Tribe, Office of Program Development. 1976. *The Navajo Nation Overall Economic Development Program, 1976 Annual Report.* Window Rock, Ariz.: Navajo Tribe, Office of Program Development.

Nehring, Richard, and Benjamin Zycher. 1976. *Coal Development and Government Regulation in the Northern Great Plains: A Preliminary Report,* R-1981-NSF/RC. Santa Monica, Calif.: The Rand Corporation.

Office of Management and Budget (OMB). N.d. *Catalog of Federal Domestic Assistance.* Washington, D.C.: Government Printing Office, issued annually.

Office of the White House Press Secreatry. 1978. "Fact Sheet: Energy Impact Assistance Program." Press Release, May 4.

Old West Regional Commission Bulletin. 1977. "Regional Impact Proposal." 4 (July 15):1.

Powerplant and Industrial Fuel Use Act of 1978, Pub. L. 95-620, 92 Stat. 3289.

Public Lands-Local Government Funds Act (1976), Pub. L. 94-565, 90 Stat. 2662, 31 U.S.C. §§1601 *et seq.*

Public Works and Economic Development Act of 1965, Pub. L. 89-136, 79 Stat. 552, 42 U.S.C. §§3121 *et seq.*

Rapp, Donald A. 1976. Western Boomtowns, Part I, Amended: *A Comparative Analysis of State Actions,* Special Report to the Governors. Denver: Western Governors' Regional Energy Policy Office.

Richards, Bill. 1976. "Western Energy Rush Taking Toll Among Boom Area Children." *Washington Post,* December 13, p. 1.

Savage, M. D. 1977. "An Overview of Strategies for Financing the Mitigation of Socio-Economic Impacts." In *Financial Strategies for Alleviation of Socioeconomic Impacts in Seven Western States,* edited by L. D. Bronder, N. Carlisle, and M. D. Savage, pp. 543–71. Denver: Western Governors' Regional Energy Policy Office.

Schneider, Richard, and John S. Gilmore. N.d. *Report of the Denver Workshop on State-Local-Federal Relationships in Socioeconomic Impact Assessment.* Denver: University of Denver, Research Institute.

Siegan, B. J. 1976. *Other People's Property.* Lexington, Mass.: D. C. Heath.

Sloan, E. A. 1975. *The Geographic Distribution of Nurses and Public Policy.* Washington, D.C.: Government Printing Office.

Stinson, T. F., and S. W. Voelker. 1978. *Coal Development in the Northern Great Plains: The Impact on Revenues of State and Local Government,* Agricultural Economic Report No. 394. Washington, D.C.: U.S. Department of Agriculture.

Strabala, Bill. 1977. "Officials Eye Coal Concerns." *Denver Post,* November 17, p. 4.

Supalla, R. J. 1976. "Land Use Planning: An Institutional Overview." *American Journal of Agricultural Economics* 58 (December): 895–901.

THK Associates, Inc. 1974. *Impact Analysis and Development Patterns Related to an Oil Shale Industry: Regional Development and Land-Use Study.* Denver: THK Associates.

Uhlmann, Julie M. 1976. *Gillette Human Services Project, Annual Report,* August 31, 1976. Laramie: University of Wyoming, Wyoming Human Services Project.

U.S. Congress, House of Representatives. 1978. Report 95-211 to accompany H.R. 10787, *Authorizing Appropriations for Activities Carried Out by the Secretary of the Interior through the Bureau of Land Management,* 95th Cong., 2d sess., May 10.

U.S. Department of Energy (DOE), Energy Impact Assistance Steering Group. 1977. *Report to the President on Energy Impact Assistance,* Draft Report. Washington, D.C.: DOE.

U.S. Department of Housing and Urban Development (HUD), Office of Community Planning and Development. 1976. *Rapid Growth From Energy Projects: Ideas for State and Local Action, A Program Guide.* Washington, D.C.: Government Printing Office.

Valeu, R. L. 1977. "Financial and Fiscal Aspects of Monitoring and Mitigation." Paper presented at the Symposium on State-of-the-Art Survey of Socioeconomic Impacts Associated with Construction/Operation of Energy Facilities, St. Louis, Missouri, January.

Wachs, M. W. 1976. "Planning for Energy Needs: 701 Helps Meet the New Challenge." *HUD Challenge,* June, pp. 18–21.

Witzenburger v. *State of Wyoming,* Wyo. 1978, 575 P. 2d 1100.

Woolfe, D. A. 1976. "Zoning Is Doing Planning In." *Practicing Planner* 6 (June):10–13.

Wyoming Industrial Development Information and Siting Act (1975), Wyoming Statutes

Wisconsin Statutes. 1971. Chapter 76. §§35-502.75 through 35-502.94.

Wyoming Industrial Siting Administration (WISA), Docket Number WISA-75-3.

Wyoming Joint Powers Act (n.d.), Wyoming Statutes §§9-18.3 through 9-18.20.

CHAPTER 9

Capital Availability

Expanding the production of western energy to meet the levels specified in our scenarios will require a level of capital investment that is unprecedented in our eight-state study area. Between 1975 and the year 2000, according to our Low-Demand scenario, energy extraction, conversion, and transportation facilities could require an investment of up to $130 billion (in 1975 dollars). Although the sum is less than 3 percent of that expected to be invested nationally during the period, there are serious questions as to whether this amount of investment capital could be made available to develop energy. Several major factors affect the availability of this capital: they include technological uncertainties associated with large-scale synfuel conversion facilities, market uncertainties related to energy pricing, the economic interests of large energy firms, and a restrictive and unstable regulatory situation. These factors can pose two types of obstacles to the development of western energy resources: financial risk and monopolized energy markets.

For capital to be available, private investors must be convinced that the proposed western energy projects will be at least as profitable and secure as other investments. In financial markets, uncertainty per se has economic consequences. Usually investors can be brought into high-risk ventures only if enticed by "risk premiums," offering a higher rate of return than do more secure investments.

A second obstacle to the development of western energy is a tendency toward monopoly or oligopoly among the large, diversified energy firms. Oil and gas companies have acquired substantial portions of other energy resources, such as coal and uranium. This raises concerns about the possibility that by controlling both raw materials and processing technologies these firms could in the future, control energy supplies and prices. On the other hand, it can be argued that these firms are the only investors with sufficient capital and expertise to commercialize the production technologies for synthetic fuels made from coal and oil shale (Sanger and Mason, 1977).

In this chapter, we identify some of the barriers to capital availability, including the risk factor and lack of competition discussed above. The next sections detail the impact of western energy development on capital markets nationally and the problems and issues surrounding capital availability. The final section identifies and evaluates some policies to make capital available for the development of energy resources in the West.

Capital-Related Consequences of Energy Development

The major questions surrounding capital availability that are addressed in this study concern, first, the total investment required for west-

Table 9-1: *New Plant Investments in Western Energy for Low-Demand Scenario (in billions of 1975 dollars)*

Energy Development Activity	1976-1980	1981-1985	1986-1990	1991-1995	1996-2000	1976-2000
Gasification	0	.7	4.9	13.8	23.7	43.1
Oil Shale	.3	1.6	3.2	7.0	16.1	28.2
Power Plants	7.6	3.8	2.5	1.2	1.2	16.3
Surface Mining	1.7	1.7	2.0	2.6	3.3	11.3
Transporation	5.1	4.1	4.8	6.5	8.7	29.2
Total	14.7	11.9	17.4	31.1	53.0	128.1
New plants nationally all industries[a]	648	770	914	1086	1289	4707
Western plants as percentage of new plants nationally	2.26%	1.54%	1.90%	2.86%	4.11%	2.72%

[a]New plant and equipment expenditures assumed to be 7.8 percent of Gross National Product (GNP), which is assumed to grow by 3.5 percent per year. The figure of 7.8 percent is the average from 1966 to 1975. (U.S. Dept. of Commerce, N.d.).

ern energy development and, second, whether capital formation nationally will be able to meet this requirement. To gauge the investment required, capital-cost estimates were made for the major facilities that would be required in the eight-state region according to our Low-Demand scenario (as described in Chapter 2).[1] This resulted in an estimated total construction cost over the first 25 years of $128 billion (Table 9-1). In terms of national investment in all industries, this is less than 3 percent of the cost of new plants and equipment projected for installation by the year 2000. These estimates indicate, therefore, that the development of western energy resources will not have much effect on the availability of financial resources nationally.

Nevertheless, the projected need for $128 billion is unprecedented in scale for the study area. By comparison, the total equipment and structures owned by private business (including farms) in Federal Region VIII (Colorado, Montana, North Dakota, South Dakota, Utah, and Wyoming) was only about $31.2 billion as of 1973.[2] Therefore, energy development can be expected to create an enormous expansion in the West's economy. Furthermore, at least in the initial stages,

most of the manufactured equipment and most of the financial resources going to western energy projects will come from outside the region.[3]

From the perspective of the national economy, money markets could easily finance the development of western energy at this scale. Therefore, in this study we have focused on the question of whether western energy projects are likely to be able to attract sufficient capital. This is largely a question of the commercial viability of particular projects. Although many investors could invest in energy development, they are not likely to do so unless they expect the venture to be at least as profitable and/or as secure as other investments.

The size of a facility is a key determinant

[1] The cost data were taken principally from Carasso et al., 1975.

[2] Federal Region VIII corresponds to our study area except that it does not include New Mexico or Arizona. Data estimated from U.S. Dept. of Commerce, Bureau of the Census, 1975, Table 675 and Section 33.

[3] The new Bucyrus-Erie plant for manufacturing draglines in Pocatello, Idaho, may be indicative of a growing westward shift of mine equipment manufacturers.

Table 9-2: *Capital Resources Required for Construction of an Energy Facility (in millions of 1975 dollars)*

Facility	Size	Materials and Equipment	Labor and Other	Interest During Construction[a]	Total
Mines					
Surface coal	12 MMtpy	64	35	30	129
Underground coal	3.4 MMtpy	35	21	20	76
Uranium[b]	1,000 mtpy	8	16	6	30
Conversion					
Power	3,000 Mwe	459	459	351	1,169
Gasification	250 MMcfd	469	369	219	1,057
Synthoil	100,000 bbl/day	1,065	649	695	2,409
Shale oil[c]	100,000 bbl/day	434	381	296	1,111

Source: Linearly scaled from data in Carasso et al., 1975.

[a] At 10 percent per year.

[b] Mining and milling to yield yellowcake (U_3O_8).

[c] Mining, retorting, and upgrading.

of how well the financial risks can be handled. Initial studies indicate that the unit cost of conversion facilities will be a decreasing function of plant size (FEA, 1974:38); this economy of scale gives an advantage to larger plants. The capital costs of various facilities are estimated in Table 9-2. These facilities are large compared to current average practices (and some do not yet exist on any commercial scale); nevertheless, these sizes are expected to be within the range of possibility during the next 20 years.

As noted in Table 9-2, each conversion facility is likely to cost over $1 billion (in 1975 dollars), requiring a capital investment beyond the reach of many would-be developers. In fact, the performance of such a facility would have a strong influence on the entire investing company and would pose a substantial financial risk. A rule of thumb for the size of individual investments is that they should be no more than 40 to 50 percent of a firm's assets. The Alcan pipeline accounts for about 50 percent of its backers' assets and is considered to be about as large a single investment as can be justified. (*Business Week,* 1977.)

Problems and Issues

In this section, two major issues affecting the availability of capital for western energy development will be discussed:

- Financial uncertainties in western energy development; and
- Potential for monopoly control of the West's resources.

Financial Uncertainties

Financial risks in energy development include three areas: technical risks, economic uncertainties, and governmental/institutional uncertainties. Technical risks—such as equipment failures, operational problems, and unforeseen problems with new technologies—all inhibit the flow of investment dollars to energy projects. Economic uncertainties include labor disputes and, in particular, uncertainties about the competitive pricing of synthetic fuels as compared to crude oil and natural gas. Governmental and institutional uncertainties include regulations and requirements that may change during a project, legal challenges by opposition groups, and the long lead times required for compliance with federal, state, and local government procedures. All three kinds of risks can reduce the potential funds available for a given energy project.

Technical, economic, and institutional uncertainties delay energy projects and, therefore, raise their costs because interest in construction loans and equipment purchases must be paid even during interruptions to a project. Among these uncertainties affecting power plant construction, one study suggests that institutional delays are responsible for about 40 percent of construction delays, economic factors are responsible for about 30 percent, and technical factors for about 25 percent (1975 Federal Power Commission study cited in Environmental Policy Institute, n.d.). Each of these factors is discussed further below, emphasizing construction of synthetic fuel facilities.

Technical Uncertainties: Technical uncertainties, of course, are magnified for the new technologies such as oil shale retorting and coal gasification. These technologies are still untested on a commercial scale, and the consequences of unforeseen problems are further compounded by the facilities' large size, needed for commercial viability. The capital intensity and the size of conversion facilities require a large capital investment, one which often exceeds the total assets of some developers.

For new energy technologies, the federal government shares development risks with the private sector by funding a large amount of energy research. The $4.25 billion budget proposed for research in this field by the DOE for fiscal year 1979 represents 15.2 percent of the total federal expenditures for all research (Greenberg, 1978). However, developers still face many risks, especially in

the later stages. Not the least of these risks is the increasingly large size of facilities—from laboratory scale units to pilot plants to commercial-scale demonstration plants—which requires larger concentrations of capital. Federal efforts currently appear to be concentrated on reducing technical uncertainties in the early stages of a technology's development. Economic and governmental uncertainties, however, are most critical at the commercialization stage.

Economic Uncertainties: The economic uncertainties facing developers of western energy resources are generally those which can alter demand conditions: the world energy market, federal and state regulatory commissions, and national policies on energy and the environment. Although each of these has strong institutional features, they are grouped together here because they are determinants of the business climate rather than of specific projects.

Prices set by OPEC for a barrel of oil are now the international yardstick. Since these prices are well above OPEC's production costs, this yardstick price could be lowered uniformly by the cartel or randomly if the cartel disintegrated. While such a decrease would benefit consumers, it could be costly for those who have invested in energy resources with high production costs, such as coal gasification and oil shale developments. The uncertainties over prices make investors reluctant to develop new energy technologies despite the apparent opportunities for profit.

As easily exploitable sources of natural gas have become scarce in recent years, pipeline companies have responded by seeking supplemental sources. These include imports by pipeline from Canada and Mexico, imports in the form of liquefied natural gas (LNG), and the production of synthetic natural gas (SNG). Because the supplemental sources come only at much higher costs than fuels from conventional domestic sources, difficult pricing issues have been raised for the Federal Energy Regulatory Commission (FERC), formerly the Federal Power Commission. The traditional pricing method had been to "roll in," or average together, the costs of all sources to a pipeline company. With the advent of expensive, supplemental sources in 1972, the commission switched to an incremental method of pricing. Under this system, users of the new gas would pay the full cost of the new gas (FPC, 1972). Although this policy has since been nullified, by this 1972 order the commission signaled its intent, first, to discourage the use of high-cost supplements; second, to give incremental users an incentive to seek alternative sources of energy; and third, to assign the costs to those who receive the benefits (*Columbia LNG Corp.* v. *FPC*, 1974).

In the case of SNG, these considerations will apply to western energy development. The FERC has jurisdiction over coal gasification facilities through its pricing of SNG. For example, when considering certification for a new pipeline, the commission may attach conditions limiting the price that can be paid to SNG sources feeding that pipeline, as was done in the certification of Transwestern Coal Gasification Company (FPC, 1975). Rolled-in pricing, in particular, could make SNG attractive since pipeline companies could spread the extra cost of the comingled gas among all their customers. Similar stimulation could make oil shale development attractive. Prices for shale oil have always been projected above market oil prices (see box).

State utility commissions have jurisdiction over intrastate markets, which are relatively more important in the case of electric utilities. Authorizing legislation has not spelled out pricing formulas in any detail, so the courts have given state commissions wide latitude. Most utility regulators seem to have concluded that shortages and blackouts are less tolerable to consumers than high prices.

The level of western coal development is affected by two other significant uncertainties. First, some recent policies have decreased the attractiveness of western coal, based on air-quality standards. The 1977 CAA Amend-

> **Shale Costs Keep Rising**
>
> *In 1970, a Cabinet task force estimated that shale oil could be produced for about $4 a barrel; in 1973, estimates were made of $5.60; in 1974 the cost was figured at $6.80; in 1975 estimates ran in the neighborhood of $15. Superior Oil Company now claims that Number 6 oil can be produced for no more than $15/barrel, but the economics depend on credits being assumed for other products being recovered from the ore. Superior wants to go ahead with a 13,000 barrel/day plant, but has been held up since 1970 by negotiations for a land swap with the Bureau of Land Management. Upon completion of Interior Department approvals, 7 to 9 years would be needed to bring the plant on line.*
> —Weidenbaum and Harnish, 1976:6; BNA, 1977a.

ments, for example, require that all coal-fired facilities, regardless of the sulfur content of the coal they burn, be equipped with the BACT for removing sulfur oxides from flue gases. As discussed in Chapter 5, this policy could, after 1990, reduce the demand for Northern Great Plains coal to only half of what had been anticipated (Krohm, Dux, and Van Kuiken, 1977). This uncertainty about the level of demand makes investment in western coal mines more risky than had previously been the case.

The second uncertainty that affects the development of western energy resources is mandatory fuel-switching. Under the Energy Supply and Environmental Coordination Act of 1974 (ESECA) and its amendments, adopted by the energy conference committee in November 1977, certain major fuel-burning installations are prohibited from using natural gas or petroleum as their primary energy source. The purpose is "to foster greater national energy self-sufficiency by requiring certain . . . installations . . . to use indigenous coal and other fuel resources in lieu of natural gas or petroleum" (*Coal Industry News,* 1977:5).

ESECA would seem to mean a potentially significant increase in the demand for coal, but after three years on the books, it actually caused very little fuel-switching (Rowe, 1977). Even though coal prices fell while oil prices rose between 1974 and 1977, oil-fired power plants were oftentimes still cheaper to own and operate than coal-fired facilities. In such cases, utilities have gone to court to oppose the conversion orders. Opposition has also come from other federal agencies. For example, at EPA's request the Federal Energy Administration[4] deleted 23 power plants from its list of 54 slated for mandatory conversion to coal (BNA, 1977b). EPA said that the conversion of those power plants "would [come in] conflict with the national commitment to protect and improve the environment and, therefore, would be inconsistent with the purposes of ESECA." Such differences of opinion among various federal agencies and the lack of a firm energy policy cause coal and uranium producers to face significant uncertainty about the future level of demand.

Governmental and Institutional Uncertainties: Even if an energy development project is technically feasible and economically sound, it still faces several governmental and institutional barriers to completion and operation. One of these is simply the number of regulatory permits that must be sought. For example, Southern California Edison and its partners spent 13 years and $22 million in acquiring water rights and preparing an EIS for the Kaiparowits project. At that point they still needed 220 additional authorizations from 42 agencies (Myhra, 1977).

Those interests opposed to development tend to favor complicated procedures for several reasons: first, such intricacies give them more time to prepare their case and more opportunities to publicize it; second, the larger the number of approvals required,

the greater the chance of a single permit being denied; and third, the mounting costs caused by a lengthy procedure may induce the developer to abandon a project. An energy specialist for the Environmental Defense Fund (EDF) has remarked, "Delaying is often the answer. You tie up the economic investment of a company, and they finally do something else" (Myhra, 1977:26). His point is sound since delay is expensive. For example, if a $1 billion project is half completed at an annual interest rate of 10 percent, each month's delay adds over $4 million in interest costs.

Many of the governmental uncertainties faced by developers of western energy resources occur at the state level, especially with regard to policies on water allocation (see Chapter 3), land reclamation (see Chapter 6), taxation, and impact mitigation (see Chapters 7 and 8). Officials in some western states are worried that the West is being viewed as an "energy colony" for the rest of the country. While it is by no means a consensus position, this concern and the proposals for dealing with it do contribute to the uncertainty with which investors and developers must deal. For example, the boomtown repercussions of energy development can be quite severe, as discussed in chapters 7 and 8. Many of the alternatives for anticipating and dealing with these problems call for either direct or indirect assistance from developers. Until the nature and extent of the developers' responsibility is established, investors and developers will not know what its effect will be on the cost of development (Gilmore, 1976; Jacobsen, 1976; ERDA, 1977).

Another policy area with a direct economic effect on energy development is the various states' severance taxes on energy minerals, which were discussed in the preceding chapter. The perception of taxes by developers, on the one hand, and by state and local governments, on the other, may diverge considerably. In most cases, the revenues are more significant for state and local governments than for the developer; that is, the proportional impact

---

[4]The program is now administered by the Economic Regulatory Administration (ERA) within DOE.

on developers and their ultimate customers is not as great as the absolute impact on the units of government that receive the revenues. Consider, for example, Montana's 30 percent severance tax. Since transportation can triple the cost of coal at its ultimate destination, such taxes represent only about 10 percent of the delivered price. But if, as the Low-Demand scenario projects, Montana is producing 525 MMtpy by the year 2000, the severance tax will yield more than $1.5 billion per year. By comparison, the state's annual revenues from all sources in 1973, prior to substantial coal revenues, were only about $700 million (fiscal 1973).

Uncertainty about governmental actions is perhaps even greater when the proposed development is on an Indian reservation. The CERT represents 22 tribes who collectively own more than half of the nation's uranium resources and 16 percent of its coal deposits. The increased aggressiveness of Indians has already resulted in renegotiation of coal leases which had originally been arranged by the BIA during the 1950s and 1960s. For example, the Crow Tribe in Montana has received a proposal for a royalty of $1.35 per ton from AMAX, Inc. compared to the 17.5 cents per ton which had been the common payment for Indian coal resources as late as 1974 (Crittenden, 1978).

Energy developers and financial institutions have worked out various strategies to deal with all these uncertainties. In fact, much of the traditional behavior of the electric and gas utilities has been oriented towards reducing their financial risks. Electric utilities, especially, have received a legal mandate to provide "all the power demanded at all times to all comers" (Maher, 1977:189). This has led the industry to exhibit several distinctive features, including an expansionist attitude — that growth is necessary, proper, and inevitable — and backward vertical integration, whereby utilities acquire ownership of their sources of supply, such as mines, trains, and storage facilities (Maher, 1977:189).

The concern with assuring the supply of electricity has also led to long-term fuel contracts and to power-pooling among utilities within a particular region. Virtually all of the electric utilities in the West have been interconnected via the Western Systems Coordinating Council, a voluntary group encouraged, at least implicitly, by the ERA (Breyer and MacAvoy, 1974:112–13). Similar pools have been established to cover most of North America. In the U.S., these regional councils are further grouped into a National Electric Reliability Council.

Gas utilities have also emphasized reliability and continuity of service, although not quite to the same extent as electric utilities. Pipeline companies have usually sought assurances on both ends that their lines would be fully used well into the future. Specifically, the pipeline companies have required that gas reserves be guaranteed to them by producers, thereby assuring supply; and they have insisted on long-term "take-or-pay" contracts[5] from the companies distributing gas to consumers (Breyer and MacAvoy, 1974:5–6). Oil pipeline companies too have rules which enable them to keep their lines full at all times (Pipeline Demurrage, 1962).

Financial institutions have started to fund selected projects which, according to a traditional analysis of the financial balance sheet, would not be considered creditworthy. In a related trend, a developer's arrangements with financial institutions and other parties now involve many investors, with banking specialists often coordinating them. Lenders also have moved increasingly toward project financing, whereby assurance is sought that a particular project has a good chance of success (Wilson, 1976:69, 72). This success is measured by the actual cash flow that will be generated from the energy project as a source of repayment rather than from the strength of the firms involved (Vickers, 1977:217). Hence, long-term contracts, utility-owned rail-

[5] Take-or-pay contracts require the buyer to pay a specified amount whether or not he has enough storage capacity to accept delivery.

road cars, and other devices have been adopted to ensure that the project is financially sound in itself. In its reliance on real assets, project financing is akin to a mortgage, but the lenders' share in the commercial risks remains: the debt is repaid only if, as, and when the energy is produced. The approach, which originated in the oil industry, has become common throughout other energy industries. It has been used especially for coal mining projects and has been largely a response to fluctuations in the coal industry which since 1971 have resulted in the average rate of return staying some 2.5 percentage points below that for other mineral industries (see Tomimatsu and Johnson, 1976: Table 9).

Electric utilities also have been trying to find ways of stretching their borrowing power. As a result of rapid growth in what is the most capital-intensive industry,[6] electric companies are relying on external sources (securities markets) for a steadily increasing share of their construction capital. Another consequence of these large and lengthy construction projects is capital exposure: $40 billion is now tied up in unfinished facilities—equivalent to 18 percent of all investments in electric plants in the U.S. Traditionally, utilities have increased their borrowing to pay the interest on construction work in progress (CWIP), putting further strains on their creditworthiness. Recently, however, many state utility commissions have allowed customers to be charged immediately for the interest on CWIP. Thus, the rate-payers are in effect made into investors, and the utility is enabled to reduce its borrowing—legally defined—by just that amount. The device has been challenged before many state utility commissions, and it is not yet a common practice (Corey, 1977; Newburger, 1977). Proposals to stimulate financing of coal gasification projects

[6] Electric utilities currently have an average of $3.50 invested in plant and equipment for each $1.00 of annual revenue. This compares with less than $1.00 invested by the steel industry and 55¢ invested by the auto industry for similar annual revenues.

would use a similar method (*Business Week,* 1978a).

Control of Western Energy Resources

Major oil companies are acquiring large holdings of coal, oil shale, and uranium reserves in the West. This has brought accusations that the large firms are trying to close off competition from other energy resources. The firms themselves hold the opposite view, claiming that only large, diversified firms have enough capital and expertise to do the job.

Increasing amounts of capital have been invested in western energy development by large, diversified energy firms. Exxon, Gulf, and Kerr-McGee are among the largest owners of domestic coal and uranium reserves, including substantial portions of these resources that are located in the West. In addition to oil companies, other firms that traditionally have been powerful in the West, such as the Burlington Northern and Union Pacific railroad, are now developing energy resources or have merged with energy companies such as Anaconda Copper has done with ARCO. While this trend is bringing unprecedented sources of capital, technical expertise, and marketing channels to the West, it also brings the danger of excessively concentrated economic power. And according to some, it is also possible that oil companies are acquiring mineral rights and patent rights for processing simply to preempt any potential competitors. There is some evidence to support this contention: for example, of 24 petroleum companies with major coal reserves, only eight had done any mining as of 1977 (FTC, 1977).

Concentration of ownership can involve mineral resources, production facilities, patents, or federal leases; the adverse consequences could include restricted production, higher prices, slower innovation, or domination of state and local governments. The diversified energy firms which have been entering western energy production maintain, however, that their participation is needed. A recent Bureau of Mines study agreed with that point of view, stating that the emerging trend of consolidated holdings should strengthen the financial structure of coal mining and aid its position in the new capital market. Furthermore, a parent or controlling company with a coal subsidiary or affiliate can readily shift capital according to a project's needs (Tomimatsu and Johnson, 1976:11). Nevertheless, there is some controversy between the major oil companies, who claim they are the only ones with enough capital and expertise to do the job, and some consumer-oriented groups, who fear that the majors are simply nipping competition in the bud.

Since 1973 the major oil firms have grown very large and extremely profitable compared to any other industry (Hayes, 1980:92); in fact, today 11 of the 20 largest industrial corporations in the country are oil and gas companies (*Fortune,* 1980:276). A more salient determinant of whether a market will be competitive, however, is the size of a firm relative to the size of the industry. In this regard, the oil industry is less concentrated than the average manufacturing industry. The four largest firms in the typical manufacturing industry account for roughly 40 percent of production (Scherer, 1970:63). In the petroleum industry as of 1974 the share of the market accounted for by the four largest firms was somewhat lower, ranging from 26.0 to 35.1 percent at various stages of processing (Markham, 1977:17).

Certain other features of the energy industry, however, tend to increase the influence of the large companies more than the foregoing figures suggest.

First, the top firms at one stage of processing, such as the production of crude, tend to be the same firms which lead the other stages, such as marketing. There are only about 20 such vertically integrated firms, but they ac-

count for 58 to 94 percent of processing at the various stages. Second, the majors "do not function as independent or competitive, but as cooperative entities at every strategic point of the industry's integrated structure" (Adams, 1977:7). They cooperate in joint ventures for bidding on offshore leases, building pipelines, developing foreign reserves, and other projects. Joint ventures "establish a community of interest among the parents and a mechanism for avoiding competition between them" (Adams, 1977:7). Finally, the oil companies have been diversifying into other energy resources.

This last trend, horizontal diversification, bears most directly on the West. Oil and gas are expected to be superseded by coal, uranium, and oil shale as the major western energy resources,[7] and the integrated energy companies have already acquired the largest single share of the reserves of these materials. As shown in Table 9-3, oil and gas companies own 41.1 percent of the nation's privately held coal reserves. In western operations, independent coal companies have become the exception rather than the rule. Similar penetration by oil and gas firms has occurred in the uranium industry, where 10 out of the 20 largest holders of uranium reserves are oil and gas companies (Greider, 1977).

Federal leasing policies will have a direct influence on the competitiveness of western energy markets. Although until recently federal coal accounted for only one percent of national production, approximately 60 percent of the western coal reserves are owned by the federal government and an additional 20 percent is dependent on the mining activity on adjacent federal land for its economic production (U.S. DOI, BLM, 1979: p. 2-1). The federal leases granted to date have resulted in an even greater concentration of control than exists for the coal industry as a whole;

[7] The one major exception to the decline of oil and gas is in western Wyoming, where recent discoveries are likely to yield the best fields since Alaska's Prudhoe Bay (Ratner, 1976).

Table 9-3: *Coal Reserves of Industrial Groups, 1976 (in millions of tons)*

Group	Reserves
Oil and Gas	
Continental (Consolidation)	13,700
Exxon (Monterey, Carter)	8,400
El Paso Natural Gas	5,200
Occidental (Island Creek)	3,570
Gulf (Pittsburg and Midway)	2,750
Mobil	2,500
Sun	2,200
ARCO	2,200
Phillips	2,000
Tenneco	1,700
Other	10,880
	55,100
Railroads	
Burlington-Northern	11,400
Union Pacific	10,000
Other	1,990
	23,390
Metals	
Newmont et al. (Peabody)	8,900
AMAX[a]	5,040
U.S. Steel	3,000
Bethlehem Steel	1,800
Other	3,200
	21,940
Independent Coal	
North American	5,100
Westmoreland	1,950
Pittston	1,700
Other	6,860
	15,610
Utilities	
Pacific Power and Light	1,700
Other	4,720
	6,420
Other Industries	11,790
All Identified Reserves	134,250

Source: NCA, 1977: Appendices B, H.

[a]AMAX might also be classified with oil and gas, as 21 percent of it is owned by Standard Oil of California.

and Continental Oil) hold 43.6 percent of the leased acreage (Bierman et al., 1977: Table 52).

Production on the federal leases has been virtually nonexistent or, at most, leisurely, suggesting the possibility that they were obtained primarily for speculative purposes. Of the 16.4 billion tons of coal under lease in the eight-state region, only 32.6 million tons (or .19 percent) were actually mined during 1976. Moreover, a BLM study found that the owners of leases covering 60 percent of these reserves have never produced any coal and have not indicated any plans for production before 1990. The average age of these leases is already 11 years. A Federal Trade Commission staff report shows that the 20 largest holders of federal coal leases are oil companies (Bierman et al., 1977:499–500).

Various measures have been adopted by the federal government to promote a more competitive market. For example, a traditional policy in leasing mineral rights has been the requirement of "diligent development" (Mineral Leasing Act, 1920). Thus, under current regulations, mining plans "must provide for the mining of all the reserves of the logical mining unit of which the lease is a part in a period of not more than forty years" (C.F.R., 1978). One of the purposes of "diligent development" is to discourage private companies from holding reserves indefinitely in anticipation of future price increases or, worse, from causing prices to rise by withholding coal reserves from production. Such legislation, however, may be inadequate. In regard to leases issued before the current moratorium,[8] the Interior Department has never cancelled a lease for failure to maintain diligent development (U.S. Congress, House Committee on Interior and Insular Affairs, 1974).

this is especially true in subregional market areas. For example, in Utah, Arizona, and New Mexico, which tend to serve coal markets to the West, four firms (Peabody, Arizona Public Service, El Paso Natural Gas,

[8]Issuance of new federal leases has been virtually halted since 1971, due to challenges by environmentalists and planning by the Interior Department for completely new procedures. Full resumption of new leasing is not expected before the mid 1980s. (See Lee and Russell, 1977; and Bagge, 1978.)

Summary of Problems and Issues

A number of risks and uncertainties inherent in the development of energy resources in the West potentially limit the availability of capital on the scale needed. These risks arise from technical, economic, and governmental and institutional factors. In addition, diversified energy companies have acquired large holdings of energy resources in the West, raising fears that these firms are closing off potential competition from the new forms of energy.

Although overall, western energy development would not be constrained by the national availability of capital, individual energy projects will be subject to a number of uncertainties that threaten secure investment. Financial difficulties for any particular facility will tend to spread to other facilities since financing tends to follow proven successes. The problems evident for individual facilities, therefore, can have a broad influence on the amount of capital that is available for western energy development generally.

Policies for Capital Availability

Given the uncertainties surrounding large-scale energy development ventures in the western U.S. (and elsewhere), the attendant financial risks will often be greater than for other investment opportunities, therby raising the cost of financing. The policy objective on which our discussion is based is to ensure the availability of sufficient investment capital for western energy development. To meet this goal, two categories of alternative policies are considered:

- To provide financial subsidies to developers; and
- To promote new sources of investment.

There are several specific alternatives within each of these categories (Table 9-4). For example, subsidies for energy projects could be

Table 9-4: *Policy Alternatives for Capital Availability*

Objective	General Alternative	Specific Alternative
Assure adequate capital for development of western energy resources	Provide financial subsidies to developers	Grant tax preferences
		Guarantee prices
	Promote new sources for investment	Encourage new participants by improving leasing system
		Require consumers to assume financial risks for new energy projects

provided either through tax incentives or by price guarantees to encourage investment in new, risky technologies.

The second category of alternatives, promoting new sources of investment, would draw different participants into the financing of western energy development. Increasing consumer rates to help pay for new facilities and streamlining the system of leasing energy resources could, along with other results, expand the amount of investment in western energy development.[9]

[9] Another category of alternatives which is available, but which is not addressed specifically in this chapter, is to improve the regulatory environment so as to reduce the delays and uncertainties in constructing energy facilities. Specific alternatives within this category might include expediting the siting process (discussed at length in Chapter 11), deregulating oil and gas prices, and permitting "rolled-in" prices for synthetic fuels. Government regulation and its impact on economic growth and technological innovation has been receiving increasing scrutiny. In fact, 1978 has been considered "the year of regulation" by some Washington observers because of the attention paid to regulation per se (see Clark, 1979:108). Recent actions include the President's appointment of a Regulatory Analysis and Review Group of the U.S. Regulatory Council and an Executive Order mandating "regulatory analyses," much like environmental impact statements, for major federal actions.

Financial Subsidies

Two types of subsidies for energy development are considered here:

- Tax preferences for energy projects; and
- Price guarantees.

Tax Preferences: The possible policies which fall into this category include allowing a larger investment tax credit for energy facilities, accelerated depreciation, or tax-exempt status for energy facility bonds. All such preferences would require both changes in the tax code and congressional action. The IRS already administers thousands of tax preferences— some designed to promote equity and others as incentives for a wide range of activities. Those with the largest impact on energy investment currently are investment tax credits, percentage depletion, accelerated depreciation, and special treatment of intangible development costs (Brannon, 1975). Tax exemption for bonds is another preferential measure that has been granted to help finance pollution control equipment, but it has not, as yet, been applied to energy facilities generally.

In this section, we will pay particular attention to the investment tax credit. As it is currently in effect, this provision allows businesses to reduce their tax liability by 10 percent of their expenditures on new equipment. Recent proposals have focused on several modifications (Samuelson, 1978:138): (a) increasing the rate; (b) making the rate permanent (it is now scheduled to revert to 7 percent eventually); (c) increasing the rate for certain industries; (d) extending the credit to buildings as well as equipment; and (e) allowing a company to offset up to 90 percent of its tax liability with the credit (compared with 50 percent now). The evaluation which follows will concentrate on increasing the credit for investments in domestic energy production.

The relative effectiveness of these options depends critically on their amount and importance to individual firms. For example, the investment tax credit is presently limited to 50 percent of a firm's net income, thus restricting many utilities in their use of this provision. It is not only the percentage of the credit, therefore, but the maximum credit allowed that determines its effectiveness in making capital available for new energy projects. An investment tax credit cannot remove much of the market-price uncertainty of an energy investment nor does it guarantee to the investor a profitable operation once the facility is functioning. Tax credits do, however, have the advantage of reducing capital exposure by allowing the developer to recapture part of its capital outlay almost immediately. Of course, without sufficient profits the tax credit would not be effective; this appears to be a problem not only for many smaller firms but also for some large electric utilities.

One factor which potentially limits the effectiveness of the investment tax credit is that the percentage of credit can be changed annually in the tax code, thus increasing uncertainty about the profitability of an investment. For example, if an investment is made in anticipation of large tax credits, the firm is gambling that the allowable percentage will not change during the construction period, which can often last eight years, or longer. The investment tax credit has recently had an erratic history, and, therefore, this uncertainty may reduce its effectiveness as a means of increasing investment in large-scale energy facilities.

The cost to the government of such subsidies depends on the extent of the subsidies offered and the range of energy facilities covered. Certain subsidies (particularly price guarantees) would be quite inefficient in the traditional economic sense, since they would support technologies that are not economical at existing market prices. The resulting disparity and liability, of course, will rest on the government and the taxpayers. Therefore, the benefits of technological improvements and increased domestic energy production must be balanced against these costs.

The investment tax credit has an unusual pattern of impact in terms of cost to the public. It shows up as a government cost—in the form of lost tax revenues—early in the life of a particular project. A 10 percent investment tax credit on a $1 billion facility will reduce the developer's taxes by $100 million, assuming the firm can take advantage of the full credit. Subsequent success or failure of the project will not affect the amount of subsidy.

In general, the fundamental feature of tax preferences is that taxpayers ultimately absorb the cost through reduced federal revenues. A major question concerning the equity of these policy alternatives is which facilities will be covered. If synthetic fuel facilities receive a benefit that, say, conventional gas producers do not, then those conventional producers will be forced to face a new source of competition. Consumers of the synthetic gas, though, will benefit by paying the market price, even though they are using gas which costs more than that to produce. This economic benefit to some consumers (since all parts of the country would probably not use synthetic gas) would be subsidized by all taxpayers. The benefit which taxpayers generally will receive is some experimentation with and refinement of new energy technologies and an increased domestic supply of energy with all of its implications (such as an improved balance of payments and a lower unemployment rate).

In considering the flexibility of financial subsidies, two aspects must be considered: the extent to which they can be tailored to specific technologies or facilities and the extent to which they can be modified over time as changing circumstances dictate. Since tax codes are changed almost on a yearly basis, the investment tax credit is very flexible in terms of allowing changes over time. It permits credits on a yearly basis and could be changed in the annual tax code revisions. However, as discussed previously, this flexibility can limit the effectiveness of tax credits since industry cannot count on uniform tax credits throughout the construction period for large energy facilities. With regard to tailoring such provisions to specific technologies or facilities, investment tax credits are not flexible; they are generally applied across the board. The IRS simply is not set up for making technical judgments based on engineering, environmental, or economic factors. Two possible solutions would be to allow certain industries an exemption from revisions for some specified period of time or to allow greater administrative discretion in expanding or reducing the tax credit. This latter option would, of course, require a revision in the tax code, setting a maximum and minimum that policymakers either could change as needed or could apply to specific industries without annual tax code revision. This could provide industry with some greater stability and yet assure taxpayers that they will not be locked into costly long-term policies.

A major problem in deciding how to provide financial subsidies to energy developers is that the policies that are attractive to industry may be opposed by other parties—taxpayers who will bear the costs or proponents of competing energy options, such as solar power. Nevertheless, generally speaking, the various financial subsidy policies, depending on the specifics of each, can probably find support among policymakers. Tax preferences, in particular, have considerable precedent and do not involve a new approach to economic policy.

On the other hand, allowing substantial administrative discretion with investment tax credits to provide increased flexibility, as discussed previously, would require major policymaking changes and would seriously challenge long-standing traditional values. Decision-making in the tax area is traditionally left to the legislative body; any increase in administrative discretion in these areas would probably be difficult to achieve.

Price Guarantees: Price guarantees would support the development of synthetic fuel and other advanced energy technologies by setting a minimum, or floor, price for the product. In the case of synthetic gas and shale oil, the guaranteed price could be as much as 50 percent above the market price of gas and oil. The purpose of guarantees is to assure that new technologies are proven on a commercial scale and that industry has sufficient experience to generate improvements and refinements in the technologies.

Price guarantees differ from tax preferences in that the guarantees would probably be more narrowly targeted and could be administered by the DOE rather than the IRS. Although guarantees resemble tax preferences as energy subsidies, the fiscal impact of tax preferences depends on how extensively firms take advantage of them. Tax preferences are not budgeted expenditures, but reduce the income available in the government budget in an *ex post* way that is often difficult to predict. Price guarantees, on the other hand, are a function of both the firm's costs and market conditions. Whatever expenditures actually are made must be budgeted and go through the usual disbursement procedures.

Price guarantees could effectively reduce investors' risks, depending on the established level of support. If a potential investor is guaranteed a profit (that is, guaranteed a sufficient price floor), much of the risk of investing would be removed. Essentially, the market price uncertainty, as well as the liability, would be transferred to the federal government. Price guarantees would also reduce some of the uncertainties associated with the costs of construction and operation. Of course, not all risks would be removed: for example, unexpected technical difficulties could raise costs above the price floor. Therefore, while price guarantees relieve investors of market-price uncertainties, they do not totally eliminate the problem of capital exposure. Thus, the developer must still raise funds at the time of construction, keep the funds tied up for a long time, and bear the potentially disastrous risks of technical problems, strikes, etc. As discussed above, the financing of a single, large facility would

nearly equal the entire fixed assets of many would-be developers. Thus, price guarantees help the ultimate profit potential of the larger firms which can get these energy projects under way, but getting started is still the major hurdle for the smaller firms.

In the case of price guarantees, the liability incurred by government to support such technologies could be quite substantial. Projections by the ERA estimate gasified coal at $5.50 per million Btu's in 1978 dollars, compared to $2.26 for domestic gas and roughly $4.20 for Alaskan imported gas (OGJ, 1978). Thus, the amount of the guarantee or subsidy could be $1.30 over the price of Alaskan gas, implying a subsidy to a 250 MMcfd gasification plant of about $110 million per year, roughly the entire annual operating cost for such a facility (less interest charges and administration).

Naturally, the total liability that the government would incur with a price guarantee would depend on both the duration of the guarantee and the relationship between the market price and guaranteed price during that period of time. In 1975, the Synfuel Interagency Task Force of the President's Energy Resources Council used a dynamic simulation model (FIST) to estimate the federal liability incurred over various time periods and market prices in supporting a shale oil plant. As Table 9-5 shows, a price guarantee could become quite expensive—depending on the rate of return, the time period of guarantees, and the real price of competing oil. For example, to guarantee a 15 percent discounted cash flow rate-of-return for 20 years when oil costs $11 per barrel (in 1975 dollars) would cost the government $157 million (1975 dollars). As market prices rise closer to the guarantee level (which is generally expected in the case of natural gas), the subsidy would decrease. If the market price rises slowly, however, or stays below the guaranteed price, the government could be supporting expensive sources of energy for many years.

A major question concerning the equity of public subsidies in general is whether or

Table 9-5: *Analysis of Price Guarantees for a Shale Oil Plant (producing the equivalent of 50,000 crude barrels per day)*

| | | Net Present Value (discounted at 10 percent) Cost to Government (millions) | | | | | |
| | Price Required for Specified Returns ($/bbl) | 2-year Termination | | 5-year Termination | | 20-year Life | |
Guaranteed Rate of Return		Oil $11/bbl	Oil $7/bbl	Oil $11/bbl	Oil $7/bbl	Oil $11/bbl	Oil $7/bbl
12 percent DCF	10.10	0	59	0	129	0	228
15 percent DCF	12.70	32	106	69	232	157	521
20 percent DCF	17.94	128	203	280	442	628	993

$/bbl = dollars per barrel. DCF = discounted cash flow, after taxes, on total capital.

Source: Synfuels Interagency Task Force, 1975: vol. 3.

not they favor the larger energy-related firms over the smaller ones. In the case of price guarantees and investment tax credits, the subsidies are definitely of more value to the larger companies. While price guarantees will encourage a large company to put up the initial capital necessary to develop a facility, they do not help the smaller firms as much: their main problem is coming up with the initial capital to start such a plant. Given the advantages of price guarantees for the larger company, it is possible that the question of equity may merge into a regional conflict. Many of the eastern coal mines are owned by smaller firms, whereas the western coal deposits are owned by large conglomerates whom the price guarantees will assist. Hence, the eastern firms may regard federal subsidies as unnecessarily shifting coal production to the West.

In contrast to the investment tax credit, price guarantees could easily be tailored to a specific technology (or even to an individual plant as part of a demonstration program), but they are not inherently flexible over time. Although a price-guarantee program can be terminated at any time, this would not retroactively affect facilities already in operation or under construction. Flexibility could be enhanced, of course, by initially designing the price guarantee so that it would only last for a particular facility for a relatively short time (such as five years). This approach, however, would also seriously limit its effectiveness in attracting investment capital.

Opposition to price supports can probably be expected, especially from individuals and groups who believe the money would be better invested in conservation efforts and in developing renewable energy resources. Price supports also cause concern among those who fear expensive price supports over an indefinite period of time. Since the various technologies have not yet been developed to a point where they are competitive in the market place, taxpayers will have to pay the difference between the actual cost and the market price. And finally, all taxpayers will be paying for some portion of the energy used by a small segment of the public.

Promotion of New Sources of Investment

The new sources of investment we will consider here include investors other than large firms and their financial backers. In selected situations these new sources might be enlisted experimentally, although none would be able to substitute completely for the traditional participants in energy development. Two specific policies discussed below are to:

- Improve the system of mineral leasing to help smaller firms; and
- Allow consumers to assume financial risk for new energy projects.

Improvement of Mineral Leasing: One approach to broadening participation would be to streamline the procedures for making energy resources on federal land available for development by the private sector. The system of distributing uranium resources, especially, can be characterized as chaotic and archaic. In this environment, competitive advantages may accrue to the larger firms who can employ staffs of geologists, lawyers, and claim investigators. Moreover, since the federal government itself does not know who holds valid mining claims (U.S. Congress, Senate Committee on Interior and Insular Affairs, 1976:671), it may be possible for a few firms to quietly gain control of large portions of the nation's uranium resources.

Other energy resources are already allocated under leasing systems; however, modifications in procedures could enhance the opportunities of smaller firms. Two particular proposals which have been considered are to replace preference-right leases on coal lands with a competitive bidding system and to institute a two-stage competitive bidding system for less well-mapped resources such as geothermal energy. The two-stage competitive bid would award an exploration lease to the firm that agreed to accept the lowest share of the revenues received if and when the resource is commercially developed. (This is called the discovery-bonus share.) In the second stage, bidding would be similar to conventional, competitive bonus-bidding methods, except that it would take place only after economically workable quantities of an energy resource were discovered (U.S. Congress, Senate Committee on Interior and Insular Affairs, 1976: 734–35). Another approach is to institute special leasing procedures to encourage activity by small firms. For example, DOI reserved

35 percent of the oil and gas tracts discovered off the coast of Massachusetts for leasing to smaller development firms (Bernstein, 1978).

If wider opportunities for investment are created, more participants and a more competitive energy market could evolve. It is also possible that streamlined leasing procedures could increase the total pace of development. For hard-rock minerals such as uranium, the current system is so chaotic that there may be seven or eight claims on a single piece of land, with many legal bases for one to challenge another. One report strongly suggests that the gaps and anomalies in the legal structure of mineral leases create confusion and vexatious litigation (U.S. Congress, Senate Committee on Interior and Insular Affairs, 1976:666). In the case of poorly mapped resources (geothermal, oil, and gas), the two-stage competitive bidding system described previously could also attract greater participation by postponing the larger cash outlays for a production lease until the exploration stage is completed.

The net effect of these forces is difficult to predict. While more companies will be able to participate under the improved leasing systems, it is possible that investment capital for western energy development will decline nevertheless, due to substantial increases in bonus bids and royalty payments made to the federal government. These costs may reduce the willingness of industry to develop federally owned resources. It may also push upward the prices that are ultimately charged to consumers for energy.

A quantitative estimate for potential revenues may be gained from examining the previous experience with coal leasing in the West. The bonus payments received in 1974, at 2.44 cents per ton, were more than three times as large as was collected in any previous year (U.S. Congress, Senate Committee on Interior and Insular Affairs, 1976: Table 9.19). Still, the market value of the coal at that time was much greater—about $10 per ton. A 12.5 percent royalty payment (which is now the Interior Department's goal) would have yielded $1.25 per ton, or more than 50 times as much as was actually collected. Even allowing for the fact that bonuses are paid immediately and royalties are deferred, collections in 1974 were probably less than one-twentieth of what they could have been under a percentage-bonus system. Multiplying by 20 the average bonus receipts of $2.09 million per year during the period between 1966 and 1971, it may be estimated that more than $40 million of potential revenue annually has not been collected. And this figure would increase if, as expected, federal coal leasing activity were to reach a faster pace than before the 1971 moratorium.

As noted above, revised leasing procedures would generate more revenue for federal and state governments, thus constituting a benefit for taxpayers. Although a prime rationale for instituting these new leasing procedures is to enhance the competitive forces which will tend to hold down energy prices, that effect may be diluted somewhat. The added cost of bonuses and/or royalties will, to some extent, be passed along by energy developers to energy consumers.

The proposed changes in mineral leasing make the process at least potentially more flexible by making it more organized. The current system leaves the government in a rather passive position, with private companies nominating coal tracts and claiming uranium resources at times and places of their own choosing. A planned leasing program would give the government a greater opportunity to guide the pace of development, as well as to concentrate it in areas which are the most environmentally acceptable. However, a basic restructuring of the mineral leasing system would require new legislation and, hence, would be virtually irreversible. Mining is currently regulated under laws enacted in 1872 (General Mining Law) and 1920 (Mineral Leasing Act); fundamental legislation in this area is not passed frequently.

Assumption of Risk by Consumers: Energy consumers are another possible source for investment capital. Utility consumers could assume a major portion of the capital cost and associated risks of new projects if project costs could be coincidentally recovered in consumer rates. This is essentially the plan that was considered by DOE to secure financing for the first proposed coal gasification facility near Beulah, North Dakota (*Business Week,* 1978a; OGJ, 1978). The DOE plan would effectively make gas customers guarantee a major portion of the development and construction costs, with a consortium of energy firms putting up the remainder. If this policy were to be employed for all synthetic fuel projects, capital availability would be a minor problem.

Even the proponents of consumer risk assumption concede that coal gasification, at its current stage of development, entails greater economic cost than do other gas sources. As noted earlier, the ERA estimated the cost of synthetic gas at $5.50 per mcf (in 1978 dollars), whereas the most expensive gas alternative (imported LNG) was only $4.27 (OGJ, 1978). Consumers would not only be paying higher energy costs, but they also would be assuming a major share of the risks in the event of unexpected technical problems.

This option could be implemented in the case of regulated industries, where the monopoly situation leaves consumers little choice but to accept special assessments in the case of project failure. Moreover, the regulatory framework facilitates other devices to improve the developers' propsects, such as "rolled-in" pricing and provisions for the recovery of interest payments during construction. In fact, these devices have been coupled with the first application before the FERC for an "all-events tariff" (OGJ, 1978). Such a regulatory package provides considerable security to the developers and may be expected to substantially increase the number of gas companies that will adopt coal gasification.

Clearly, consumer assumption of risk implies that gas consumers will bear what are expected to be the higher costs of synthesized gas. Under the first "all-events tariff" that has been proposed, outside lenders would be guaranteed a complete return of their investment. This would be achieved mainly by assessing consumers with a "cost-of-service tariff" sufficient to cover all costs of construction and operation. The pipeline companies would probably receive a somewhat lower rate of return on their equity (which represents 25 percent of the total investment) since the ERA has recommended that "appropriate consideration [be given] to the degree to which risks associated with this project are being preguaranteed by consumers" (OGJ, 1978). In all other respects, consumers would bear any additional costs.

As noted previously, compared to other "supplemental" sources of gas, synthetic natural gas from coal will cost an additional $1.23 per thousand cubic feet (in 1978 dollars). Thus, a plant of 125 MMcfd capacity would entail an added cost of $50.5 million per year, or about $6.30 per year for each of the participating companies's customers. This is a minimal estimate, since the cost of new technologies tends to rise in the development stage, and even cheaper alternatives may become available. For example, a major gas discovery in Alberta has recently cast doubt on the economic viability of the Alaskan gas pipeline (*Business Week,* 1978b).

The general public would gain some benefits from a policy of consumer assumption of risk. Probably the most important is the reliability of supply, as compared to imports and to limited domestic reserves. In addition, experience with the new conversion processes may well lead to technological breakthroughs and, eventually, to lower costs. Moreover, the policy can be applied on a plant-by-plant basis. Many of the benefits, however, will be national in scope, whereas the costs will be paid by a relatively small number of energy consumers. Thus, utility customers and utility commissions may oppose this approach.

Summary and Comparison of Alternatives to Increase Capital Availability

The effectiveness of these alternatives depends critically on the particular method of implementation. In general, subsidies could be very effective because they bring the financial resources of the federal government directly to bear (see Table 9-6).

Very different kinds of firms would be involved in each of these alternatives. The "new sources" strategy reduces the risk of financial failure and, hence, could attract small firms or those facing financial constraints (for example, some utilities). In fact, some options within this overall strategy could be deliberately designed to aid the smaller firm. By contrast, subsidies do not greatly reduce capital exposure—the amount of capital tied up for long periods and subject to possible loss. Thus, while expected profitability is increased statistically, substantial risks remain. Hence, participation under the subsidy plans would tend to be limited to well-capitalized firms, such as integrated oil companies.

Most of the options considered here are inefficient in the economic sense, inasmuch as they promote the use of resources and technologies whose cost is greater than that of available alternatives, such as imported oil. The primary rationale for adopting these technologies is the need to balance economic efficiency with reduced dependence on foreign sources and to encourage faster technological advancement. Additional economic costs are borne by taxpayers, in the case of subsidies, and by energy consumers, in the case of the consumer-assumption-of-risk alternative. If this latter plan is used primarily as a tool to develop new technologies, then a very small segment of energy consumers would be paying for a program with broad national benefits. In the case of improved leasing systems, the costs and benefits and their distribution is uncertain, but taxpayers generally should benefit through higher government revenues paid

Table 9-6: *Summary and Comparison of Alternatives to Increase Capital Availability*

Criteria	Provide Financial Subsidies	Promote New Sources of Investment
Effectiveness: Achievement of Policy Objective	Potentially very effective as they can bring to bear the large financial resources of the federal government. Effectiveness of tax credits depends on stability. Tax credits reduce capital exposure but large risks remain. Price guarantees reduce risks somewhat but do not influence initial capital requirements.	Leasing options will have an uncertain effect; more firms, on the one hand, but more competition and higher costs in the form of government payments, on the other. Consumer risk assumption essentially reduces all risks. Could be applied on a case-by-case basis.
Efficiency: Costs, Risks, and Benefits	All subsidy options are inefficient in the sense that they promote resources that are more expensive than other available options. If selectively applied to promote new technological developments, total costs are not large and domestic energy production may be increased.	Leasing options would increase revenues to government. Competition would be enhanced, but higher leasing costs might be passed on to consumers. Consumer risk assumption would increase costs relative to other sources. However, could be applied selectively.
Equity: Distribution of Costs, Risks, and Benefits	Taxpayers bear subsidy costs; competitive energy sources and conservation may suffer. Energy developers and consumers receive the benefits. Also, taxpayers would receive the national benefits of decreased oil imports. Small firms may not be able to take advantage of subsidies.	Energy consumers would generally bear increased costs of both specific alternatives. Smaller firms might benefit from improved leasing procedures.
Flexibility: Adaptability to Changes	Tax preferences tend to be administered inflexibly. Price guarantees could be applied selectively.	Leasing options would tend to be permanent, but would give government more administrative flexibility. Consumer risk assumption could be administered flexibly by FERC.
Implementability: Adoptability and Acceptability	Subsidies have considerable precedent in nonenergy fields; also taxation policies do not require disbursement from the federal budget. Likely opposition from groups who believe the money would be better invested in conservation and renewable energy sources.	Most leasing options would require new legislation. Consumer risk assumption would require a change in philosophy by FERC and may be opposed by the utilities customers.

for the right to develop federally owned resources.

Implementation may come most readily for the various subsidy strategies, especially tax incentives. They have considerable prece-

dent and politically may be acceptable since they do not require any government expenditures, only the reduction of revenues. Improving the leasing system, on the other hand, strongly affects well-defined groups, such as uranium and coal mining firms. Finally, strategies such as consumers' assumption of risk go against the traditional orientation of such agencies as the FERC which would have to implement the plans, although they may be experimented with on a case-by-case basis.

Conclusion

Expanding western energy production will require an unprecedented level of capital investment in our eight-state western region. In order to provide the capital necessary for extraction, conversion, and transportation facilities, several uncertainties will have to be reduced or resolved; these include technical, economic, regulatory, and institutional factors. Perhaps the problem of capital avail-

ability is best demonstrated by the fact that despite the 1973 Arab oil embargo and subsequent efforts by the federal government to encourage the development and use of domestic energy resources, not one commercial oil shale or coal-based synthetic fuel plant has been built. Such facilities require an investment on the order of $1 billion per plant, equivalent to the entire fixed investment of many large energy firms. So far, no company has been willing to accept such a risk in the face of the technical, economic, and regulatory uncertainties involved. If the federal government wishes to increase domestic energy production by using these newer technologies, then it appears that some type of government programs, such as those discussed in this chapter, will be necessary to reduce the risks involved.

REFERENCES

Adams, Walter. 1977. "Horizontal Divestiture in the Petroleum Industry: An Affirmative Case." In *Horizontal Divestiture,* edited by W. S. Moore. Washington, D.C.: American Enterprise Institute for Public Policy Research.

Bagge, Carl. 1978. "Federal Leasing? In the 1980's Perhaps." *Coal Mining and Processing* 15 (September):49.

Bernstein, Peter J. 1978. "System to Aid Small Firms." *Denver Post,* January 18.

Bierman, Sheldon L., et al. 1977. *Innovation versus Monopoly: Geothermal Energy in the West,* Final Report, Report No. DGE/3036-1. Washington, D.C.: Energy Research and Development Administration, Division of Geothermal Energy.

Brannon, Gerard M. 1975. "Existing Tax Differentials and Subsidies Relating to the Energy Industries." In *Studies in Energy Tax Policy,* A Report to the Energy Policy Project of the Ford Foundation, edited by Gerard M. Brannon, pp. 3–40. Cambridge, Mass.: Ballinger.

Breyer, Stephen G., and Paul W. MacAvoy. 1974. *Energy Regulation by the Federal Power Commission.* Washington, D.C.: Brookings Institution.

Bureau of National Affairs. 1977a. "Interior Policy, EIS Delay New Oil Shale Demonstration Plant." *Energy Users Report,* Current Report No. 190 (March 31):16.

Bureau of National Affairs. 1977b. "New England Utilities to Challenge FEA Conversion Order in Federal Court." *Energy Users Report,* Current Report No. 194 (April 18):9–11.

Business Week. 1977. "Financing: The Real Test for Alcan Pipeline." November 28, pp. 102 4.

Business Week. 1978a. "How to Finance Gas Produced from Coal." June 19, pp. 33–36.

Business Week. 1978b. "A Halt to the Alaska Gas Line?" September 25, pp. 155, 158.

Carasso, M., et al. 1975. *The Energy Supply Planning Model,* 2 vols. San Francisco: Bechtel Corporation.

Clark, Timothy B. 1979. "The Year of

Regulation." *National Journal* 11 (January 20):108.

Clean Air Act Amendments of 1977, Pub. L. 95-95, 91 Stat. 685.

Coal Industry News. 1977. "Energy Act Coal Conversion Proposals: A Summary." 1 (December 12):5.

Code of Federal Regulations (C.F.R.). 1978. Title 30, §211.10(c)(6)(ii).

Columbia LNG Corp. v. *FPC,* 491 F. 2d 651 (1974).

Corey, Gorden. 1977. "Further Observations on the Taxation of Regulated Utilities." *Public Utilities Fortnightly* 100 (November 10): 16–18.

Crittenden, Ann. 1978. "Coal: The Last Chance for the Crow." *New York Times,* January 8, sec. 3, p. 1.

Energy Research and Development Administration (ERDA). 1977. *Assistance from Energy Developers: A Negotiating Guide for Communities.* Washington, D.C.: ERDA, Office of Planning, Analysis and Evaluation.

Energy Supply and Environmental Coordination Act (1974), Pub. L. 93-319, 88 Stat. 246, as amended by Pub. L. 94-163, 89 Stat. 871 (1977).

Environmental Policy Institute. N.d. *The Need for Energy Facility Sites in the U.S.* Washington, D.C.: Environmental Policy Institute.

Federal Energy Administration (FEA). 1974. *Project Independence Blueprint, Final Task Force Report: Synthetic Fuels from Coal.* Washington, D.C.: Government Printing Office.

Federal Power Commission (FPC) Order No. 622, June 28, 1972.

Federal Power Commission (FPC) Opinion No. 728, April 21, 1975.

Federal Trade Commission (FTC). 1977. Survey noted in "U.S. Oil Industry Stakes Out Role for the Future," by William Greider. *Washington Post,* May 22, p. A1.

Fortune. 1980. "The Fortune Directory of the 500 Largest U.S. Industrial Corporations." 101 (May 5):274–301.

General Mining Law of 1872, 17 Stat. 91.

Gilmore, John S. 1976. "Boom Towns May Hinder Energy Resource Development." *Science* 191 (February 13):535–40.

Greenberg, Daniel. 1978. "New R&D Budget:

A Boost But No Bonanza." *Science and Government Report* 8 (February 1):1–3.

Greider, William. 1977. "U.S. Oil Industry Stakes Out Role for the Future." *Washington Post,* May 22, p. A1.

Hayes, Linda Snyder. 1980. "Twenty-Five Years of Change in the Fortune 500." *Fortune* 101 (May 5):88–96.

Jacobsen, Larry G. 1976. "Coping with Growth in the Modern Boom Town." *Personnel Journal* 55 (June):288–89, 303.

Krohm, G. C., C. D. Dux, and J. C. Van Kuiken. 1977. *Effect on Regional Coal Markets of the "Best Available Control Technology" Policy for Sulfur Emission,* National Coal Utilization Assessment. Argonne, Ill.: Argonne National Laboratory.

Lee, L. Courtland, and David C. Russell. 1977. "Whatever Happened to Federal Coal Leasing?" *Coal Mining and Processing* 14 (June):60–63, 112.

Maher, Ellen. 1977. "The Dynamics of Growth in the U.S. Electric Power Industry." In *Values in the Electric Power Industry,* edited by Kenneth Sayre. South Bend, Ind.: University of Notre Dame Press.

Markham, Jesse. 1977. "Market Structure and Horizontal Divestiture of the Energy Companies." In *Horizontal Divestiture,* edited by W. S. Moore. Washington, D.C.: American Enterprise Institute for Public Policy Research.

Mineral Leasing Act of 1920, Pub. L. 66-146, 41 Stat. 437.

Myhra, David. 1977. "Fossil Projects Need Siting Help Too." *Public Utilities Fortnightly* 99 (September 29):24–28.

National Coal Association (NCA). 1977. "Implications of Investments in the Coal Industry by Firms from Other Energy Industries." Washington, D.C.: NCA.

Newburger, David. 1977. "Electric Power— Who Pays for Expansion?" *Environment* 19 (June):50–52.

Oil and Gas Journal (OGJ). 1978. "DOE Backs First Coal Gasification Plant." 76 (June 12): 22.

Pipeline Demurrage and Shipment Rule on Propane, 315 I.C.C. 443 (1962).

Ratner, Steven. 1976. "Geologically Weird 'Overthrust Belt' Excites Oil Drillers." *New York Times,* December 27, p. D1.

Rowe, James L., Jr. 1977. "Conversion to Coal

Still Is Mostly Talk." *Washington Post.* June 19, p. L1.

Samuelson, Robert J. 1978. "Carter's Tax Tightrope." *National Journal* 10 (January 28): 133–38.

Sanger, Herbert, Jr., and William E. Mason. 1977. *The Structure of the Energy Markets: A Report of TVA's Antitrust Investigation of the Coal and Uranium Industries.* 3 vols. Knoxville, Tenn.: Tennessee Valley Authority.

Scherer, F. M. 1970. *Industrial Market Structure and Economic Performance.* Chicago: Rand McNally.

Synfuels Interagency Task Force. 1975. *Recommendations for a Synthetic Fuels Commercialization Program.* Report Submitted to the President's Energy Resource Council, 4 vols. Washington, D.C.: Government Printing Office.

Tomimatsu, Tommy, and Robert E. Johnson. 1976. *The State of the U.S. Coal Industry: A Financial Analysis of Selected Coal Producing Companies with Observations on Industry Structure,* Bureau of Mines Circular 8707. Washington, D.C.: U.S. Department of the Interior, Bureau of Mines.

U.S. Congress, House of Representatives, Committee on Interior and Insular Affairs. 1974. *Federal Coal Leasing. Hearings* before the Subcommittee on Mines and Mining, 93d Cong., 2d sess., July 25–August 15.

U.S. Congress, Senate, Committee on Interior and Insular Affairs. 1976. *Report to the Federal Trade Commission on Federal Energy Land Policy: Efficiency, Revenue, and Competition.* Committee Print, by the Federal Trade Commission, Bureau of Competition and Economics. Washington, D.C.: Government Printing Office.

U.S. Department of Commerce. N.d. "New Plant and Equipment Expenditures." *Survey of Current Business,* various issues, 1966–1975.

U.S. Department of Commerce, Bureau of the Census. 1975. *The Statistical Abstract of the United States.* Washington, D.C.: Government Printing Office.

U.S. Department of the Interior (DOI), Bureau of Land Management (BLM). 1979. *Final Environmental Statement: Federal Coal Management Program.* Washington, D.C.: Government Printing Office.

Vickers, Edward L. 1977. In *Southwest Energy-Minerals Conference Proceedings,* U.S. Department of the Interior, Bureau of Land Management, vol. 2, pp. 209–39. Santa Fe, N. Mex.: Bureau of Land Management.

Weidenbaum, Murray L., and Reno Harnish. 1976. *Government Credit Subsidies for Energy Development.* Washington, D.C.: American Enterprise Institute for Public Policy Research.

Wilson, Wallace W. 1976. "Capital for Coal Mine Development." *Coal Mining and Processing* 13 (January):68ff.

CHAPTER 10

Transporting Energy

The demand for western energy resources comes largely from urban areas outside the West, and some projections estimate that approximately half the expected production of western coal may be exported in raw form. As a result, existing transportation facilities both for raw resources (such as coal) and for converted forms of energy (such as electricity) will have to be expanded dramatically. Many of the benefits associated with such an expansion will be received outside the West, while the adverse effects will be confined largely to the western states and to other states along the transportation routes. These disparities between expected costs and benefits have made transportation issues central to political conflicts in our eight-state region.

Four modes of transportation—railroads, coal-slurry pipelines, oil and gas pipelines, and high voltage transmission lines (HVTL)—are critical to the movement of energy from the West. In transporting coal, three types of trains are used: conventional, unit, and dedicated. When conventional trains are used, the cars carrying coal are treated (for purposes of regulation) like any any other cars. Unit trains, by contrast, are made up entirely of cars carrying coal. Dedicated railroads are those lines used exclusively for transporting coal (as with a rail line linking a mine to a single-source user). The coal-slurry pipeline is used to move pulverized coal suspended in water. Of these various modes, railroads, coal-slurry pipelines, and high voltage trans-

mission lines are involved in the problems and policy issues of western energy development.

This chapter focuses on policy issues and obstacles to energy transportation that affect the development of energy resources in the West. Following a brief outline of the most significant factors, four major transportation issues are discussed. Finally, a variety of policy options for addressing these problems and issues are considered.

Factors Affecting the Impact of Energy Transportation

The expansion of networks for transporting energy raises difficult political questions because of the conflicts caused by the regulatory arrangements, disagreements about the ecological and economic consequences, and the direct competition between various methods of transporting energy.

Institutional Factors

The movement of energy from the West has already begun to generate conflicts among the various parties interested in western energy development; in many cases, this is due to the way each system of transportation is regulated. There has never been a single, coherent federal transportation policy for energy resources; instead, there is a patchwork of

269

policies based on the functions of the specific modes of transportation. Thus, federal responsibility for the transportation of coal is divided between the major roles of the Department of Transportation (DOT) and the Interstate Commerce Commission (ICC) and the secondary roles of the Department of Commerce (DOC) and the DOI. This regulatory fragmentation is a key factor in the rapid expansion of coal production, as specified in the objectives of the National Energy Plan and other major federal energy projections. If coal production does reach or even approach the 1.1 billion tons projected for 1985 by the National Energy Plan, significant demands will be made on the nation's various transportation facilities. Railroads, which already carry about two-thirds of the coal used by utilities nationally, would be called upon to increase coal tonnage by an average of 8 percent per year until 1985. A National Academy of Engineering study has forecast that by 1985 this country could need 60 new eastern rail-barge systems (each 100 to 200 miles in lenght with a 2 million ton annual capacity), 70 new western barge systems (each 1,000 to 2,000 miles in length with a 3 million ton annual capacity), and four coal-slurry pipelines (each 1,000 miles in length with a 25 million ton annual capacity) to meet these demands (Rittenhouse, 1977: 49). While most studies are optimistic that the nation will have sufficient coal-carrying capabilities, the margin is narrow and leaves little room for error. That is why the institutional uncertainties surrounding energy transportation are important: a breakdown in any aspect of transportation regulation could hinder both national and regional energy policies (see Diamond, 1977:38–47).

In addition to regulatory fragmentation, most transportation agencies are functionally independent of the organizations setting energy policy. With at least four powerful federal agencies delegated broad authority in the transportation arena, any coordination with the national goals and objectives of an energy policy is extremely difficult. As a result, there are major areas of conflict within the federal government and across federal-state jurisdictions.

In terms of federal interagency policy, two key areas of difficulty have been the rate structures for freight and the environmental restrictions on coal use. Federal rate policies for freight are a source of uncertainty in the area of coal transportation. This is particularly the case with the setting of unit-train rates by the ICC. And the efforts of environmental protection agencies to reduce air pollution or to improve the quality of water sources have had serious implications for planners: they need some assurances that attempts to develop rail networks will not be undercut by environmental policies which rapidly shift the demand for coal with particular characteristics. (See Chapter 5 for a discussion of the potential effects of a BACT policy on the demand for western coal.)

Federal, state, and local conflicts in the transportation sector are also significant. State agencies may severely obstruct national transportation policies by controlling such factors as rights-of-way and eminent domain. This form of institutional uncertainty has already impeded the use of coal-slurry pipelines, many of which have been delayed by the lack of powers of eminent domain. Another controversial federal-state factor in the West, water use, presents perhaps the greatest limitation to the use of slurries. An additional problem between federal and state transportation agencies is the impact of large coal trucks on highway systems.

Local governments may also present serious obstacles to the transportation and distribution of coal. The most obvious are local ordinances which restrict railroad or motor vehicle performance (speed limits, for example). In many western towns, local ordinances require trains to slow to five miles per hour at grade crossings (O'Hara, 1977:4).

Environmental and Economic Factors

Other significant factors center on the human and physical effects of the demands generated by each mode of transporation. The extent of impact depends on the assumed level of energy production. The Low-Demand scenario, which amounts to a level of western energy development just over 42 Q's by the year 2000 (see Chapter 2), results in the capital and land requirements for energy transportation facilities shown in Table 10-1. Coal-slurry pipelines, as a new method for shipping raw coal, would require large amounts of capital for construction (up to $14.2 billion) and relatively large commitments of land. (Although once the pipeline is buried, the land can be used for other, limited purposes such as grazing.) Unit trains, on the other hand, use only moderate capital resources and make very limited additional claims on land since many of the necessary rail lines already exist. And electric transmission lines would involve considerably lower capital costs (in the neighborhood of $2.2 billion) but would consume the greatest quantities of land for their rights-of-way (up to 205,000 acres).

The environmental impact of railroads primarily includes noise, fragmentation of wildlife habitats, and disruption of highway traffic in towns through which trains pass. The last of these is considered to be the most significant, particularly as the use of unit trains increases. Unit trains of around 100 cars moving through small western communities can be especially disruptive and create a potential for accidents because grade separations (overpasses and underpasses) are not common. Of course, these local effects have been present for a long time, but they would increase in intensity and frequency as energy development expands. Slurry pipelines have as their major environmental effect the consumption of water, estimated at 788 acre-feet per million tons of coal, which would come from water-scarce western locations. In addition, coal "fines" (very small particles of coal) produce a sludge at the receiving end of a slurry line. Thus, both rails and slurries have costs and advantages in terms of environmental impact.

High voltage electric transmission lines affect large quantities of land, since right-of-way corridors are required. As the voltage is increased, wider corridors are necessary; on the other hand, fewer lines (and corridors) are needed at the higher voltages because of their larger carrying capacity. Thus, to cut both construction and land costs, and because of the increasing size of power plants, utilities have been moving toward the higher voltage lines. In addition to the amount of land required, HVTL have been associated with electric shocks, interference with electronic communications, and biological effects on both humans and animals. These problems, which may be minor or nonexistent at low voltages, become of more concern as voltages are increased.

Of the methods of transporting energy that

Table 10-1: *Increased Requirements by 2000 for Three Methods of Transportation (Low-Demand Scenario)*

Mode	Annual Energy Transport Level[a]	Capital (billions of 1975 dollars)	Land (thousands of acres)
Coal-Slurry Pipeline	350 billion ton-miles	14.2	184
Unit Trains	306 billion ton-miles	9.9	67
Electric Transmission	29.6 million megawatt-miles	2.2	205

[a]Energy transportation requirements for the scenario are expressed by multiplying the amount of energy transported by the distance of transport.

are considered in this study, oil and gas pipelines produce the fewest physical alterations. The construction of oil, gas, or coal-slurry pipelines will incur some short-term environmental damage due to trenching and clearing rights-of-way; however, much of the natural environment is restored after this phase. The exception is that trees must be kept off pipeline corridors.

Problems and Issues

Direct competition between the various modes of transporting energy is the result of a combination of these regulatory, environmental, and economic factors and has been a major element affecting western energy development. Because railroads were the means of transportation that activated the growth of the West during the latter part of the nineteenth century, conflicts have arisen as newer forms of transportation challenge the dominance of rail interests. For instance, disputes have developed between the railroad and trucking industries because federal funds provided significant support for the construction of the interstate highway system but not for the improvement and expansion of the rail network. One recent response has been the proposed Railroad Deregulation Act, which would encourage innovation in the rail industry by making it easier for railroads to dispose of unprofitable services or tracks and so concentrate their efforts on profitable ventures (Linsley, 1979:6–11). Proponents of railroads and of coal-slurry pipelines have engaged in a hotly contested debate over the past few years about the future method of moving western coal resources. In order to protect their future economic stake in carrying coal, railroads have bitterly fought any legislation that would provide slurry pipelines with access to railway rights-of-way (Kirschten, 1979:1112–16).

Environmental, economic, and political factors must be taken into account in any attempt to resolve these conflicts. For example, regarding the direct competition between rail

and slurry interests, railroads appear to have advantages when: (1) variable quantities are shipped over time; (2) there are multiple points of origination and destinations; and (3) established mainlines can be used as routes. Slurry pipelines appear to be superior when: (1) large volumes of coal are shipped over long distances; (2) high rates of inflation occur; (3) coal mines are big and are close together; (4) there is a substantial, secure market in relatively few locations; (5) adequate supplies of water are available; and (6) the terrain favors pipeline construction. The cost structures of railroads and slurries also differ considerably. Pipelines must be built from scratch, requiring a large construction force but few permanent employees. Railroads, on the other hand, need a significant number of workers to operate. Thus, there is considerable opposition to slurry pipelines because they threaten to reduce (or at least reduce the future growth of) these labor forces.

Energy transportation is a very complex policy area, in which various routes and modes are selected by developers in the same manner as energy-facility siting decisions are made (see Chapter 11). Historically, transportation has been the focus of political conflicts, but as greater quantities of coal and other energy resources are extracted from the West, energy transportation problems and issues are expected to become even more central to the national debate on energy policies. Four issues appear to be particularly significant for the future development of western energy resources:

- Competition between railroads and coal-slurry pipelines;
- The uncertainty surrounding the potential capacity of the coal industry to carry increasingly heavy traffic in the face of financial, health, and safety problems;
- The potential hazards from electric transmission lines and the resultant public opposition to such facilities;
- The water consumption of coal-slurry

Unit Train Near Colstrip, Montana

pipelines originating in the water-short West.

These and other issues raised in the context of how best to transport western energy can rarely be resolved at any single level of government. Federal, state, and local governments, as well as farmers, ranchers, railroads, utilities, and other participants, all have an interest. Taken together, these problems contribute to an overall uncertainty about the future of western energy development. The sections which follow discuss each of these policy issues in turn.

Railroads vs Slurry Pipelines

Railroads have long had a virtual monopoly on coal traffic. Coal tends to be considerably more profitable than shipments of other commodities because it is sent in high-volume, long-distance shipments. Slurry pipelines provide an alternative to railroads for coal traffic, but they are strongly opposed by the powerful railroad interests in the West.

Rail development in the West during the 1800s was subsidized primarily by federal land grants totaling approximately 54 million acres within our eight-state study area. Although the initial grants generally did not include mineral rights, the railroads often mortgaged or sold these lands in order to purchase prime agricultural lands and lands with mineral rights and timber (Mossman and Morton, 1957:38–39). Because the railroads often sold package-deals of transportation and land to new settlers, the railroads controlled productive lands along their routes and greatly influenced the pattern of urban and rural development and, consequently, of economic activity within the region.

Hauling Prices Escalate

When City Public Service Company of San Antonio, Texas, started building its coal-fired plant in 1973, Burlington Northern and Southern Pacific estimated coal shipment costs of $7.90 a ton, roughly equal to the cost of coal at the mine in Wyoming. Later the price went to $10.93, then $11.94. When the railroads requested ICC approval of $18.23 a ton, San Antonio protested. San Antonio officials contend, and DOE agrees, that at any price over $15.64 a ton, it would be cheaper to burn oil. The ICC granted an increase of rates to $16.12 a ton. The city could pay $1.4 million per year more for coal than for oil under this new rate. —Drew, 1978; Ray, 1978; BNA, 1978b.

In an early attempt to regulate rail rates, state railroad commissions established rate ceilings under which railroads had to operate (Locklin, 1972:211–14). Federal action followed in the Act to Regulate Commerce (commonly known as the Interstate Commerce Act of 1887), which established a basis for the federal regulation of prices in interstate transportation and set a precedent for future federal intervention into virtually every aspect of rail and pipeline commerce. This early federal and state regulation of rates highlighted conflicts of interest among various groups in the West. For example, the banking industry was often politically aligned with railroad interest in their battles against cattlemen (Richards, 1977). The railroad industry also developed an extensive national lobbying effort, characterized by the early rate-making associations and more recently by the Association of American Railroads (AAR). The former were composed of representatives of the industry who met to determine rates among competing rail lines so that "rate wars" and other such conflicts could be avoided.

Tight governmental regulation of the railroads has benefited some lines financially while contributing to the failure of others. In part this is because the high fixed costs of railroads (debt services on roadbed and rolling stock, maintenance, and so forth) force them to operate at nearly peak capacity. Thus, railroad consolidation is frequently promoted as a way to achieve greater efficiency by scaling up the size of railroads and reducing competition.

In addition to regulatory problems, railroads face a new economic issue in the form of competition from coal-slurry pipelines for exporting coal from the region. Slurries would provide a transportation system made for electric utilities and other large, constant users, based on long-term contracts often at lower rates than the railroads could match. The Slurry Transport Association argues that slurry pipelines would end rail monopolies, thereby lowering transportation costs, and would enhance the reliability of energy supplies (Federation of Rocky Mountain States, 1976: Appendix B).

Railroad companies generally say that coal-slurry pipelines could threaten the solvency of rail companies by forcing them to lower freight rates to remain competitive (Federation of Rocky Mountain States, 1976:Appendix B). Slurry pipelines, they argue, would cut into the railroads' most profitable and convenient hauls: transporting large volumes of materials over long distances. In general, railroad firms believe that the pipeline concept represents a "redundant system" that would undermine efforts to rebuild America's rail network while at the same time consuming scarce western water resources in an inefficient manner. William Dempsey, president of the AAR, told Congress in April 1978 that there was no way that slurry pipelines and railroads could operate together (BNA, 1978a:25 [see box]).

One issue entering into the picture is the different regulatory situation in which slurry pipeline proponents find themselves. Although coal-slurry pipelines are "common carriers" in legal terms, they are "contract carriers" by the nature of slurry technology, and pipeline financing required long-term contracts. While the development of slurry pipelines would certainly contribute to the use of western coal, some technological uncertainty does exist because at the moment there is no experience with slurries of between 600 and 1,500 miles, the length now being proposed for many lives originating in the West. (Guccione, 1978).

If slurries do succeed as a major transportation alternative, it has been estimated that they could account for over 25 percent of the total cost of moving coal in the West by the year 2000. This market penetration could reduce the revenues of western railroads by as much as $628 million and reduce rail employment by more than 6 percent (U.S. Congress, OTA, 1978:74–77). Any offsetting pipeline employment would be only short-term—during the construction phase of development—and, thus, would tend to exacerbate boomtown conditions in some localities as a 400-person construction force moved along each pipeline at about 40 miles per month (Freudenthal, Ricciardelli, and York, 1974:29).

The railroads have attempted to block construction of slurries primarily by not allowing them to cross railroad rights-of-way. Without the right of eminent domain, slurries may have difficulty, for example, in crossing the 49 sets of railroad tracks between Wyoming and Arkansas. However, the railroads, which during their expansion were granted eminent domain at the state level as well as receiving extensive land grants, have not yet been very successful in attempting to restrict slurries from their rights-of-way. So far, slurry interests have won every case brought to court disputing their right to cross railroad rights-of-way, although several decisions have been appealed to higher courts by the railroads. And the Carter administration has moved to support federal legislation on eminent domain that would enable pipeline promoters to construct their slurries beneath the tracks of coal-transporting railroads (Kirschten, 1979: 1112).

Railroad Capacity

There is some question whether the railroad industry will be able to provide adequate transportation capacity for energy resources. The issue of capacity is a result of problems surrounding the adequacy of railroads' rolling stock, track conditions, and financing capabilities.

The issue of rail capacity largely revolves around the ability of the industry to provide adequate transport facilities to meet coal demands while responding to the national goal of increasing the efficiency and reliability of railroad service. Central to the question of capacity are the projected needs to add up to 38,000 new coal hopper cars plus replace all existing rolling stock by the year 2000. However, a great deal of uncertainty surrounds these estimates; the railroads and shippers (who own a majority of the new coal cars) may, in fact, be able to lease or otherwise obtain a sufficient number of coal hoppers and locomotives, even though significant backlogs of orders have existed (Manalytics, 1977:6; and *Wall Street Journal,* 1978a).

Moreover, it is unclear whether existing trackage can accommodate the probable number of trains traveling between points of coal production and consumption. If projections of future coal shipments are even reasonably accurate, numerous stretches of track will be overloaded (Manalytics, 1977:6). Such overloading would result in greater expense and more time required for transport because of the necessity to use longer routes that avoid bottlenecks. For example, because of routing, coal shipments from Wyoming to utilities in eastern Oklahoma have taken three days longer than the railroads initially expected (Clock, 1978). But it may be possible to reduce these bottlenecks by changing the signalling systems and by adding some bypass sections to a single-line track. In this way track capacity might be doubled (U.S. Congress, 1977:Vol. 3, pp. 462–63).

```
┌─────────────────────────────────────────┐
│          Low Return on Investment         │
│                                           │
│   Despite expanding shipments of coal,    │
│   railroads have comparatively low return │
│   on investment. Burlington Northern and  │
│   Southern Pacific Transportation for ex- │
│   ample, have 2.1 percent and 2.5 percent │
│   rates of return on investment, respec-  │
│   tively, which are so low that they claim │
│   that they can't raise needed equity capital │
│   without increasing freight rates.—Drew, │
│   1978.                                   │
└─────────────────────────────────────────┘
```

Table 10-2: *Applications for Track Abandonment 1971-1977*

State	Approved		Pending	
	No.	Miles	No.	Miles
Arizona	5	79	3	72
Colorado	9	110	4	50
Montana	6	142	2	69
New Mexico	3	58	1	14
N. Dakota	4	145	3	28
S. Dakota	13	461	5	322
Utah	7	44	1	4
Wyoming	5	86	0	0

Source: McAvoy and Snow, 1977:152-53.

Finally, the railroads' ability to secure financing is unclear. The costs of expanding rail capacity are roughly proportional to the increased costs of producing coal. And uncertainty about the demand for western coal brings the necessary investment into question. One report estimates that railroads might be required to invest about $40 billion by 1985 to meet capacity and maintenance needs, of which coal-related investments could conservatively require 10 percent (Richard J. Barber Associates, 1977:37). This would amount to twice the railroads' recent annual capital outlay.

Railroad profits, therefore, may need to improve significantly to attract the additional equity (see box, "Low Return on Investment"). Alternatively, help with financing may have to be provided by the federal government, customers, and other participants. Rate structures that permit railroads enough profit to expand for coal transport may make potential alternatives (such as slurries) more attractive to shippers (U.S. Congress, 1977:Vol. 3, p. 465).

All of the capacity problems discussed above depend on the economic health of the railroads, which in turn is related to various restrictions on the rail industry—"common carrier" requirements and rules on entering markets and abandoning lines.

"Common carriage" refers to the requirements, mandated by the 1887 Interstate Commerce Act, that a means of transportation must carry, at published rates, the goods of all shippers who request the specified service. Common carriers cannot treat shippers in either a discriminatory or a preferential manner, a restriction that has led to some uncertainty regarding unit trains as a way to transport coal. For example, even if a long-term shipping arrangement exists, railroads may have to "bump" old shippers or to reduce service to utilities in favor of new shippers who have an equal claim to the railroad's limited capacity.

Entry into new market areas and the abandonment of lines can take place only after the granting of a certificate of public convenience and necessity by the ICC. This requirement poses a particular problem if a railroad wishes to eliminate some classes of traffic or to abandon such traffic to avoid financial loss. Low-use routes are more common in the West than elsewhere in the country; there some 4.7 percent of the carloads move on "potentially uneconomic light density lines," compared with 1.2 percent in four other regions (U.S. DOT, 1977:160). Although track abandonment in the West was less than the national rate of 8.2 percent from 1970 to 1976, applications were filed and/or granted for abandonment of 23 percent of the track in South Dakota during the same period (see Table 10-2). Thus, rail capacity and rail use generally are decreasing on all but the most heavily traveled routes. But because of the restrictions on abandonment of low-use routes, rates must be high enough on the high-volume lines to subsidize the others and to maintain overall solvency. This may not be possible where there are alternative modes of transportation for shippers of high-volume commodities such as coal.

In November 1979 when the ICC removed its restrictions on long-term contracts between railroads and coal customers, a significant obstacle to rail transportation of coal was eliminated. Under the old system of regulation, the ICC required that rates be negotiated annually—a practice that left the utilities and railroads uncertain as to future policies and prices. In addition, this approach had given slurry pipelines a major advantage in entering into long-term contracts with large users of coal.

Perhaps the most controversial aspect of transporting coal by railroads results from unit-train movements through the small communities of the region. Unit trains usually consist of about 100 cars, each holding 100 tons of coal, and are pulled by five or more diesel locomotives. Except for stops to change personnel and to refuel, these trains move nonstop from origin to destination (Glover, et al., 1970:1 [see box, "Concerns of Increased Coal Train Traffic"]). The Sierra Club has filed suit in the U.S. District Court in Washington in an attempt to force the Burlington Northern and Chicago and North Western railroads to reassess the impact of a proposed 113-mile link through the Wyoming coal fields. This new route is projected to allow 30 to 48 unit trains daily to pass through a number of small towns. In addition, these trains would make 306 crossings of county highways in the area each day.

In addition to disrupting traffic, railroads can also produce serious noise pollution, which may be a major consequence in terms of daily and long-term human irritation and hearing loss (Lang, 1975:108–115). The primary source of noise from trains is the locomotive, and the noise level is essentially independent of a train's speed. Levels of engine noise will typically vary between 90 and

---

> **Concerns of Increased Coal Train Traffic**
>
> *The effects on small communities can be particularly disruptive. The mayor of Lusk, Wyoming explains: "We get one freight a day through here now and that ties up traffic coming in and out of town." In anticipating the increased train movement, Mayor Hammond summed up the worries of his constituency: "What do we do if there is a fire or if someone has a heart attack on the other side of the track when these unit trains start running? We just don't know what's going to happen."*
> —Richards, 1976.

---

100 decibels A-weighted (dBA); the horns used at grade crossings are also quite noisy and have been measured at 110 dBA at 100 feet (URS Company, 1976:V-8). Current railroad noise-emission standards limit the maximum sound level from locomotives to 96 dBA at 100 feet from the track, with the standard decreasing to 90 dBA for locomotives manufactured after 1979. These criteria are based on the current best practical technology and economic feasibility. Apparently it is difficult to restrict the level of engine noise below 90 dBA.

The loss of livestock due to train traffic is also an economic issue. Ruth Rice, a rancher and a member of the Sheridan County (Wyoming) Board of County Commissioners, has lost livestock to unit trains; she and other westerners are ready to take action to deal with these problems. She has said: "We are finding out that we are all citizens and we have the same rights as the railroads and big mining companies. We may have created a monster in these trains. And when you deal with a monster you have to make sure you protect yourself" (Richards, 1976:A4).

Opposition to HVTL

The public resistance to electric trans-

mission lines has increased dramatically in recent years, and opposition has focused on a wider range of issues than merely the question of land use. Concerns have focused on the aesthetic impact of power lines, the problems of interference with communications, and particularly on the possible adverse biological effects, on both humans and livestock, of HVTL.

It is not yet clear whether HVTL will continue to be passively accepted in the West or elsewhere. In several cases during the past few years, transmission-line projects have met extensive opposition, especially from farmers and ranchers (see Gerlack, 1978:22–37). Their objections emphasize several factors, including the negative aesthetic impact of the lines, annoyance with radio and television interference, and a range of health and safety issues. The unattractiveness of HVTL has not become a salient policy problem in the West; however, landowners in other regions have raised this objection as a significant factor in reducing property values (Katz,

High-Voltage Transmission Lines

Table 10-3: *Effects of Alternating Current (AC) Transmission Lines on the Environment*

Effects of Construction and Maintenance
   Clearing of land within right-of-way
   Clearing and maintenance of access roads
   Soil compaction and other physical damage
      due to construction activities
   Interference with agriculture due to towers
   Aesthetic impact of towers and lines

Hazard of Electric Shock from Contact with Lines

Effects of Corona
   Radio and television interference
   Audible noise
   Production of ozone and oxides of nitrogen

Effects of Electric and Magnetic Fields
   Fuel ignition by spark discharges
   Induced electric shocks
   Biological effects of electric and magnetic
      fields
   Interference with cardiac pacemakers

Source: Miller and Kaufman, 1978:8.

1977). Similarly, interference with communications is not yet a major obstacle, although such annoyances have begun to be voiced (see *Business Week,* 1977:27).

Human safety is currently the focal point for protests against HVTL in our eight-state study area. Farmers and ranchers have experienced electric shocks while operating farm equipment, and residents have become aware of the findings of recent research that concluded, "Biological effects will probably be induced in humans exposed to overhead lines . . . and such effects may be harmful" (Bernstein, 1978; see also Young, 1973). It must be noted, however, that the conclusions of other research contradict these findings (see Miller and Kaufman, 1978:6–15). Nevertheless, a number of potential consequences of HVTL have been raised as policy issues. The possible effects of constructing

Cross the Northern Great Plains

kilovolt (kV) line as it can be transmitted for 10 miles over a 138 kV line (U.S. Congress, 1977:Vol. 1, p. 357). However, each mile of transmission line can require from 9 to 27 acres of land depending on the voltage which affects the width required for the right-of-way. This demand for land, the competition for other land uses, and the economics of scale offered by larger transmission lines explain why utilites have increasingly moved to higher voltages. Table 10-4 compares the costs for the various alternating current (AC) voltages and illustrates the lower total cost of higher-voltage options. Utilities have taken advantage of reductions in the right-of-way requirements and the fewer lines and towers mandated by HVTL. Although not shown in Table 10-4, overall energy efficiency (for AC lines) is about the same at any voltage, but the total cost decreases with the use of HVTL, because the number of lines needed decreases as voltage increases.

From the perspective of the utilities, there is an urgent need to build new high voltage facilities. As a spokesman for the American Electric Power Company has said, "Without these high voltage lines, utilities may simply not be able to meet the growth in power demand over the next decade" (*Business Week,* 1977:27).

Legislation has been enacted in a number of states in recent years to attempt to avoid conflicts by bringing diverse groups into the planning process for electricity transmission. However, these "open" procedures have not resolved the difficulties. The public outcry in both North Dakota and Minnesota exemplifies the continuing problem (for example, see *Denver Post,* 1978). Both these states have siting commissions that use special procedures for siting HVTL, and in both cases public participation is an integral part of the proceedings. But even after extensive public debate and discussion, evaluating the environmental, social, and economic consequences of HVTL, conflict still arose in both states when the actual construction of transmission lines began (Bernstein, 1978).

and operating transmission lines are summarized in Table 10-3.

In addition to aesthetics, health, and safety, there are also questions about the economics of HVTL, particularly in terms of the choice of converting coal to electricity within the West and transporting the energy by HVTL or shipping coal out of the region by rail or pipeline for conversion elsewhere. It is already more economical in many cases to transmit electricity from power plants at the mouth of a mine than to ship coal out of the West, especially to regions where air-quality requirements place restrictions on facility siting (see Federation of Rocky Mountain States, 1974:86).

It appears that HVTL can achieve even greater economies of scale than they have to date. For instance, electricity can be transmitted as effectively for 300 miles over a 765

Table 10-4: Cost Comparison for Alternating Current Transmission Systems at Various Voltages (for transmission of 51,000 MWe for 1,000 miles)

kV	Number of Line Systems[a]	Line Cost[b] (thousands of dollars/mile)	Station Cost Per Line[b] (thousands of dollars/mile)	Land Used Per Line[c] (Acre/Mile)	Total Land Cost ($4,000/Acre) Over 1,000 Miles (billions of dollars)	Terminal Equipment Cost Per Line[d] (thousands of dollars)	Total Equipment Cost[d] (billions of dollars)	Total Cost[e] (billions of dollars)
230	241	85	19	9.1	8.8	3	25.07	33.87
345	108	109	42	11.0	4.75	4	16.32	21.07
500	51	155	82	15.2	3.10	9	12.08	15.18
765	22	217	179	19.4	2.41	29	8.72	11.13
1,100	11	310	349	21.8	0.96	74	7.26	8.22
1,500	6	411	775	26.7	0.64	155	7.14	7.78

[a]Number of line systems = total capacity (51,000 MW)/system carrying capability.

[b]Compensation station. See Cirillo et al., 1977:225, Table 5.40.

[c]Total land used = land used per line x number of line systems (Cirillo et al., 1977:225, Table 5.41).

[d]Equipment cost = line cost + compensation station cost + terminal cost.

[e]Land & equipment.

Direct current (DC) transmission of electricity, which is primarily applicable to long-distance transmission, does not create the electric field, corona, and shock hazards associated with AC lines, thus permitting use of a smaller right-of-way and creating fewer land-related problems. However, the intermediate tapping of DC lines is costly. Furthermore, DC lines are grounded through the soil and metallic structures with unknown effects—something that AC lines (which are grounded in an additional overhead line) do not cause (Montana DNRC, 1974:Vol. 4, pp. 20–25). These difficulties and the convenient use of electricity at intermediate points between the originating power plant and the ultimate destination have menat that AC is the predominant form of HVTL in this country.

Slurry Pipelines and Water Use

The principal detrimental factor for coal-slurry lines is their consumption of water from arid western sites. Agricultural and ranching interests in the West fear that slurries would threaten the regional water supply in the long run, even if new water sources are developed.

The principal conflict over coal-slurry pipelines concerns their consumption of water from water-short western coal areas. Some agricultural and ranching participants in our eight-state study area believe that the water requirements for coal-slurry lines would place an unnecessary burden on the region's water supply. Several national environmental groups also support this view.

Legislators representing the West have divergent viewpoints on the use of western water to ship western coal. Proponents of pipeline transportation view the technology as a way to avoid the degradation of the western environment by shipping coal out of the region rather than converting it on the site. Opponents, however, fear that slurry pipelines may abridge the rights of western states to regulate the use of their water. In the most recent defeat of federal legislation designed to promote the development of slurries, western congressmen voted for the measure (HR 1609) by nearly a two-to-one margin (Congressional Quarterly, Inc., 1979:98-A).

Compared with many other energy conversion and transportation alternatives, however, coal slurries are not particularly water-intensive. For example, the consumption of western water by slurry pipelines would be much lower, on a per-unit-of-energy basis, than for most forms of coal conversion carried out at the mouth of the mine. The water requirements for generating electric power on a per-unit-of-energy basis is far greater than for any other method of converting or transporting coal, more than twice that of a slurry pipeline (see Figure 2, Chapter 3). Thus, shipment of coal from the West by slurries would consume less water than converting the coal within the region. On the other hand, if coal is to be exported from the region, railroads use negligible amounts of water compared to slurry pipelines.

Slurries are able to use low-quality water, such as highly saline water that is unfit for agricultural, industrial, or domestic purposes. But in the Powder River Basin, the origin of many of the anticipated coal-slurry pipelines, most available water is also suitable for other functions. This means that competition with existing users remains a critical issue throughout the West (see Strain, 1977). The problems and issues of water allocation are discussed in Chapter 3.

Summary of Problems and Issues

Railroads, slurry pipelines, HVTL, and trucks are the means of transportation most likely to raise problems and issues regarding the transportation of energy resources within and from the eight-state study area. Each of these forms of transportation has already produced conflicts among interested parties over the economic and physical effects.

Economic conflicts are underscored by the history and operating characteristics of railroads. The early federal and state regulation of the rail industry was a response to monopoly conditions that still exist to a large degree, especially with regard to transporting energy in bulk forms such as coal. The development of an alternative technology—the coal-slurry pipeline—has been held back by railroad opposition and by fragmented national energy policies. Coal slurries are feasible in both a technological and an economic sense; but some citizens and government officials have expressed concerns that the water required for slurry pipelines will deplete the region's already scarce resources and that pipelines will adversely affect the economic health of the railroad industry. Finally, the entire issue of developing and transporting coal has become entangled in federal-state-local debates about the appropriate level of government control of any future energy policy.

Conflict also arises over the physical impact of each of the methods of transportation. New corridors in the West will have an environmental impact due to the removal of vegetation and the disruption of wildlife habitats. Some of the effects will be of limited duration during construction, while others will continue during the operational phase. Health and safety questions have also been raised, particularly for unit trains which produce noise pollution and introduce the risk of accidents for highway traffic. Small towns will be especially vulnerable to disruption from unit trains since the railroad routes within our study area tend to bisect these communities. The increased use of unit trains prolongs the amount of time these towns are effectively divided and riases concerns about accidents at rail crossings. HVTL have met with strong resistance because of their aesthetic effects, their effects on land values, and their potential safety and health hazards.

Alternative Policies for Energy Transportation

Increasing domestic energy production and protecting and enhancing the environment are both national policy objectives. As emphasized throughout this study, the West is expected to contribute significantly to increases in the national supply of energy, most of which will be transported for use in other regions of the country. The primary policy objective in the area of energy transportation is to provide an dequate capacity for moving the energy resources produced in the West while minimizing the undesirable environmental and socioeconomic effects. In this section, four categories of transportation policies are discussed, each focusing on a specific aspect of this policy objective. The four categories of policies are as follows:

- Enhance the coal-carrying capacity of railroads;
- Mitigate the adverse effects of unit trains;
- Promote the use of coal-slurry pipelines; and
- Minimize the impact of HVTL.

Table 10-5: *Transportation Policy Options*

General Alternative	Specific Alternative
Enhance coal-carrying capacity of railroads	Expand railroads in energy areas of the West.
	Increase rail rates.
	Assure economic and regulatory stability.
Mitigate adverse impact of unit trains	Improve safety and mobility for vehicles crossing rail routes.
	Reduce train noise in communities.
Promote coal-slurry pipelines	Grant eminent domain to coal-slurry pipelines.
Minimize impact of high voltage transmission lines	Limit voltage.
	Increase research on technological alternatives.

Table 10-5 lists a number of specific policies in each of these four categories. It should be noted that broad concerns about the form of transported energy—such as the choice between exporting raw coal or converting it first to electricity—are not considered here.

Policies to Enhance the Capacity of Railroads

We will briefly discuss three specific alternatives to enhance the capacity of railroads:

- Expanding the railroad network into the new energy-producing areas of the West;
- Increasing the railroads' rate of return on energy transportation; and
- Assuring the economic and political stability of the railroad industry.

Expanding Rail Lines into the West's New Energy Regions: Increasing the capacity of railroads is, to some extent, a matter of increasing the rail industry's share of the energy-transportation markets in new areas of development. There are a number of cases where rail routes currently are not serving some important energy-producing areas, such as the San Juan Basin in northwestern New Mexico and the nearby cities of Aztec, Bloomfield, Farmington, and Shiprock. Given the existing financial problems of the rail industry, however, it is unlikely that such rail service will be provided without substantial subsidies from the public sector. In the past, the railroads have been active in seeking preferential tax credits from government, and a significant historical precedent exists for a variety of such supports. But these policies are receiving a more critical assessment in the light of growing tax-consciousness by politicians and the general public alike. It is possible, therefore, that continued subsidies of the railroads will become less acceptable in the future.

One factor working in the railroads' favor

in financing these capacity increases is that rail facilities can be added more quickly than a new mine can be opened or a new power plant built. Thus, uncertainty over the general financial requirements for upgrading or adding new routes will not be critical since investment capital can be sought as it is needed. (Richard J. Barber Associates, 1977:463). Thus, despite the fact that the new 110-mile rail line to coal fields near Gillette, Wyoming, is the first major extension of national rail service in 70 years, there is optimism that at least a few additional routes will soon be opened in the West. For example, burlington Northern and Chicago and North Western are planning a 113-mile line from Gillette to Douglas, Wyoming, connecting the main routes of those two railroad companies (U.S. DOI, BLM, et al., 1974).

Increasing the Economic Return on Coal Transported by Rail: Even though railroads now are allowed to enter into long-term contracts with energy customers, the price of transporting coal by rail is still affected by ICC's rate-making activities and by the difficulties surrounding the calculation of transportation costs and "reasonable rates." Under the terms of the 1976 Railroad Revitalization and Regulatory Reform Act (the so-called 4-R Act), the ICC's mandate calls for the reduction of regulatory interference and emphasizes the carriers' financial needs. This new federal flexibility has allowed railroads to increase their rates for coal transportation; expanding future capacity may well be enhanced by further deregulation of the industry. Thus the proposed Railroad Deregulation Act is seen as a key to improving the industry's financial health. According to former Secretary of Transportation Brock Adams, the new proposals "will give rail operators freedom to provide a variety of rates and services, tailored to the needs of specific shippers"—that is, freeing railroad managers from many of the existing restraints on pricing, market entry, and market exit (Linsley, 1979:8–9). If such policies were adopted,

beginning in 1980 the ICC would gradually surrender its control over rate increases over a five-year period. But significant aspects of the ICC's regulation of "economic fair play" would be continued in the reform plan. The agency's authority over "discrimination," such as allowing special rates for larger shipments, would be maintained, as would its responsibility for overseeing the obligations of common carriers.

One important factor underlying any attempt to evaluate and implement such reforms is the difficulty in obtaining a full understanding of how to calculate the actual costs of various transportation options and how to compare these costs with reasonable rates. Tariff data may be collected annually or during other time periods; costs differ according to volume, points of origin, size of car, type of train, and delivery mechanism (docks, ports, or utilities). Defining cost, therefore, is a complex task for regulators.

Assuring the Economic and Political Stability of Rail Companies: A third alternative that would safeguard the economic stability of railroads would be to use federal subsidies and other types of supports. Such policies might include federal loans, federal guarantees for repayment of obligations such as bonds, or the outright purchase of stock by a government fund (as provided for in the "4-R Act").

The investment needed by most railroads in the West clearly will be greater than their recent annual level of capital outlays. But a transportation strategy that incorporates federal subsidies to provide needed capital must take into account several complex and controversial factors.

First, subsidies may benefit a broad range of western economic interests outside the transportation sector. There are, for example, direct economic links between the coal-hauling activities of rialroads and the economy of the West. Railroads employ about 4.8 employees per million tons of coal per 100 miles and a total of nearly 30,000 workers in the western states alone. Railroads also carry assessed property valuations of 20 cents per annual ton of capacity along rail routes; these tax revenues accrue mostly to county governments.

Second, subsidizing the rail industry also benefits shippers and consumers of noncoal commodities. The implementation strategies that diminish the burden of shippers (e.g., electric utilities) in favor of government support (including loans, grants, and tax incentives) would spread the costs over a larger number of citizens. If the increased use of western coal is a national priority benefiting the general public, then government subsidies may distribute the increased costs more equitably than do high rates for shipping coal as a means to subsidize other goods (such as industrial or agricultural commodities) and routes. It must be noted, however, that this type of public policy might come into conflict with the railroads' interest in becoming more independent (U.S. Congress, 1977:Vol. 3, p. 466).

And third, any use of government supports for rail interests must take into account the dependency of western agriculture, forestry, and mining on the capacity of railroads. Track availability, route restrictions, and rate structures are highly salient issues to these users throughout our study area. Any subsidies that encourage better all-around service protect these interests, but government programs that enable railroads to reduce service, as with the abandonment of routes, are generally resisted by these key western industries.

Federal loans have been a major source of funds for the rail companies, especially for financially ailing lines that need improved equipment. In addition, loan guarantees have been made to railroads in western states, both for acquiring equipment and for rehabilitating the tracks (*Wall Street Journal,* 1978b). However, the relatively prosperous western lines, such as Burlington Northern and Union Pacific, are generally less eligible for these programs than are the financially weaker eastern railroads.

In addition to federal loans, other government supports could be used to enhance rail capacity. Tax incentives on capital investment are a flexible policy option that does not commit the railroads to any fixed plan or set of services. Likewise, grants can be used to respond to specific short- or mid-term needs. However, major commitments of this type could tend to become a nearly irreversible form of public support for a significant part of the private sector.

Policies to Mitigate the Adverse Impact of Unit Trains

There are a number of possible options that could reduce the negative consequences of increasing the transportation of coal by rail for the communities located along the way. These include:

- Improving the safety for and mobility of vehicles crossing rail routes; and
- Reducing the noise from unit trains.

Improving the Safety and Access Aspects of Rail Transportation: Transporting coal by railroad generates many safety and access risks that can be reduced by improving the design and operation of rail lines. These modifications include installing or improving grade-crossing safety signals, building grade separations, and constructing by-pass rail lines around some localities. Safety signals at certain crossings might reduce the risk of accidents, especially in towns where traffic is heavy. In general, rural signalization is not considered as a serious option because of the cost. In addition to safety, most communities will need to have access across railroad tracks both for emergency vehicles and for general traffic; this requirement will increase as the number of coal shipments by rail escalates. The construction of grade separations in larger towns is already a major issue, and more underpasses and overpasses may be demanded in smaller towns as a result of the development of western energy resources. Finally, by-pass lines around larger cities would largely avoid the crossing problem and would permit higher train speeds. (By-pass lines also help solve the noise problem, as discussed below.)

It is very difficult to generalize about the cost of implementing the measures just outlined; the cost will be highly dependent on site-specific factors. It has been estimated that grade-crossing signalization can cost up to $70,000 per crossing, while the construction of grade separations generally ranges from $750,000 to $1.5 million, depending on conditions at the specific location. To get some idea of the total magnitude of such projections, it is only necessary to note that there are 167 grade crossings in urban areas in Colorado alone. To construct grade separations for each of these would cost from $125 to $250 million (URS Company, 1976:VII-20 and VII-22).

The cost of constructing by-pass lines around communities will depend very much on what routes are available and what land values near the community are. One source indicated that typically the cost for constructing new double-tracks is $500,000 per mile (U.S. Congress, OTA, 1978). Thus, to build a by-pass of 15 miles could be expected to cost some $7.5 million. Besides these direct costs, however, such a plan would also entail a number of other important costs and benefits for the community. Land values along the new route, land-use plans, and patterns of community growth could all be substantially altered by the construction of a new rail line. Conversely, if an existing rail line were abandoned, it could have a disruptive effect on some businesses and require the relocation of some rail-dependent industries.

The method (or combination of methods) for financing these various measures is, of course, dependent upon how the program is structured. There are four general sources of funds to carry out such a program: state highway or general revenue funds, railroad revenues, local government funds, and federal funds.

Most states currently spend some state highway funds to upgrade highway/railroad crossings, but the amounts are probably inadequate to address the problem effectively. For example, in Colorado the Public Utilities Commission's Railroad-Highway Grade Crossing Protection Fund receives $20,000 per month from the Highway User Trust Fund to be used for upgrading or installing grade-crossing protection. This fund is used only in cases where no federal funding is available; the railroad must pay a minimum of 10 percent of the total cost, based on the benefit to the railroad, with the remaining 90 percent of the cost divided between the county or city that owns the roadway and the fund itself (URS Company, 1976:VIII-12). Therefore, not only is the total dollar amount in this fund quite limited, but it cannot be used for the full spectrum of measures we have considered here.

A potentially attractive means for financing such programs designed to mitigate the impact of unit trains, in states with substantial coal development is to use some portion of the state's revenue from severance taxes on coal production. This is an equitable approach since, in effect, it passes the costs on to energy consumers. Of course, this solution would not be available to those noncoal-producing states that lie on the major routes for transporting coal from the West to the coal-consuming regions.

It would seem that equity would also be served if the railroads themselves were required to pay for some portion of the costs of reducing the adverse effects of unit trains. However, there appears to be little flexibility in most existing state laws that would allow mandatory participation by the industry.

Finally, as with many transportation issues, the federal government might play a role in sharing the cost of reducing the impact of unit trains. Federal funding could take the form of a special appropriation; one such bill has already been introduced in Congress (see box, "Grade Crossing Bill Makes Progress"). Among the many groups that might oppose

ENERGY FROM THE WEST

such a federal commitment, however, are proponents of slurry pipelines who may view this kind of public expenditure as inequitable.

Reducing Railroad Noise Emissions: Options to reduce train noise include improving the engineering design of locomotives, coal cars, or tracks; reducing train speeds; and constructing noise barriers. Engineering design modifications must be accompanies by research and development efforts, some of which can come from the private sector but most of which are likely to result from public financing. As a former director of the National Bureau of Standards has noted, "Government funding is necessary to clarify for the public that there are engineering solutions or alternative system solutions to noise problems" (Branscomb, 1975:16).

Reducing the speed of trains is another possibility for controlling noise from the wheel-track interface, but in most communities, speeds are already controlled at levels between 20 and 30 miles per hour. And speed reductions do nothing to control the major source of noise, the locomotive. Moreover, lower speed limits would exacerbate the already serious problem of traffic congestion in small communities that are bisected by railways.

The third option is to construct noise barriers. But to be effective (that is, to lower noise levels below 80 dBA for locomotives

at 100 feet), one study has demonstrated that barriers as high or higher than five-story buildings would be required (Wyle Laboratories, 1973). Building such barriers or lowering the tracks to an equivalent depth would not only be expensive, but it also might be physically impossible within the existing rights-of-way. Alternatively, thick stands of trees will provide some barrier effect. But this too would be difficult to achieve in many of the arid western regions, and again, the width of rights-of-way is insufficient in many localities. Given these obstacles, one analysis has concluded that there is little that can be done to reduce noise pollution in sensitive areas near railroad tracks other than relocating either the tracks or the people (URS Company, 1976:VII-49).

Policy to Promote the Use of Coal-Slurry Pipelines

Promoting the increased use of coal-slurry pipelines is a category of transportation policy that can be aided by granting eminent domain to slurries. However, slurries could still be restricted by an inability to acquire water rights in some western states. If a federal coal-slurry pipeline bill, such as the one that has been proposed by the Carter administration, is passed, several objectives would be accomplished. Perhaps the most significant would be the attainment of a higher overall level of economic efficiency in transporting western coal. At least theoretically, a viable choice of means of transportation should provide increased economic competition and a reduction in the degree to which coal consumers are held captive by the railroads. In the words of officials at DOE, steps must be taken to ensure "that coal users do not bear a disproportionate share of the cost of upgrading and maintaining railroads" and that "proposals for railroad deregulation emphasize protection for captive shippers," such as electric utilities, that depend upon a single coal carrier (Kirschten, 1979:1113). The evolution of a mixed system, with each method of transportation sharing the energy market, would be greatly facilitated by legislation granting eminent domain to the slurries.

Although even the extensive use of coal-slurry pipelines probably would not significantly modify national patterns of coal consumption over the short term, pipeline technology could increase the rate of development and use of western coal. Slurries, which are the most economical means of transportation for long distance shipments, would likely expand the areas outside the region to which western coal could be economically moved. And slurry pipelines could encourage a more rapid development of western energy resources by removing any bottlenecks at the transportation stage.

An additional, but less significant, effect of the passage of legislation promoting coal-slurry pipelines would be to build greater diversity and reliability into the system of transportation for western coal. There would be fewer disruptions in service, whether the result of a railroad strike or severe weather, if alternative means of transportation were in operation and available to take on emergency shipments.

Although there is a precedent for granting eminent domain to slurries in the form of the 1947 Amendment to the Natural Gas Act (which gave similar authority to gas pipelines), the development of coal-slurry pipelines in the West would result in reduced revenues for railroads in the region and increased competition. Thus, as noted earlier, railroads have provided the major source of opposition to slurries. But the rail interests have been joined by agricultural lobbies and by national labor unions in this stance. For example, the United Mine Workers opposes slurries because they may make it easier to transport nonunion western coal into the eastern markets during labor strikes (Corrigan, 1976: 295). And many western politicians have serious reservations about slurry pipelines; in particular, state officials have focused on

the issue of water. Governor Ed Herschler of Wyoming has testified against slurry pipelines before the Senate Energy Committee, arguing that "if Wyoming is to continue to play a substantial role in fulfilling the nation's energy demands, then our water must be kept available for uses in the state of Wyoming, and not shipped to other states" (BNA, 1978c). And in 1977 the Colorado General Assembly approved legislation to temporarily prohibit the export of water from the state through slurry pipelines.

Policies to Minimize HVTL Impacts

In the following sections, we will briefly consider two policy options that would reduce the impact of high voltage transmission of electricity:

- Limiting the voltage of HVTL; and
- Increasing research on alternatives to AC transmission.

Limiting the HVTL Voltage: Since conflicts over the hazards of high voltages have seriously limited the use of HVTL technology, instituting a voltage ceiling, below which certain adverse effects would be minimal (e.g., 350 to 400 kV), is a possible option. A regulatory policy that restricted the voltage level to a maximum of 600 kV, for example, could reduce the danger from the corona and electric shocks to humans and wildlife in the vicinity. However, this advantage would be countered by the increased economic cost and by the corresponding need for a greater number of HVTL in order to transmit the same amount of power. Thus, setting a maximum limit on HVTL voltage would directly affect the cost of AC transmission: lower voltage means not only more transmission lines but also higher costs for land, materials, and labor. These higher costs would be borne by consumers. And a large portion of the benefits in terms of reduced health and safety risks also would be countered by the increased environmental damage associated with expanded rights-of-way and construction activities.

Increasing Research on DC Transmission Alternatives: A second policy that would minimize the impact of HVTL would be to expand research on alternatives to AC transmission, especially DC transmission. Since DC technology has few of the electric-field effects of AC facilities, research could be undertaken to make DC options feasible and competitive in both an engineering and an economic sense. Today DC is not an alternative for short-distance, distribution-oriented transmission, but an expanded, accelerated research program could reduce the cost of DC terminal facilities and could bring about enough improvements to match the superiority of DC technology in long-distance, point-to-point transmission.

To remove existing technological and economic uncertainties about DC transmission will require considerable expenditures both by the electric power industry and by the federal government. For example, in 1977 the federal government spent approximately $11 million for underground and overhead DC transmission research, and the Electric Power Research Institute (EPRI) allocated $9.7 million for research on overhead DC lines and $7.2 million for underground lines (Kash, et al., 1976). If the research is successful, electric utilities and their customers will benefit from the availability of the more efficient options; residents along the lines will benefit since many side-effects of DC transmission would be less troublesome than with AC. Some effects, however, may be worse than with AC options. For example, telephone interference from DC power lines would be a more serious problem than it would be with comparable AC technology.

Summary and Comparison of Alternatives

The central policy objective in addressing issues of energy transportation is to provide an adequate capacity for transporting western energy resources while minimizing the associated undesirable environmental and socioeconomic effects. In examining railroads, slurry pipelines, and HVTL, the policy issues and alternatives we have discussed are different for each; Table 10-6 summarizes some of the key features of the four categories of policy alternatives that were considered.

In order to use the supply of coal in the West, the coal-carrying capacity of railroads must be increased. While railroads should be able to physically meet the increased demand for track and rolling stock, the requirements for investment capital will be considerably larger than has recently been needed. Thus, a range of government subsidies will probably be necessary. Such government supports are an equitable method of spreading the cost of investment and operation among the beneficiaries of rail service, but the future climate for federal subsidies is highly questionable if public concern about budgetary policy continues to grow.

Some of the effects of unit trains on western communities could be mitigated through improved crossings, although there is a serious question as to whether sufficient funding can be made available. Little can be done to reduce the other consequences on communities (especially noise) unless railroads are entirely removed from the immediate vicinity.

Coal-slurry pipelines are technically and economically feasible ways to transport coal. The biggest obstacle is whether enough western water is available—in both a physical and a political sense. Extension of the right of eminent domain to slurries would help remove the second major obstacle. If slurries are built, they would promote competition between the various coal carriers, resulting in a more efficient system of transporting coal from the West to the centers of demand. However, slurries are relatively inflexible; they must be operated at a more or less constant volume and at a very high capacity between a fixed point of supply and a fixed destination. Thus, under no circumstances

Table 10-6: *Summary Evaluation of Transportation Policy Alternatives*

Criteria	Enhance Rail Capacity	Mitigate Impact of Unit Trains	Coal-Slurry Pipelines	Mitigate Impact of HVTL
Effectiveness: Achievement of policy objective	May be required to assume needed expansion of transportation capacity.	Grade separations and rerouting would help access. Noise will be difficult to reduce without rerouting.	Slurries represent an alternative coal transport mode and could allow increased coal production. Concerns about water use could block slurries.	No direct effects on capacity are evident. Increased R&D on DC transmission is a long-term option.
Efficiency: Costs, risks, and benefits	Rates may rise or government subsidies increase. Financial stability of railroads should increase. Employment benefits, due to personnel requirements for operation. Capacity improvements will be available for nonenergy shippers.	Cost of approximately $70,000 per crossing for signalization; $750,000 to $1.5 million per grade separation; and $500,000 per mile for new track. Rerouting would have other land-use effects and could hurt existing rail-depending businesses.	Slurry costs are lower than rail in some cases. Railroads' financial strength will be lessened. A mixed system of slurries and rail would provide a more reliable transportation system.	Voltage limits reduce some effects, but at a considerable increase in cost and land use, due to increased number of lines. Risk that increased R&D on DC transmission may not pay off. Utilities, consumers, and residents could benefit from R&D.
Equity: Distribution of costs, risks, and benefits	Coal rates could subsidize other commodities. Increased flexibility could result in loss of service for some communities.	State revenues from severance taxes would pass costs to energy consumers. Federal funding would probably be required for construction of grade separations.	Rail profits may decline.	Benefits of lower voltage accrue to those in vicinity of HVTL; costs will be borne by consumers and utilities who must finance construction of more lines.
Flexibility:	Would help railroads' competitiveness with other modes. Subsidy programs are often difficult to change once started.	Could be implemented flexibly at the state level.	Slurries are a less flexible means of transportation in terms of routing and commodities. Federal legislation could preempt state decisions on eminent domain.	Voltage regulation would not be flexible. R&D would increase options available for HVTL.
Implementability	Subsidies are increasingly unpopular, due to desire to balance federal budget. Rail abandonment would be strongly resisted by some interests.	State programs would have relatively few procedural obstacles, but adequate funding will be the major constraint.	Legal precedent exists for federal control. Considerable opposition to slurries from states, railroads, and agriculture threatens slurries as a widely used alternative.	Utilities would oppose voltage limits, but would probably favor increased R&D. Consumer groups might be opposed to voltage limits because of the higher electric bills.

R & D = research and development

alternate

R & D = research and development.

would slurry pipelines totally substitute for rail transport. And the diversity of a "mixed" rail-slurry transportation system may be seen as an advantage, encouraging the future development of energy resources in the West.

The adverse effects of HVTL can be reduced by promoting DC transmission through an expanded national commitment to research on this still uncertain technology. Limiting the maximum voltage for HVTL would minimize some of the consequences of AC transmission while exacerbating others. States may wish to pursue this approach; but since the trade-offs are unclear, it probably should be done on a flexible, case-by-case basis.

REFERENCES

Bernstein, Peter. 1978. "High Voltage Power Lines Protested." *Denver Post,* February 8.

Branscomb, Lewis M. 1975. "Noise Control for the Future." *Noise Control Engineering* 4 (January/February):13–16.

Bureau of National Affairs (BNA). 1978a. "ICC Seeks Slurry Pipeline Jurisdiction as Railroad Industry Reaffirms Opposition." *Energy Users Report,* Current Report No. 244 (April 13): 25.

Bureau of National Affairs (BNA). 1978b. "Increased Railroad Hauling Rate Could Make Oil Use Cheaper Than Coal." *Energy Users Report,* Current Report No. 272 (October 26):17.

Bureau of National Affairs (BNA). 1978c. "DOE Assures State Rights in Coal Slurry Lines; Bill Goes to Floor." *Energy Users Report,* Current Report No. 249 (May 18):21.

Business Week. 1977. "The New Opposition to High Voltage Lines." November 7, p. 27.

Cirillo, Richard R., et al. 1977. *An Evaluation of Regional Trends in Power Plant Siting and Energy Transportation.* Argonne, Ill.: Argonne National Laboratory.

Clock, Greg. 1978. "Coal Boom Causes Rail Growing Pains." *Daily Oklahoman,* September 24, p. 1.

Congressional Quarterly, Inc. 1979. *Energy Policy.* Washington, D.C.: Congressional Quarterly.

Corrigan, Richard. 1976. "Railroads Versus Coal Pipelines—New Showdown in the West." *National Journal* 8 (March 6):290–95.

Denver Post. 1978. "Troopers Arrest Eight in Power Protest." January 13.

Diamond, S. Lynn. 1977. "Regulatory Update:

Leaders, Policies, and Trends." *Traffic Management* 16 (November):38–47.

Drew, Christopher. 1978. "Texas Battle of Coal Hauling Rates Risks U.S. Energy Plan, Efforts to Lift Rail Net." *Wall Street Journal,* June 16, p. 16.

Federation of Rocky Mountain States, Inc. 1974. *Energy Development in the Rocky Mountain Region: Goals and Concerns.* Denver: Federation of Rocky Mountain States.

Federation of Rocky Mountain States, Inc. 1976. *A Comparison of Unit Trains and Slurry Pipeline Transportation Costs.* Denver: Federation of Rocky Mountain States.

Freudenthal, David D., Peter Ricciardelli, and Michael N. York. 1974. *Coal Development Alternatives: An Assessment of Water Use and Economic Implications.* Cheyenne, Wyo.: Wyoming Department of Economic Planning and Development.

Gerlach, Luther P. 1978. "The Great Energy Standoff." *Natural History* 87 (January):22–37.

Glover, T. O., et al. 1970. *Unit Train Transportation of Coal,* Bureau of Mines Information Circular 8444. Washington, D.C.: U.S. Department of the Interior, Bureau of Mines.

Guccione, E. 1978. "Railroads vs. Pipelines: There's Room for Both." *Coal Mining and Processing* 15 (February):48–51.

Interstate Commerce Act of 1887, 24 Stat. 379 (also known as the Act to Regulate Commerce).

Kash, Don E., et al. 1976. *Our Energy Future: The Role of Research, Development, and Demonstration in Reaching a National Consensus on Energy Supply.* Norman: University of

Oklahoma Press.

Katz, Barbara J. 1977. "Reaching for the Switch: Maryland Neighborhoods Fight Super Power Line." *Washington Post,* December 11.

Kirschten, Dick. 1979. "Railroads v. Pipelines— Cashing in on Coal." *National Journal* 11 (July 7):1112–16.

Lang, William W. 1975. "The Status of Noise Control Regulations in the USA." *Noise Control Engineering* 4 (November/December): 108–15.

Linsley, Clyde, Jr. 1979. "Can We Keep the Railroads Alive?" *Transportation USA* 5 (Summer):6–11.

Locklin, D. Phillip. 1972. *Economics of Transportation.* Homewood, Ill.: Richard D. Irwin.

MacAvoy, Paul, and John Snow, eds. 1977. *Railroad Revitalization and Regulatory Reform.* Washington, D.C.: American Enterprise Institute for Public Policy Research.

Manlytics, Inc. 1977. *Coal Transportation Capacity of the Existing Rail and Barge Network, 1985 and Beyond.* Palo Alto, Calif.: Electric Power Research Institute. As cited in U.S. Congress, Senate Committee on Energy and Natural Resources, and Committee on Commerce, Science, and Transportation. *National Energy Transportation,* vol. 3: *Issues and Problems,* Committee Print by the Congressional Research Service. Washington, D.C.: Government Printing Office, 1978, p. 461.

Miller, Morton, and Gary Kaufman. 1978. "High Voltage Overhead." *Environment* 20 (January):6–15, 32–36.

Montana Department of Natural Resources and Conservation (DNRC), Energy Planning Division. 1974. *Draft Environmental Impact Statement on Colstrip Electric Generating Units 3 and 4, 500 Kilovolt Transmission Lines and Associated Facilities.* Helena, Mont.: Montana DNRC.

Mossman, Frank H., and Newton Morton. 1957. *Principles of Transportation.* New York: Ronald Press.

Natural Gas Act Amendment of 1947, Pub. L. 80-245, 61 Stat. 459.

O'Hara, Edward. 1977. "Moving Coal: Transportation Is Critical to Expand Production." *Transportation USA* 4 (Fall):2–5.

Parmenter, Cindy. 1978. "House Unit Passes Rail Overpass Bill." *Denver Post,* May 5.

Railroad Revitalization and Regulatory Reform Act of 1976, Pub. L. 94-210, 90 Stat. 31.

Ray, Mel. 1978. "ICC Comes under Fire." *Electrical World* 190 (December 1):4.

Richard J. Barber Associates. 1977. *The Railroads, Coal and the National Energy Plan: An Assessment of the Issue.* Washington, D.C.: Richard J. Barber Associates. As cited in U.S. Congress, Senate Committee on Energy and Natural Resources, and Committee on Commerce, Science, and Transportation. *National Energy Transportation,* vol. 3: *Issues and Problems,* Committee Print, by the Congressional Research Service. Washington, D.C.: Government Printing Office, 1978, p. 465.

Richards, Bill. 1976. "Paying the Price for Western Energy." *Washington Post,* December 13, p. A4.

Richards, Bill. 1977. "Changes Sweep into Wyoming." *Washington Post,* June 27.

Rittenhouse, R. C. 1977. "Fuel Transportation: Meeting the Growing Demand." *Power Engineering* 81 (July):48–56.

Strain, Peggy. 1977. "Water, Land, Life—It's All One in Valley Pipeline Debate." *Denver Post,* November 13.

U.S. Congress, Office of Technology Assessment (OTA), Coal Slurry Pipeline Project Staff. 1978. *A Technology Assessment of Coal Slurry Pipelines,* Prepublication Draft. Washington, D.C.: OTA.

U.S. Congress, Senate Committee on Energy and Natural Resources, and Committee on Commerce, Science, and Transportation. 1977. *National Energy Transportation,* vol. 1: *Current Systems and Movements;* vol. 3: *Issues and Problems,* Committee Print by the Congressional Research Service. Washington, D.C.: Government Printing Office.

U.S. Department of the Interior (DOI), Bureau of Land Management (BLM), et al. 1974. *Final Environmental Impact Statement for the Proposed Development of Coal Resources in the Eastern Powder River Coal Basin of Wyoming,* 6 vols. Cheyenne, Wyo.: BLM.

U.S. Department of Transportation (DOT). 1977. "Rail Abandonments and Their Impacts." In *Railroad Revitalization and Regulatory Reform,* edited by Paul MacAvoy and John Snow. Washington, D.C.: American Enterprise Institute for Public Policy Research.

URS Company. 1976. *Coal Train Assessment,* Final Report for Colorado Department of

Highways. Denver: URS Company.

Wall Street Journal, 1978a. "Freight Car Orders Climbed 65% in May." June 12.

Wall Street Journal, 1978b. "U.S. to Back Railroad Loan." May 31, p. 19.

Wyle Laboratories. 1973. *Assessment of Noise Environments Around Railroad Operations,* WCR 73-5. N.p. As Cited in URS Company. *Coal Train Assessment,* Final Report for Colorado Department of Highways. Denver: URS Company, 1976, p. VII–49.

Young, Louise B. 1973. *Power Over People.* New York: Oxford University Press.

CHAPTER 11

Siting Energy Facilities

Many of the most troublesome energy issues involve finding acceptable sites for new energy facilities. The process of siting matches a particular technology for supplying energy, such as a power plant or mine, with a specific geographical location. Since the impact of energy development is heavily dependent on the type of technology chosen and the location at which that technology is deployed, siting policies will largely determine the actual impact of developing western energy resources. For example, siting water-intensive facilities (such as steam-electric power plants) in a water-short area (like the UCRB) is likely to intensify water availability problems and spur conflicts over water rights. Likewise, siting energy conversion facilities near national parks or other clean-air areas will increase the possibility of conflicts over air quality.

Because energy development begins with choosing a site for a new facility, siting decisions often become the focus of a broader controversy over how the costs and benefits of development should be distributed. Recent decisions to site energy conversion facilities in the West and to develop western resources have already led to a rivalry between this energy-rich region and the energy-consuming eastern states (Hall, White and Ballard, 1978: 197–98).

In this chapter we will address the obstacles to and constraints on siting energy facilities in the West. Following a brief discussion of the major factors underlying these controver-

sies, we will outline the four most significant siting issues and then describe and evaluate a range of alternative policies for improving the process of siting energy facilities.

Factors Affecting the Siting of Energy Facilities

A number of problems associated with siting energy facilities in the West arise because there now exists no comprehensive national siting strategy. Decisionmaking authority at the highest level of government is often fragmented, parochial, and reactive in nature. As a result, policymakers often lack the capability to balance economic, energy, and environmental trade-offs in selecting sites for new energy facilities. Nor is it easy, given the fragmented policy system, to determine which sites should be avoided or which factors are critical in terms of impact. Moreover, the siting process has inherent redundancies which, when coupled with the intervention of interest groups, can delay the siting of new facilities for very parochial reasons. The siting process is characterized by the involvement of many different interests and authorities; each represents a different viewpoint and often each possesses extremely limited information.

Although the initial determination of the need for and the location of new facilities is usually made by energy industries, their decisions are shaped and modified by public policy. As Russell E. Train, the former administrator of the EPA noted, the present siting process is a purely "reactive" system in which the industry selects the sites and state and federal regulatory agencies respond to requests for permits (Carter, 1978a:671). This reactive tendency is clearly illustrated by the typical process of planning energy facilities.

Utilities and other energy companies engage in system-wide planning which constantly assesses their need to replace or add facilities. Important aspects of such planning that influence the siting decisions are: load forecasting, generator selection, reliability analyses, territorial considerations, corporate policy, and basic economic policies (Cirillo et al., 1976:5). Government agencies and public interest groups generally are not active in the early stages of planning for new energy facilities. Once the data indicate that additional demands for energy will develop, several options are open to the company in addition to acquiring sites for the construction of new facilities. For example, an electric utility may be able to expand or replace units on existing sites, or it may choose to purchase power from other sources.

Considerable controversy surrounds the determination of the need for new energy supplies, and it is at this point in the siting process that governmental and other interested participants normally become involved. Opponents of utilities maintain that these firms purposely engage in projects to increase capacity for self-serving reasons. On the other hand, utilities claim that questioning the need for more energy has been a tactic used by obstructionists to delay new energy projects. In any case, whether deliberate or not, delays do result; Robert I. Hanfling, the acting director of utility projects for the DOE, notes that conflicts over the need for new energy facilities is one of the major problems in getting projects completed on schedule (Cavanaugh, 1977:89). Indeed, Montana Power Company's application for two additional power plants at Colstrip, Montana, was denied by the state

Department of Natural Resources and Conservation largely because the need for new facilities in Montana had not been adequately established. Most of the electricity generated at the proposed Colstrip units was to have been exported to the Pacific coast states (see Western Interstate Nuclear Board, 1977:52–53).

Once the need for new energy facilities has been determined, and assuming that the utility decides to locate these facilities on new sites, a screening process usually identifies several prospective sites for additional analysis. The most significant criteria used in evaluating these sites are economic, engineering, safety, environmental, and institutional considerations. Although the utility's ongoing planning and site-selection processes must consider how the proposal conforms to federal, state, and local requirements, the utility's plans are largely proprietary and not subject to public inspection. Historically, once the utility persuaded the state utility commission and the Federal Power Commission (FPC) that a new facility was needed and that the project was economically sound, there was little public debate in the siting process itself. The utility would apply to the state utility regulatory commission for a certificate of public convenience and necessity; if this application was approved, the utility would acquire the site using its right of eminent domain and would then build the facility after securing the necessary permits. However, a number of changes have occurred during the past decade to transform this closed and orderly siting process into one of the most controversial aspects of developing domestic energy resources.

Problems and Issues

In the absence of any systematic federal or regional siting policy, few states in the West have attempted to develop an overall siting process on their own. Thus, siting decisions usually are made piecemeal, are not coordi-

Coal-Fired Power Plant with Wet-Cooling Towers, Colstrip, Montana

nated with related choices about energy development, and are subject to long delays. Furthermore, inadequate attention is given to the combination of technology and location which can be used to reduce many of the adverse effects of energy development. Among the most important issues which have created this disjointed, highly uncertain situation are:

- Federal and state environmental legislation;
- Increased public participation;
- Interregional conflict; and
- Intergovernmental jurisdictional disputes.

These problems and issues, along with the inherent importance of siting, have created a complex system of policymaking which is essentially beyond the control of developers, of federal agencies, or of the state and local interests which face the most direct consequences of energy development. The section which follows discusses each of these issues in more detail.

Federal and State Environmental Legislation

Extensive federal and state environmental legislation restricts the areas in which development can occur and helps to define how new facilities will be constructed. However, the comprehensive nature of these regulations requires expensive and lengthy review processes that add considerably to the financial risk of new projects.

As environmental concerns have increased during the past decade, numerous laws and regulations which affect the siting of energy facilities have been enacted at both the federal and state levels. The earliest federal environmental legislation affecting sites for energy facilities stems from the Rivers and Harbors Act (1899) which gave the Army Corps of Engineers power to issue permits for the use of navigable waters (see Beatty, 1973:497). The federal role was expanded by the FWPCA Amendments (1972) which gave EPA the authority to issue permits for energy facilities that discharge effluents (see Ward, 1977: 2059). This act also controls the indirect impact on water quality from expanded local sewage treatment made necessary by the siting of new energy facilities.

The most far-reaching environmental law affecting siting is the 1969 National Environmental Policy Act (NEPA) which mandates the preparation and review of an EIS for all proposed projects which require federal action and which will significantly affect the human environment. The EIS must include a discussion and evaluation of: the environmental setting or existing conditions, the proposed facility, the environmental effects of development, the socioeconomic effects, possible policy alternatives, and unavoidable adverse effects. The EIS process also requires a review by a wide range of federal, state, regional, and local agencies and citizens groups (see Baram, 1976:3–39). Because of these extensive review guarantees, the EIS is a major avenue for public participation in siting decisions.

Another major step toward federal control of the siting of new facilities came in 1970 when the CAA was passed (see Rosenbaum, 1977:144–57). This legislation, among other things, extended federal regulatory authority to include stationary air pollution sources, such as energy facilities. Since 1970, federal air-quality controls have been subject to almost continuous modification. As described briefly in Chapter 5, the 1977 CAA Amendments, established regulations for the PSD of air quality. Under the terms of the 1977 amendments, those areas designated Class I (national parks and other scenic spots) are to be most vigorously protected. Class II areas have more liberal PSD restrictions, and Class III will allow the most degradation in air quality (Kirschten, 1977a:1261–63). Class II and III areas may be upgraded to Class I by the administrator of EPA; the Northern Cheyenne Reservation in Montana is the first such redesignated area in the country. Thus, some parts of the West are effectively precluded as sites for energy development in order to maintain their currently pristine air quality.

Wildlife preservation legislation constitutes another constraint on siting decisions. Conservationist and ecological interests have strongly supported legislation to protect both wildlife and wildlife habitats. The Endangered Species Conservation Act (1969) provided federal protection for threatened plants and animals. The Endangered Species Preservation Act (1973) extended this authority, prohibiting the critical habitat of endangered fish, wildlife, and plants as possible locations for new energy facilities. A list of endangered species compiled in March 1967 included 72 native species. As of the end of 1976, some 170 native species were listed as endangered. Although the locations of the critical habitat of several known species have been identified, this information forms only a partial basis for excluding potential sites. Many new species will probably be added to the endangered list, and the EIS required for any new energy facility may also identify new endangered species (see Carter, 1978b:628).

Federal legislation also protects historically or culturally significant places. This authority stems from laws enacted as early as 1906, but the most recent legislation is the Historic Preservation Act (1966). In 1970, under pressure from the National Trust for Historic Preservation, Congress extended the 1966 act and increased the funding for preservation efforts. Basically, these laws mandate the initiation of a federal-state advisory process to eliminate or mitigate any effects of projects on cultural properties that are, or may be, listed in the National Register of Historic Places. These requirements are over and above the NEPA mandate, thus adding another set of procedures to the siting process. Because of the ubiquity of potential "historical places," some energy developers now employ staff members or consultants to examine proposed facility sites for evidence of historical or cultural significance (Holden, 1977:1070–72).

Increased Public Participation

Increased public participation has opened siting decisions to more diverse interests. Environmental interest groups have been the most visible, but agricultural groups, coalitions of ranchers, Indians, and other interested parties have also become important participants in siting decisions. The major conflicts arise over the question of how to develop mechanisms for public participation which are meaningful but do not cause unreasonable delays for energy projects.

Individual citizens, organized interest groups, and governmental agencies have used the legal avenues provided by environmental legislation to participate in siting decisions. Public participation in decisions that were once considered to be primarily in the domain of private industry has been greatly facilitated by legislation such as NEPA and FWPCA. Greater citizen involvement in choosing sites for energy facilities is based on a need to make the political system more responsive to the general public by incorporating more diverse interests in decisions on policy formulation and implementation (Nelkin, 1977:13). In the West, the participants include environmental and consumer interest groups, economic interests such as farmers and ranchers and cultural and ethnic groups such as Indians.

Environmental and consumer interests have often been the most vocal opponents of sites for energy facilities. Groups such as the Sierra Club, the Audubon Society, the Friends of the Earth, and the Environmental Defense Fund are nationally recognized in part because of the number of times they have intervened in cases to halt or modify siting proposals. Such opposition is usually seen by industry spokesmen as a major roadblock to increasing the domestic supply of energy. According to Richard F. Walker, president of the Public Service Company of Colorado, delays occasioned by environmental interest groups have been a major cause of increases in the price of energy over recent years. But characterizing all environmental and public interest groups as opponents of energy projects is simply inaccurate. Significant differences exist between the energy-policy positions of various groups representing consumers and environmentalists. Thus, while the more aggressively antinuclear environmental groups, such as Ralph Nader's organizations, have opposed the construction of new nuclear power plants as part of a "low energy growth" position, other groups, such as Common Cause, have supported energy development through "mixed market" strategies (McFarland, 1976:47–51).

The most intense environmentally oriented public opposition in the West is voiced over the site-specific concerns of spoiling a unique environment. Opposition to the proposed Kaiparowits power plant in southern Utah, for example, centered on the impact of the facilities at the site. Many of the people resisting this project did so because they valued the region's unique heritage rather than because of any environmental or physical risks associated with development (*Wall Street Journal,* 1976:1). In contrast, proposals to site similar fossil-fuel facilities in western North Dakota did not generate such widespread environmental opposition, probably because the region is not perceived to be as physically and culturally attractive.

Economic interests intervene on both sides of siting questions, and local reactions are difficult to generalize. Whereas those who favor energy development in the West tend to see economic benefits as outweighing environmental and socioeconomic costs, groups

---

Rancher Fearful of Development

A rancher outside Colstrip, Montana, articulately expressed his fear of new energy facilities. "To me, I just don't think you can put a numerical value on what we have right here. I think it's our responsibility to do with it what we can and turn it over to our children in good shape. . . . Then you get conversion plants, and you think, okay, what's this going to do with the water? How much water is going to come out of the Yellowstone? What's a 200 or 1,000 percent increase of the population of my hometown going to do? . . . to the tax base? . . . to the sociological undercurrent? . . . to my community?"— (Johnson, 1975)

---

in opposition to such plans usually perceive local development as producing more costs than benefits (see box). Ranchers in Colorado, for example, have voiced their concern that the siting of Public Service of Colorado's Pawnee power plant would create emissions that would harm pastures by decreasing their productivity. In another case, the North Dakota Farmers Union joined forces with the Friends of the Earth to demand strict reclamation procedures for coal mining so that agricultural production would not be sacrificed to energy development (see Fradkin, 1977:120).

Those who support local siting of a facility because of economic benefits may include local landowners, realtors, bankers, chambers of commerce, or wholesale/retail merchants. For most residents in southern Utah, the prospect of some 35,000 new jobs, an additional annual payroll of $100 million, and yearly tax revenues of $28 million overrode environmental concerns as local citizens largely supported the Kaiparowits proposal (Myhra, 1977:26).

Indian tribes have also increased their participation in site selection for energy facilities. The Indian tribes control between 10 and 16 percent of the nation's coal (one-fifth of the strippable western reserves), 50 percent

of the country's uranium (one-sixth of the recoverable uranium), and about 3 percent of the nation's remaining oil and gas (Crittenden, 1978:1). Although Indians do not have a united position on siting energy facilities, the newly created Council of Energy Resource Tribes (CERT) is attempting to establish a common set of rules on prices, production controls, and environmental protection (Greider, 1977). The Bureau of Mines has assisted in this effort by helping Indian tribes evaluate the financial implications of different negotiating positions.

Efforts by Indians to gain greater control over their energy resources have the potential to generate conflicts in the siting process. Some tribes, most notably the Apaches, maintain that federal and state siting regulations are not applicable on Indian lands. This contention, if upheld by the courts, opens the possibility that in the future, sovereign tribes could operate energy facilities under their own environmental standards. The successful court battles waged by Indians over land claims and fishing rights have already created a significant white backlash among ranchers and sportsmen in the West (*Business Week,* 1977). Nevertheless, Indian tribes appear committed to taking tougher bargaining positions with energy developers: for example, the Navajo recently refused to allow Western Gasification Company (WESCO) to build a coal gasification plant on the reservation or to grant a right-of-way for a site just off the reservation (*Coal Mining and Processing,* 1978).

Interregional Conflict

Increased reliance on domestic energy resources has forced many utilities from other regions to build energy facilities in the West. This pattern of development has generated concerns in the West about regional exploitation and neocolonialism.

Interregional conflicts are likely to arise because many energy facilities sited in the

West will serve the demands of other regions. The provisions of the Second National Energy Plan and other Carter administration proposals (such as the synthetic fuels program announced in July 1979) place a high priority on developing western energy resources both to help decrease dependence on foreign imports and to make it possible to maintain national economic growth. Developing western resources may cause the West to bear a disproportionate share of the impact of achieving these goals, although the region will also enjoy the associated economic benefits (see Lamm, 1976). Thus, building energy facilities in the West will result in trade-offs between economic and environmental costs and benefits as well as trade-offs between regional autonomy and regional interdependence.

It is particularly disturbing to many westerners that other regions will receive fuel but will not experience the adverse effects of the energy facilities needed to produce that fuel. For instance, officials in Montana have announced that they do not intend to be subjected to outside "exploitation" and have therefore passed tough energy-facility siting laws.

On the other hand, some westerners see the development of energy resources as a boon and are fighting to retain the advantages of using western energy on a long-term basis. Wyoming's Sen. Clifford P. Hansen, for example, was outraged by the Carter administration's plan to encourage the use of local coal to serve local markets (see box). This plan was formulated in response to a concern that coal-rich areas in the East which have high unemployment might be bypassed in favor of the West; the plan reflected the views of easterners like Ohio's Sen. Howard M. Metzenbaum, who argued, "Do we mean that we will require coal to be transported across the entire country when there is plenty of coal available in the very regions . . . the utilities operate in?" (Kirschten, 1977b:1683).

On a larger scale, a war of words exists between such regional organizations as the Western Governors' Policy Office (WESPO) and the Coalition of Northeastern Governors

---

Administration Favors Use of Local Coal

EPA administrator Douglas M. Costle justified the position of requiring stack gas scrubbers on new coal-burning facilities whether they needed them or not. This policy, he said, was intended to "encourage power plants to use locally mined high- and medium-sulfur coal instead of bringing in low-sulfur coal from other regions. This will avoid much of the regional unemployment and economic disruption that would result from greater reliance on low-sulfur coal than on control technology."—(Kirschten, 1977b:1683)

(CONEG). The western governors have argued that the West should not be forced to bear the ecological and socioeconomic costs of large-scale energy projects to subsidize the energy needs of the industrial states. CONEG governors argue, first, that their states are already suffering the negative effects of existing industries that serve the entire nation and, second, that their states' federal taxes have subsidized the development of the West (Peirce, 1976:1699).

Intergovernmental Jurisdictional Disputes

Laws and regulations governing the siting of facilities often are uncoordinated and overlap among and between levels of government. At the federal level, no single agency has the responsibility for looking at the overall need for a major energy facility from the viewpoint of the long-range public interest, and federal siting activities are often inconsistent with state programs.

The federal government has little direct authority over the siting of nonnuclear energy facilities; most siting responsibility resides in state and local governments. The disposition of public lands is one area in which the federal government does have direct authority over the location of energy facilities. However, indirect federal control of siting is the predominant situation. Through such mechanisms as the EIS requirements of NEPA and legislation such as the CAA and FWPCA, the federal government is able to intervene in the siting process. Since passage of the Federal Power Act (1935), the federal government has also assumed responsibility for overseeing the use and operation of interstate gas- and electrical-supply systems.

The consideration of more comprehensive federal siting legislation has been complicated by a continuing debate over two very different approaches to siting policy. On the one hand, there is pressure for a generic approach, national in scope, which would emphasize comprehensive land-use planning as a prerequisite for narrower choices regarding the need for energy, the type of fuel and source, the general location and type of facility, the actual site selection, the plant's design and construction, and its operating guidelines. The proposed Land Resources Planning Act of 1975, for example, included the siting of facilities as only one component of overall land-use planning. Title III of this act, the Energy Facilities Planning Act, would have required public participation and consideration of regional and national energy needs (U.S. Congress, Committee on Interior and Insular Affairs, 1977:1383–516). But there are also pressures at work for a more flexible, site-specific approach which would then aggregate the particulars of siting into a national land-use plan. Thus, the Energy Facilities Planning and Development Act, also proposed in 1975, relegated land-use planning to a function of the site-selection process. This legislation, however, would have required each state to design an energy-facility program and would have given the federal government extensive data collection and forecasting responsibilities. Many state officials were concerned that these provisions could preempt state authority. Thus, Gov. Richard D. Lamm of Colorado

testified against the proposal in Congress, and other governors expressed similar fears that the legislation threatened states' rights (The Energy Center, Inc., 1977:376).

Both of these 1975 proposals were ultimately defeated, as was the Electric Utility Rate Reform and Regulatory Improvement Act of 1976. Title V of that bill would have required energy firms to report to the FPC on their prospective sites, their efforts to protect the environment, and their provisions for involving land-use agencies in the planning process. Under the terms of the bill the FPC would have been responsible for preparing national energy-facility siting plans. And the FPC would have taken the lead role as the sole legal authority for licensing energy facilities in a federal one-stop siting process, featuring a single, composite utility application.

Many of these reforms have been proposed because the number of state permits and approvals needed to site an energy facility (most of which are over and above the federal requirements) have added significantly to the complexity of the siting process. The licensing process for a coal-fired power plant may involve up to 50 separate permits, each requiring different public hearings and legal findings (Calvert, 1978:34). For example, Table 11-1 lists the number and types of major permits and approvals that must be obtained before a coal gasification facility in North Dakota can be constructed and operated. In addition, judicial reviews of agency decisionmaking add another level of intervention (and delay) to this process.

The process of siting energy facilities in most states is extremely decentralized; numerous single-purpose laws influence the final decision. The simultaneous imposition of energy, environmental, land-use, and consumer protection laws, each of which may apply to distinct elements of a new energy facility, creates an extraordinarily complex process (Mills et al., 1975:91).

State and local authority governing energy facilities is based on the "police-power" control over state-owned lands and on the au-

Table 11-1: *Permits, Approvals, and Certifications Required for Construction and Operation of a North Dakota Coal Gasification Facility*

Agency	Permit and/or Approval
Federal	
U.S. Army Corps of Engineers	Easement for Water Intake, Pipeline, and Access Road; Section 10 Permits for Water Intake and Pipeline Crossings of Major Streams; Section 404 Permits for Wetland Disturbance
Environmental Protection Agency	New Source Performance and Air Quality Significant Deterioration Review: Deep Well Disposal Review
Federal Power Commission	Certificate of Public Convenience and Necessity
Federal Aeronautical Administration	Application for and Notice of Proposed Construction for Structure over Regulated Heights
U.S. Bureau of Reclamation	Water Service Contract, Environmental Impact Statement
State (North Dakota)	
Public Service Commission	Plant Certificate of Site Compatibility, Water Pipeline, Certificate of Site Compatibility; Water Pipeline Transmission Facility Route Permit; Mining Plan
Department of Health	License for Radioactive Measuring Device Operations; Hazardous Waste Control Plan; Wells for Temporary Water Supply; Sewage Treatment Plant
Environmental Engineering Division	Permit to Construct (Air Pollution Control Permit); Permit to Operate (Air Pollution Control Permit)
Water Supply and Pollution Control Division	NPDES* Permit for Deep Well Disposal; Solid Waste Disposal Permit
State Highway Department	Rail Siding Crossing; Pipeline Construction on Highway Right-of-way
State Water Commission	Appropriation of Underground Water, North Dakota State Water Permit (conditional permit obtained)
Secretary of State	Certificate of Authority for Foreign Corporation to Transact Business
Unemployment Compensation, Division of Employment, Security Bureau	Application for Coverage by ANGCGC*
Workman's Compensation Bureau	Covered by ANGCGC
Mercer County	
Board of Commissioners	Petition for Access to County Roads; Petition for Vacating County Road and Closing Section Lines; Certificate of Zoning Compliance; Plantsite Rezoning; Conditional Use Permit
Soil Conservation District	Erosion and Sediment Control Plan

*NPDES = National Pollutant Discharge Elimination System. ANGCGC = American Natural Gas Coal Gasification Company.

Source: U.S. DOI, BuRec, Upper Missouri Region, 1977:1-9, 1-10.

thority to implement federal legislation. The police power of state and local governments includes zoning laws, building codes, subdivision regulations, permit requirements, and performance standards. Traditionally, state public utility commissions have used the criterion of "public convenience and necessity" to evaluate siting proposals. Issuance of this certificate demonstrates that the state accepts the utility's forecast of demand and evaluation of the impact of the additional facility on retail rate structures. But during the past several years

Table 11-2: *Facilities Covered by State Siting Laws in the West*

State	Processing	Transmission
Arizona	Thermal, nuclear, or hydroelectric plants with capacity ≥100 megawatts; associated infrastructure	HVTL ≥115 kV
Montana	Plants with capacity ≥50 megawatts; processing of gas ≥25,000,000 cubic feet/day; processing of hydrocarbons ≥25,000 bbl/day; "yellowcake" enrichment whose plant costs ≥$250,000; any facility capable of utilizing, refining, or converting 500,000 tons coal/year	HVTL associated with facilities ≥69 kV (except HVTL ≤230 kV if less than 10 miles length); pipelines (gas, water, liquid hydrocarbons) from mentioned facilities, or geothermal plants, *in situ* gasification, subsurface conversion
New Mexico	Oil, gas, uranium, geothermal plants with generation capacity ≥300 megawatts	HVTL ≥230 kV
North Dakota	Plant capacity 50 megawatts; manufacture or refinement of 100 million cubic feet of gas/day; manufacture or refinement of 50,000 bbl liquid hydrocarbons/day; enrichment of uranium materials	HVTL 200 kV HVTL 69 kV if lines do not follow ¼ or whole section lines or paths of other transport modes; gas or liquid pipelines (coal, gas, water, liquid hydrocarbons) to or from facilities mentioned herein
Wyoming	Plant capacity ≥100 megawatts; facility capable of producing 100 million cubic feet synthetic gas/day; 50,000 bbl liquid hydrocarbons/day; "yellowcake" enrichment capacity 2,500 lbs/day; any industrial facility with an estimated cost of $50 million	Not specified

HVTL = high voltage transmission lines. ≥ = is greater than or equal to. kV = kilovolts. ≤ = is less than or equal to. bbl = barrels.

some western states have passed explicit bills to coordinate, supervise, and control such siting decisions. Such laws are a major step toward coordinating state land-use, health, building, and other single-interest agencies involved in siting decisions, and the laws simplify the process of evaluating the trade-offs involved in locating a particular facility (see Verity, Lacy, and Geraud, 1974).

Five of the states in our study area—Arizona, New Mexico, North Dakota, Montana, and Wyoming—have such siting laws; their characteristics are summarized in Tables 11-2 and 11-3. As indicated, the siting laws of North Dakota, Montana, and Wyoming cover the widest variety of energy processing and transmission facilities.

The siting agencies are generally charged

with the responsibility for viewing siting proposals holistically rather than following the single-purpose criteria of the older state regulatory bodies that cover health, safety, and welfare. In practice, however, the policymaking process for siting energy facilities is still fragmented. In Wyoming, for example, a relatively rigorous siting law is still subject to the decisions of the State Engineer and the Public Service Commission. But despite claims that these fragmented state laws contribute to delays in the licensing of facilities, the time required to site energy facilities in states with and in states without state regulations does not differ widely (Calvert, 1978).

State control of siting energy facilities is also bolstered by state implementation of federal environmental legislation, especially air-

Table 11-3: *Provisions of State Power-Plant Siting Laws in the West*

State	Power-Plant Siting Law	Lead Agency	Site Certification Authority	Size and Composition of Site Panel	Method of Acquisition	Application Fee	Annual Utility Forecast
Arizona	yes	yes	Arizona Power Plant Siting Committee	11 state officials, 7 others	Certification of environment compatibility	new site: $10,000 expansion: $7,500	10-year-plan
Colorado	no						
Montana	yes	yes	Board of Natural Resources		application to site authority	based on facility cost	10-year-plan
New Mexico	yes	yes	Public Utilities Commission		eminent domain	none	none
North Dakota	yes	yes	Public Service Commission		certification of compatibility	$150,000 maximum	10-year-plan
South Dakota	no						
Utah	no						
Wyoming	yes	yes	Industrial Siting Council	7 government appointees	permit	$100,000 maximum	5-year-plan updated annually

Source: Southern Interstate Nuclear Board, 1976:viii-x.

and water-quality laws and regulations through "state environmental policy acts" (SEPAs). Six of the eight western states have some sort of SEPA; Utah and New Mexico do not (CEQ, 1976:133–35). The SEPAs in Arizona, Montana, and South Dakota require an EIS that involves detailed consideration of any environmental effects prior to the development of a site. North Dakota and Wyoming have SEPAs that stress the enforcement of certain criteria after the facility is in operation. Colorado has a coordinating and planning arm that is only an advisory adjunct to the state's environmental policymaking mechanism.

Municipalities can also affect the location of energy facilities through local zoning ordinances, building codes, health and sanitation standards, and local taxation policies. Using zoning ordinances, the community has full authority to determine the siting of energy facilities within its jurisdiction. However, these tools to control land use have not been widely used in the West: there, municipalities tend to be small, and most new energy facilities are located beyond municipal boundaries. As a result, localities have been forced to "decide limited issues, play a reactive role only, and make policy by default" (Aron, 1975:344).

Although efforts have been made at both the state and federal levels to streamline the siting process and to reduce overlaps and inconsistencies in procedures for issuing site permits, the process is still complicated; overall policy is made more on the basis of state and local interests than on regional or national needs. While state siting commissions and EISs do consider the broader implications of a proposed facility, they generally focus on the impact of the new facility on the particular surrounding area and do not attempt to plan interrelated strategies for siting. Because all major electrical production facilities are tied together in a regional power pool, siting new facilities affects the reliability and efficiency of the entire regional energy system. Moreover, generalized effects on the area could be minimized if appropriate sites were chosen beforehand, rather than evaluated *ex post*

facto. However, state siting commissions pay little attention to a facility's implications for the interstate power systems, despite the fact that regional planning institutions do exist in the West. For instance, the Western Systems Coordinating Council, serving eleven western states, is more concerned with system reliability and information exchange than it is with plans for siting energy facilities in its region. An exception to the lack of regional planning in the West is Montana's Major Facility Siting Act (1975); it lists the relationship of new energy facilities to the regional grid-distribution system as one criterion to be considered in siting.

Summary

The extensive array of laws and regulations that compose the present process for siting energy facilities and the manner in which actual siting policy is made is not an effective way to weigh the alternatives and select the best sites. At present, the initial selection of sites is made by industry, based on proprietary criteria and methodologies for evaluation. The general public often becomes aware of siting choices only after a site has been selected and the developer is financially committed to the project. Interest groups can participate in the process only after plans are publicly disclosed. Accordingly, they are more likely to influence the design of the facility than the selection of the most suitable site.

The existing system for siting energy facilities leads to delays in locating new facilities and is not well suited for incorporating broad participation in determining the trade-offs among different sites. Conflicts often occur since the criteria for evaluating sites differ from group to group and from stage to stage in the siting process. An attractive site from the developers' point of view may be a poor choice in the eyes of environmental, economic, or cultural groups who become involved at a later stage. Environmental groups may be more concerned with ecological effects; local ranchers and farmers are interested in the project's impacts on agricultural productivity; and Indians may focus on possible changes in lifestyle. For the utility, however, the overriding criteria are economic efficiency and energy-system reliability.

The principal concerns of this chapter are the goals of using the best sites for new energy facilities and of getting needed energy facilities into operation expeditiously. To accomplish these goals it is essential to modify the policymaking processes by providing more accessibility to the siting decision for all major interests. Also, an overall strategy for developing western energy resources while keeping environmental degradation to a minimum probably requires a regional perspective and the early identification of changing energy needs.

Table 11-4: *Alternative Policies for Siting Energy Facilities*

General Alternative	Specific Alternative
Undertake technology-site evaluations	Designate acceptable technology-location combinations.
	Site-screening and site-banking techniques.
Employ impact-assistance methods	Require impact mitigation by developers, state, or federal governments.
	Compensate state and local governments for adverse effects.
Increase citizen involvement in site selection	Provide for the exchange of information between citizens and agencies.
	Encourage administrative interaction between agencies and citizens.
	Allow direct citizen input into agency decisions.
Provide support for participation	Provide technical or financial assistance for interested groups.

Policies for Improving the Siting of Energy Facilities

In combination, the problems and issues discussed above make siting one of the most crucial aspects of energy development in the West. These problems suggest two objectives:

- To site energy facilities in locations that will avoid unacceptable costs and risks; and
- To provide the means for interested parties to participate in site selection without causing unreasonable delays.

Each of these objectives can be attained in a variety of ways. As outlined in Table 11-4, two categories of policies respond to the first objective of siting energy facilities so as to avoid unacceptable costs and risks. These categories are to conduct site evaluations in the light of a specific energy technology and to plan methods to mitigate any adverse effects. Likewise, the second objective, providing the means for interested parties to participate without causing unreasonable delays, may be accomplished by increasing citizen in-

volvement in the actual selection of sites or by providing support for citizen participation.

Policies to Undertake Technology-Site Evaluations

As outlined in Table 11-4, this category includes two specific policy alternatives:

- To designate acceptable technology-location combinations;
- To implement site-screening and site-banking techniques.

Both alternatives assume that some areas will be excluded from energy development, and both represent an extension of current zoning practices on a larger scale.

Designating Acceptable Technology-Location Combinations: Certain combinations

Table 11-5: *Areas Excluded as Sites for New Energy Conversion Facilities in North Dakota*

Exclusion Areas:
   Designated or registered: national parks; national historic sites and landmarks; national historic districts; national monuments; national wilderness areas; national wildlife areas; national wild, scenic, or recreational rivers; national wildlife refuges; and national grasslands.

   Designated or registered: state parks; state forests; state forest management lands; state historical sites; state monuments; state historical markers; state archaeological sites; state grasslands; state wild, scenic, or recreational rivers; state game refuges; state game-management areas; and state nature preserves.

   County parks and recreational areas; municipal parks; parks owned or administered by other governmental subdivisions; hardwood draws; and enrolled woodlands.

   Areas critical to the lifestages of threatened or endangered animal or plant species.

   Areas where animal or plant species that are unique or rare to this state would be irreversibly damaged.

   Prime farm land and unique farm land, as defined by the Land Inventory and Monitoring Division of the Soil Conservation Service, United States Department of Agriculture.

   Irrigated land.

Avoidance Areas:
   Areas of historical, scenic, recreational, archaeological, or paleontological significance which are not designated as exclusion areas.

   Areas where surface drainage patterns and groundwater flow patterns will be adversely affected.

   Within the city limits of a city or the boundaries of a military installation.

   Areas within known floodplains as defined by the geographical boundaries of the 100-year flood.

   Areas that are geologically unstable.

   Woodlands and wetlands.

Source: North Dakota Energy Conversion and Transmission Facility Siting Act, 1975.

of technology and location could be encouraged with procedures similar to eminent domain or discouraged by the condemnation of areas for a particular land use. The approach of discouraging or prohibiting certain areas as sites for energy facilities rests, however, on more limited and more recent experience with such schemes as air-quality classifications. These mechanisms could be extended to include other criteria—the availability of water, socioeconomic costs, or the consumption of agricultural land. As an example of how the acceptability of locations may be determined, Table 11-5 shows some types of sites excluded under North Dakota's Energy Conversion and Transmission Facility Siting Act (1975).

For the most part, the evaluation of certain combinations of technology and location relies on formal, governmental regulatory and economic policies designed to influence the

decisions of private industry. Thus, acceptable technologies and locations may be designated directly, through federal or state laws, or indirectly, by taxation or subsidies. But the determination of acceptable and unacceptable sites also could be made by a combination of participants from government agencies, industry, and interest groups, employing a strategy similar to the "rule of reason" used by the National Coal Policy Project. The principles guiding that project emphasized sharing all relevant data, simplifying complex concepts, isolating subjective factors and value judgments, and unanimously agreeing to avoid dogmatic negotiating tactics (Carter, 1977; see also Murray, 1978).

This option is based, first, on the knowledge that some combinations of technology and location will result in more undesirable effects than others and, second, on the agreement that those combinations which result in greater adverse effects should be excluded. The remaining sites, then, are defined as acceptable locations for energy development facilities. To the extent that a sufficient number of sites can be identified to meet projected energy needs, the option of technology-site evaluation can be effective.

However, a number of problems severely limit the usefulness of zoning approaches or classification schemes. Most significantly, technology-site evaluation depends on the elimination of certain areas as possible locations for energy facilities, based on several objective criteria. Each stage of the evaluation, therefore, could result in the elimination of those sites that failed to meet a particular standard. When all the unacceptable locations for each criterion were identified, the remaining sites would be regarded as favorable by all interested parties. However, it is possible that such an evaluation could exclude most (or all) areas from future energy development. That is, since each possible combination of technology and location will have some adverse effects, no particular configuration may be acceptable if all the evaluation standards are invoked simultaneously. For example, a process which eliminated all national parks, wilderness areas, and wildlife refuges as a first step and then eliminated areas with severe water problems, regions with air-dispersal problems, and locations that form the critical habitat of endangered or threatened species might leave only a very few sites for energy development. Thus, in order for this to be an effective policy option, policymakers will have to choose evaluation criteria which generate sufficient sites to meet projected energy needs.

Technology-location evaluations should make the siting of facilities more equitable so that no locality will have to bear an excessively large share of the impact of energy development. If the sites that would create the largest costs and greatest risks are eliminated through the use of some evaluative technique, then facilities sited in the remaining locations should impose fewer ecological and socioeconomic burdens on residents. But this repre-

sents only a gross improvement in the distribution of costs and risks. The state-of-the-art in technology-site evaluation does little to determine the gains and losses for specific groups; about the best that can be done is to "make tradeoffs between costs and benefits so as to improve the equity of the decision without sacrificing unduly the interests of the majority of the affected parties" (Spangler, 1974:24). The identification of the "winners" and the "losers" in the selection of a particular site, through consideration of public values and attitudes, is still at a primitive stage. Standards weighing the importance of energy development to the "public interest" or to a particular interest group or individual have yet to be developed. But an evaluation of the distribution of costs is highly relative. The displacement of even 20 families in a sparsely populated area of the West may have dramatic and far-reaching economic and social consequences; technology-location evaluations remain quite insensitive to these kinds of effects of energy development.

Technology-site evaluation is not a highly flexible alternative. Evaluative criteria are based on social, political, technological, and financial conditions which change rapidly. Unless all areas are reassessed periodically to determine the appropriateness of their status, certain locations may be incorrectly included in or excluded from development. Sophisticated technology-site evaluations may require several iterations in which different (updated) and/or additional criteria are used.

The implementation of technology-location evaluations will have to facilitate federal-state relationships. Certain combinations of technology and location can be encouraged or discouraged through federal and/or state actions. Many federal laws, for example, already lead to the explicit or implicit exclusion of parks, monuments, wilderness lands, and wildlife areas as sites for new energy facilities.

Site-Screening and Site-Banking Techniques: Methods more flexible than the evaluations described above for determining the

Table 11-6: *Examples of Criteria Used in Site-Screening Process*

Issue	Consideration	Measure	Criteria for Inclusion
Health and safety	Flooding	Height above nearest water source	Area must be above primary floodplain
	Surface faulting	Distance from fault	Areas >5 miles from capable or unclassified faults >12 miles in length
	Radiation exposure	Distance from populated areas	Areas >3 miles from populated places >2500 Areas >1 mile from populated places <2500
Environmental effects	Thermal pollution	Average low flow	Rivers or reservoirs yielding 7 day average, 10 year frequency low flow >50 cfs
	Sensitive or protected environments	Location with respect to ecological areas	Areas outside designated protected ecological areas
Socioeconomic effects	Tourism and recreation	Location with respect to designated scenic and recreational areas	Areas outside designated scenic and recreational areas
System cost and reliability	Routine and emergency water supply and source characteristics	Cost of cooling water acquisition	Rivers or reservoirs yielding 7 day average, 10 year frequency low flow>50 cfs
		Cost of pumping water	Areas <10 miles from water supply Areas <800 feet above water supply
	Delivery of major plant components	Cost of providing access for major plant components	Areas within 25 miles of navigable waterways

> = greater than. < = less than. cfs = cubic feet per second.

Source: Keeney and Nair, 1977:224.

acceptability of sites would rely on site-screening to examine each individual configuration of technology and location. Thus a power plant using wet cooling might be screened out of a particular area although other energy technologies would be quite acceptable at the site. Table 11-6 outlines some examples of criteria which might be used in screening sites. After acceptable technology-location configurations have been agreed upon, a number of techniques are available for improving and acquiring sites. One such technique is "site banking," which would provide for state authority to approve sites (according to screen-

ing or some other device for determining acceptability) as much as ten years before they are needed. By joining a preapproved site with an acceptable technology, developers could bypass many of the lengthy individualized plant reviews now necessitated by today's "custom-designed" facilities (see Metz, 1977).

Site screening and banking may be implemented comprehensively by state or federal legislation or through more flexible site-selection and purchasing policies, but there is also the option of the "rule of reason" mentioned above.

Although the use of site screening or other

approaches could increase some costs of energy development by eliminating certain locations on noneconomic (chiefly environmental) grounds, an early determination of acceptable sites should reduce other costs by minimizing the likelihood that energy facilities will be proposed in areas where widespread public opposition exists. Thus, although some economically attractive locations will be excluded from consideration, if technology-site evaluation is implemented judiciously it could reduce the costs of siting new energy facilities by removing "troublesome" sites at an early stage, thus eliminating costly delays. For instance, early site screening might have ruled out the Kaiparowits site in southern Utah as an area for new energy facilities. Because the site has eight national parks within a 200-mile radius, a technology-site evaluation might have ruled such an environmentally sensitive region unacceptable for energy development. Since the sponsors had spent more than $20 million by the time the project was cancelled, the use of technology-site evaluation could have been an efficient option (Myhra, 1977:26).

Of course, the value of screening sites is highly dependent on the availability of adequate information concerning the criteria to be used. For instance, water availability, predominant land uses (intensive agriculture, specialty crops, etc.), known habitats of threatened species, population densities, and land ownership patterns are examples of phenomena for which there may be adequate data and which can be used as a basis for excluding certain impact-prone regions as sites for energy facilities. But broad-scale evaluations are no substitute for an in-depth analysis of the specific consequences which may arise from the interaction of a particular energy technology with its singular location. Unfortunately, much of the information needed to determine which combinations of technology and location are acceptable is lacking. Baseline data on many energy technologies (such as synthetic fuel facilities) are still unavailable, and it is not yet possible to determine reliably the level of residuals from commer-

cially operating facilities. In addition, models which predict the various effects of energy facilities, such as air pollution dispersion models, are subject to high levels of uncertainty.

The implementation of techniques such as site screening is hindered by opposition from several sectors. First, resistance can be expected to develop against any technology-site evaluation which runs counter to local and state interests and values. Thus, any site selection that is conducted by the federal government is particularly vulnerable to local concerns about "outside" studies. This opposition will likely focus on such factors as land speculation following the identification of acceptable areas. Second, technology-site evaluation may be difficult for utilities or other energy developers to accept, given the lack of precision involved in many of the cost-benefit data. Sponsors of energy projects may also object to the inclusion of noneconomic criteria in the evaluative process (Muntzing, 1976:6–7). Finally, there is a widespread perception among citizens and groups who participate in the siting process that technology-site evaluations are "administrative charades," characterized by "pseudoscientific cost-benefit jargon" (see Joskow, 1974:320–22).

In many instances, federal legislation and regulatory authority already implicitly screen sites for energy facilities. An example is the 1977 CAA amendments which exclude certain sites from consideration—international parks of any size, national parks over 6,000 acres, and national wilderness areas in excess of 5,000 acres have been designated Class I PSD areas. As discussed in Chapter 5, the West has many Class I PSD areas; and although the delineation of rigid clean-air buffer zones is not defined within the CAA amendments, implementation of the act will result in the development of implicit exclusion zones, surrounding Class I areas, for certain types of facilities. Permission would be denied for a site that, according to air-quality modeling, could be expected to violate Class I PSD increments in nearby Class I areas. This is one example of federal institutions already in place that could mandate or prohibit certain combinations of technology and location. And in some cases, this authority has been exercised. The difficulty with these policies is that direct use of such evaluations by the federal government has preempted some state efforts in selecting suitable areas. Officials in Utah, for example, have expressed dissatisfaction with the preemptive implications of the CAA amendments which, according to the state view, stymie efforts to develop local energy resources. Thus, when implemented at the federal level, technology-site evaluation may not reduce the lead time in bringing new facilities into operation because of state resistance.

Policies to Employ
Impact-Assistance Methods

The use of impact-assistance methods includes two specific policy alternatives:

- Require impact mitigation by developers, state or federal governments;
- Compensate state and local governments.

These options respond to the variety of consequences that result from siting energy technologies. Given the fact that "technological fixes" may have little effect on some aspects of energy development, particularly socioeconomic factors, policymakers may be faced with the choice of either attempting to moderate these consequences (mitigation) or making an effort to "buy off" the affected groups (compensation).

Requiring Impact Mitigation by Developers, State, or Federal Governments: Efforts by energy developers to mitigate adverse effects (such as addressing the socioeconomic impact) currently are voluntary in most states. Although energy developers have made some attempts to ameliorate the social and economic consequences of rapid population growth,

these voluntary actions tend to leave a number of factors outstanding; and many of these unmitigated effects are not susceptible to being solved by a quick infusion of federal or state funds.

As described earlier in this chapter, siting laws in several states do require certain conditions to be met before a permit is issued. These prior conditions can include lessening the social and economic impact (for example, by providing services or funding to communities) or constructing needed public facilities. The best example in the West of this type of response is the requirement that the WISA issue a permit only after studies of local community needs (Monaghan, 1977). Under this authority, developers have agreed to provide housing units and other services as part of the siting process for an energy facility.

Even without specific siting legislation, industry could be induced to take a greater role in mitigating the impact of developing energy resources. A tax strategy which allowed deductions for such efforts could encourage developers to accept more responsibility for the local impacts. Should such incentives prove ineffective, punative taxes could be levied on energy development to fund mitigation efforts.

As discussed in chapters 7 and 8, state or federal action requires adequate funding and appropriate mechanisms for distributing these funds to the municipalities which must supply the majority of public services. For example, programs could be designed to allow the state to advance funds to a community, thus providing assistance to prepare for and combat the predicted effects. Severance taxes, royalty payments, and license fees can all be structured to cover these costs. States in the West with severance taxes currently allocate only a relatively small portion of their revenues to such areas; this portion would have to be increased substantially to achieve effective results. In states without severance taxes, some other source of state revenue would have to be developed. For example, state revision of property taxes could provide funds for local governmental units where they are most needed;

property taxes on energy facilities could be collected at the state level (including prepayment of taxes by industry to provide funds when they are needed) and redistributed to affected communities.

The success of these efforts depends on the satisfactory resolution of intergovernmental jurisdictional disputes. However, the lack of coordination and bottlenecks which now characterize federal, state, and local financing and management of impact-assistance programs could cancel any gains achieved by linking the siting process to such programs. There is general agreement that melioratory efforts are best monitored and administered at the local level: first, localities are more sensitive to the costs, benefits, and risks of energy development for particular segments of the population; and second, "local government has a primary responsibility to see that the new growth and development in a community pays a fair share of the community support systems in a manner which will not place an unfair burden upon original local residents" (Monaghan, 1977:609). But most local governments are poorly equipped to monitor and manage the consequences of new energy facilities. Federal and state constraints, both legislative and regulatory, limit the capabilities of local communities to finance mitigation programs; and states have been reluctant to cede local jurisdictions sufficient authority (such as revenue-sharing powers) to respond. Moreover, states themselves have been slow to develop institutions that could assist in ameliorating the adverse consequences of siting decisions. Finally, federal programs often have been characterized by inadequate funding, an inability to anticipate or acknowledge adverse conditions, and overly complex and time-consuming procedures to apply for financial aid (Valeu, 1977:12).

It should be emphasized that reducing the burden on affected citizens can eliminate some of the costs and risks of siting energy facilities. But to be effective, mitigation strategies must provide states and localities: (1) anticipatory support before adverse effects arise; (2) rev-

enue-generating approaches applicable to energy facilities on federal lands (to which property taxes do not apply); (3) planning capabilities, including tools to collect and analyze information; and (4) risk-sharing options which provide for "bust" as well as "boom" scenarios (Heintz, 1977:4–6).

Adopting mitigation alternatives is a costly commitment. The federal government has established a $1.2 billion grant-and-loan fund solely as part of the Coastal Energy Impact Program. Impact assistance at a single site can require a substantial financial investment. For example, at the Colstrip power project in Montana, such assistance included a housing investment of about $2 million (White, 1977: 4–6). Given the relatively ineffective governmental apparatus discussed above, the administration of programs designed to alleviate the burdens of energy development may be inefficient. But evaluating the administrative efficiency of impact-assistance efforts is too narrow a focus—the true test is whether the cost of making these investments is outweighed by the reduced risks of opposition to energy projects.

Impact-assistance strategies should be designed to reduce the local inequities which flow from the siting of new energy facilities. Otherwise, some citizens will be required to shoulder an unequal share of the burden of providing energy to the general population. The crucial criterion for employing impact-assistance methods, therefore, is the degree to which the costs and risks are distributed in an equitable manner among the affected parties. Overall, mitigation efforts provide an equitable solution to the siting dilemma. Impact aid is perhaps the only policy alternative which can decisively and equitably site major facilities despite the local costs that can be expected to result. But some serious caveats are in order regarding the equity of mitigation measures. First, equity does not imply equality —no impact-assistance strategy will be able to reduce the costs and risks of development without some participants being overprotected or overexposed to risks. Second, even ap-

proaching an equitable diffusion of the negative effects of facility siting requires an extremely sensitive set of political and administrative institutions. The ultimate success or
failure of mitigation schemes, which require
substantial lead time, rests on being able to
identify secondary and future costs, such as
rising insurance premiums or declining land
values, as well as those citizens who will be
affected (Heintz, 1977:6-8).

**Compensating State and Local Govern-
ments for Adverse Effects:** Compensation
does not attempt to reduce or eliminate either
the consequences of energy development or
their causes. Rather, it offers payments to
those who will be adversely affected. Compensation can counter local opposition to energy development by determining what price
people are willing to accept to offset the adverse effects. This alternative can address a
single problem, through measures such as a
pollution tax that provides compensation for
damages to air or water quality, or multiple
problems, through bidding procedures. Thus,
a community would establish the acceptable
level of compensation required of a developer
to site an energy facility within its boundaries.
The bid could include a range of mitigation
costs based on estimates of local hardships.
If several communities bid, the developer
could then include compensation costs in the
budget for selecting the best site. After compensation had been agreed upon, development
could take place without opposition from the
compensated groups (O'Hare, 1977).

The major constraints on compensation
policies are the difficulties involved in identifying the affected citizens and the problems
of administering these complex procedures beyond the local level. Identifying those groups
burdened by energy development is a difficult
task at best, but for national or regional consequences, such as widespread air or water
pollution, the process of determining who
would be affected may be impossible. Even
if such a determination could be made, redis-
tributing federal or state funds to localities

or chanelling payments from the developer to
affected groups is a complex procedure that
may only serve to subsidize the least desirable
combinations of technology and location.

Like mitigation measures, compensation alternatives are efficient to the degree that they
reduce the risks of opposition to energy facilities. Thus, in the case of making compensation payments to affected citizens, "such payments may not be worth their administrative
costs on [the] ground of fairness alone, but if
their omission means that a valuable project
is cancelled entirely for want of a community
willing to accept it, a strong efficiency argument is applicable" (O'Hare, 1977:429). If
local communities could be reimbursed for the
identifiable socioeconomic and ecological side-
effects of energy development, many of the
production costs resulting from project delays
or cancellations could be decreased. And, impact assistance could reduce the time and
money spent in the litigation procedures
which today are still the major compensatory
avenues for those affected by siting decisions
(see Horowitz, 1977).

As the federal government's response to
the consequences of energy development has
evolved toward the use of multiple impact-
assistance tools and as state and local governments have come to appreciate the diversity
of costs and risks associated with siting energy
facilities, compensation alternatives have become more feasible. An example of how compensation could be made more adaptable to
changing conditions and times would be by
forming an "impact team." As used in Sweetwater County, Wyoming, and Moffat County,
Colorado, these teams of officials from local,
state, and federal governments and private industry have been assembled to monitor energy
development and its impact. Recent experience indicates that "the functioning of such
impact teams does a great deal to head off
conflict and reduce citizen antagonims over
resource development" (Monaghan, 1977).

Nevertheless, several obstacles stand in the
way of implementing compensation programs.
First, no matter how detailed the underlying

planning process, compensation measures are notorious for attracting "free riders" who, in order to share in the benefits of impact assistance, "join in the clamor of opposition to a project, making it appear that the numbers of people to be compensated in order to secure timely energy development are quite large" (Heintz, 1977:6–7). Second, compensation does little to resolve the problem of opposition to siting from interests outside the local community. Although there is at least the possibility that compensation measures might deprive "outside" environmentalists of allies within the affected community, compensation schemes are not likely to overcome resistance to energy development by organized environmental advocates.

In addition, the current structure for providing federal impact-assistance is ill-equipped to provide compensatory aid as a siting strategy. The federal government has a number of ongoing impact-assistance programs, including sharing profits from royalties or other development revenues with the states, or can target assistance to the specific problem itself (as in the 1976 Coastal Zone Management Act amendments). Also, some existing energy programs have been modified to provide impact assistance. However, the federal government basically is limited to a complementary role of providing back-up loans, loan guarantees, or grants. This elaborate and essential support system is highly fragmented; no single federal agency has been designated as the triggering mechanism. As a result, at least four cabinet-level departments (Commerce, Energy, Interior, and Housing and Urban Development) and various other regulatory and coordinating agencies have major responsibilities in the area of energy-related impact assistance. This administrative compartmentalization restricts the planning process that must precede impact-assistance programs; and it can restrict agency decisionmaking by leading to bureaucratic stalling or "buck-passing" in the face of citizen opposition to a siting proposal.

An example of the problems confronting effective impact assistance in the West is the difficulty encountered with the 1976 Federal Land Policy and Management Act. Under its terms, state and local governments were allowed to borrow against projected mineral-lease revenues to respond to the socioeconomic consequences of energy development. However, a number of western states, including Colorado, Utah, and Wyoming, have constitutional provisions that prohibit their participation in loan guarantee programs that present the possibility of fiscal overextension by using future revenues before they are actually generated. Conservative restrictions of this type are understandable, given the fact that in many cases, federal support may actually increase the uncertainties and risks for states and localities of siting energy facilities (Monaghan, 1977).

Policies to Increase Citizen Involvement in Site Selection

As outlined in Table 11-4, three specific alternatives for increasing public involvement in siting decisions will be considered. They are to:

- Provide for the exchange of information between citizens and agencies;
- Encourage administrative interaction between agencies and citizens; and
- Allow direct citizen input into agency decisions.

These specific alternatives were chosen for analysis because they incorporate three principles underlying effective public participation in developing energy resources: first, public involvement in site selection is facilitated by a two-way flow of relevant data; second, the resolution of administrative and regulatory conflicts in the siting process can be achieved through the use of both citizen-initiated and citizen-reaction techniques; and third, full public participation involves citizens actually sharing decisionmaking power (Bishop, McKee and Hansen, 1978:8–10).

Providing for Information Exchange between Citizens and Agencies: At present, the exchange of credible and reliable information about proposed energy facilities between industry, government agencies, and potentially affected parties is generally lacking. If confusion and conflict are to be avoided, industry must be able to communicate to the public the basic facts and information regarding siting alternatives and consequences. At the same time, efforts should be made to discover public values and needs related to the specific configuration of technology and location (see Frauenglass, 1971). Thus, a successful exchange of information must include both data dissemination and collection.

Information could be exchanged by such techniques as participation audits or by improving the notification procedures, settings, and information for public hearings. A participation audit is designed to determine whether a developer has made a reasonable attempt to involve affected citizens and relevant agencies in the siting process. Prior to issuing a siting permit, the permitting agency would require submission of a participation statement by the developer. This document would outline the nature and level of potential adverse consequences, identify those citizens to be affected, and detail the degree to which all concerned parties had been given an opportunity to become involved in preliminary plans and decisions. A site permit would not be granted until the developer could demonstrate that consideration had been given to the interests of those who would be affected by the project. By requiring such an audit of public involvement in the earliest stages of site selection, a dialogue could be created between citizens and agencies. But, conducting participation audits is constrained by the problems involved in determining the impact criteria and identifying affected or duly concerned citizens. Identification is particularly difficult in the preliminary phase of site selection when there have been few formal actions and where citizen interest tends to be low.

Public hearings are the most common mechanisms in use today for exchanging information about sites for energy facilities. Largely as a result of environmental protection legislation such as NEPA, hearings are now required for the licensing of almost every energy facility. Not only must requirements for hearings on an EIS be met, but a host of other construction and operation rules now require public hearings as standard practice. For example, Table 11-7 outlines the legislative and administrative regulations concerning public participation in energy decisions in the two western states, Colorado and Montana, with the most comprehensive procedures for hearings.

Unfortunately, the proliferation of ritualistic public hearings has become one of the major obstacles to the development of a workable siting policy for energy facilities. Neither the government nor the public is well served by a procedure for which there is often no citizen demand or which results in a one-way, agency-dominated communication process (Heberlein, 1976). There are several reasons for the failure of hearings to encourage public participation: inadequate notification, overly formal and highly structured settings, and excessive emphasis on highly technical data. Improving the staging of hearings may therefore require reforming notification methods to better identify constituencies and to use more innovative media techniques, modifying the setting in which hearings take place to include smaller, more informal groups, and describing alternatives and consequences in language the layman can understand.

Exchanges of information between the public and siting agencies can reduce the delays involved in data searches and could provide greater bases for building a consensus among citizens' groups. And this method of increasing public participation can be moderately efficient because administrative costs would be low—and in some cases, would be borne by participating agencies or groups. An early consensus on site choices would decrease the possibility of lengthy delays and thus would reduce the costs of producing energy. However,

Table 11-7: *Public Participation Regulations in Colorado and Montana*

State	Public Participation Regulations
Colorado	Sunshine Law: requires opening all meetings of public officials ("any board, committee, commission, or other policy-making or rule-making body of any state agency or authority or of the legislature at which any public business is discussed"). Public notice of meeting is required; minutes must be taken and made available to public. (C.R.S. 24-6-401, 402)
	Rule-making hearings: opportunity for public to submit views to rule-making body. (C.R.S. 24-4-103)
	Agency shall notify public of proposed rule change. (C.R.S. 24-4-103)
	Availability to public of rules and changes in rules; the agency shall supply copies upon request.
	Licensing hearings are nearly identical with NRC procedures: notification of hearings to public; holding of hearings at convenient time for parties and public; participation of parties (full rights) and limited participation of members of public; people who feel "affected or aggrieved" may petition to be a party to the hearing, and Commission decides legitimacy of interests; transcripts of hearings are made and available to the public. (C.R.S. 24-4-105; General procedures, C.R.S. 40-6-109)
	For utilities' policy- and rule-making; notice of all proceedings, applications, petitions, and orders instituting investigations or inquiries shall be given to all persons, firms or corporations who, in the opinion of the Commission, are interested in, or who would be affected by, the granting or denial of any such application, petition, or other proceeding. (C.R.S. 40-6-108)
Montana	Publication and distribution of booklet entitled "A Citizen's Guide to Participation in Government," compiled by the Montana Energy Advisory Council. This booklet describes administrative procedures in layman's terms and specifically identifies points in process where public participation occurs (elections, petitions, public notices, hearings, protests). It also lists and describes all advisory committees.
	Public has the right to examine public documents. (Constitution of Montana, Article II-9)
	Public has the right to observe deliberations of all public bodies or agencies of the state. (Constitution of Montana, Article II-8)
	Notice must be given for all proposed rules or rule changes and for agency actions with significant impact on public, such as siting a utility. (R.C.M. 82-4209)
	An applicant to construct an energy facility must prove that notice of intent has been given to persons residing in the affected municipalities. The notice, involving a summary of the application, must be published on or about the day the application is filed and must appear "in those newspapers as will serve substantially to inform those persons of the application." (R.C.M. 70-806-4)
	Parties to a hearing for certification of environmental compatibility and public need include: "any person residing in a municipality entitled to receive service of a copy of the application under subsection R.C.M. 70-806-4; any nonprofit organization, formed in whole or in part to promote conservation of natural beauty, to protect the environment, personal health or other biological values, to preserve historical sites, to promote consumer interests, to represent commercial and industrial groups, or to promote the orderly development of the areas in which the facility is to be located; or any other interested person." (R.C.M. 70-808-1-c)
	Each utility, and each person contemplating the construction of a facility in the ensuing 10 years must annually submit a long-range plan for the construction and operation of facilities. "The plan shall be made available to the public by the department.... Citizen environmental protection and resource planning groups, and other interested persons may obtain a copy of the plan by written request and payment to the department." (R.C.M. 70-814)

C.R.S. = Colorado Revised Statutes. NRC = Nuclear Regulatory Commission. R.C.M. = Revised Codes of Montana.

Source: Nelkin and Fallows, 1976.

the need to satisfy a wide range of interested parties could also result in siting choices that are more expensive than would otherwise be the case. Moreover, low administrative costs might well be offset by the cost of data exchanges. Whether data are purchased by the government, generated by agencies, or provided by the participants themselves, they are expensive to collect, monitor, and assess (see Ingram, 1973). It is also possible that exchanging data will lead to an "information overload" for both the agencies and the public involved, since the information which can be used efficiently by any decisionmaker is limited both in quality and in quantity (Carver, 1976: 280). This is particularly the case for legal or highly technical data of the type generated through public hearings on site selection (see Heberlein, 1976). Interest groups with little experience in participating in scientific and technological decisions may be overwhelmed even by procedures designed to make hearings meaningful.

However, used in conjunction with support techniques (discussed below), the exchange of information offers an equitable way to increase citizen involvement. Equal access to data in public hearings, for example, can remove many of the participatory advantages enjoyed by well-represented groups which may use information strategically to transfer the risks of energy development to poorly represented groups (see Frauenglass, 1971:495–96).

The exchange of information can be implemented more easily than most other options. Demands for public involvement in siting decisions for energy facilities have made information exchanges almost an automatic governmental response. Hearings are found at all levels of government, with functions ranging from general data collecting through providing scientific and technical expertise, to conducting special studies for particular groups (Brown, 1972).

Encouraging Administrative Interaction between Agencies and Citizens: Conflicts

over sites for energy facilities could be reduced by encouraging administrative interaction between citizens and agencies in the form of an advisory committee with broad community representation that would consider each proposed site before a permit was sought. Committees composed of industry and environmental groups, local government officials, and residents could provide suggestions on siting problems and issues, act as a sounding board for community attitudes toward agency policies, and disseminate siting information to their constituencies (see Bishop, McKee, and Hansen, 1978:59–62). But for any significant contribution to public participation in siting decisions, the advisory committees must be included in the earliest evaluations of sites and technologies. If such participation and advice were secured prior to applications for permits, many developers might avoid alienating concerned groups and could enlist local support to help obtain a more timely siting decision (see O'Riordan, 1976).

The effectiveness of these devices to achieve interaction is limited by their inability to do more than act as "trial balloons"; such committees have no binding decisionmaking authority. While they do establish a major channel of communication between professional administrators and the public, they possess no real standing or sanctions by which to ensure representation, despite their widespread acceptance by government.

Two fundamental problems plague advisory committees. First, some interest groups have complained that limited resources inhibit their ability to participate actively through these avenues. (This deficiency has led to much of the demand for financial and technical support; these options are discussed below.) Second, even if the groups were able to make their interests known through such policy forums, there is absolutely nothing which guarantees that recommendations will be heeded (Rycroft, 1979; see also Doerksen and Pierce, 1975).

Allowing Direct Citizen Input into Agency

Decisions: The third alternative for increasing public involvement in site selections, direct input into the siting process, enables citizens to share in the actual decisions about technology and location. A variety of strategies may be used, but the two options with the most direct impact on the final decision involve designating citizen review boards for each siting agency and guaranteeing citizens a legal standing in the courts after other participatory alternatives have been pursued. In contrast to advisory committees and other plans to encourage administrative interaction, citizen review boards can be given full authority to make siting decisions. At present, they include representatives of the broad spectrum of interested parties and have authority to review agency plans or policies. The creation of open and credible review boards to help site energy facilities could reduce the level and intensity of conflicts during the early stages of site selection and impact evaluation. And the realization that public participants will have a role in determining the final outcome of the site itself could serve as a safety valve, reducing conflict and helping to shorten the lead time required before the actual construction of a new facility begins. Moreover, establishing review boards within the regulatory agencies could serve to legitimize the agencies' roles and help build public trust in their programs (see Arnstein, 1972).

Lawsuits have been one of the primary avenues of citizen participation in energy policymaking for some time. This has especially been the case for environmental interest groups, who have relied on the courts as a "sort of secondary fall-back mechanism" when the legislative and administrative processes have failed to provide adequate access to energy decisions (Garvey, 1975). But court challenges by citizens' groups may also contribute to delays in the licensing of facilities, in part because adjudication often becomes part of the politics of desperation for citizens who are opposed to the location of an energy project. However, the practice of using litigation as the last line of resistance can be reduced by

providing meaningful alternatives such as those discussed above. In particular, better performance by the federal and state regulatory agencies with regard to public participation could contribute to less reliance on the judicial system for resolving siting disputes. Lawsuits could still be a viable coercive tool and a supplement to the "volunteerism" of the other participatory techniques if their availability were guaranteed by well-defined agency procedures. One such guarantee might involve the assurance of judicial recourse after other participatory options had been pursued. This strategy might resolve many siting difficulties before adjudication became necessary. And the ultimate threat of legal action by citizens' groups could be maintained to give the individuals and communities directly affected additional bargaining power in the location and operation of energy facilities (Wolpert, 1976).

Allowing citizens direct access to siting decisions is a relatively effective means of reducing some of the costs associated with energy development. Much more accurate and meaningful signals of public preferences can be sent to policymakers in less time through the use of citizen review boards or citizens' lawsuits than with other participatory options. But precisely because review boards involve sharing power there is a high degree of entrenched opposition on the part of those in government (see Arnstein, 1972).

The costs of litigation have become one of the major obstacles to the establishment of a rational national siting policy for energy facilities. Lawsuits involve securing and paying for scientific and technical expertise, and it is a lengthy process to carry legal actions through to their conclusion. Moreover, judicial decisionmaking may be uneven, since cases are not chosen or decided on the basis of establishing a systematic approach to the problem of siting energy facilities. In addition, lawsuits are decided in adversary proceedings, featuring examination in a confrontation atmosphere —a process not noted for its efficient resolution of issues (Sive, 1971). On the other hand, litigation continues to be a relatively easy strategy to implement. The role of the courts in siting facilities has expanded rapidly as the judiciary has moved from a focus on the protection of the rights of affected parties to a more comprehensive interest in the problems of direct public participation and the more general public-interest issues of locating energy facilities (see Wolpert, 1976; and Horowitz, 1977).

Policies to Support Citizen Participation

In this study we have considered a single alternative for supporting public participation: providing technical or financial assistance for citizens involved in the siting process.

This support attempts to provide interested parties with access to the best technical capabilities for making decisions about siting energy facilities. To accomplish this, interest groups can be supplied with technical expertise and/or funding. Both options are applicable to the full range of participatory alternatives discussed above.

Many interest groups find it difficult to enter into technological debates with federal or state agencies because technical details are often inaccessible to the public. And even if technical data are available, they are usually expressed in too complex a fashion to be useful to the typical citizens' group. Because technological expertise is increasingly a requisite for effective involvement in energy policy, fees for attorneys and expert witnesses, along with other reasonable costs of developing a technically competent response, could be awarded to citizens' groups. The Carter administration has already supported legislation which would provide federal funding to qualified citizens' groups acting as intervenors in hearings on siting nuclear facilities (Metz, 1977:590). A strategy for providing technical and financial support could expand such assistance to include all federal regulatory processes for all types of energy facilities and cover the range of participatory programs (see Davis, 1976).

Support for citizen involvement in the siting process is difficult to assess by itself as a means to reduce costs, risks, and delays because technical and/or financial assistance is intended to be used in conjunction with other policy alternatives. Thus, citizens' groups desiring access to the siting process through the avenue of public hearings could be provided with intervenor funding or technical staffing (see Davis, 1976). Support for public participation, therefore, may make the other alternatives more efficient and effective, depending on the circumstances in which they are used. However, some direct consequences of this kind of support can be identified. Providing technical expertise to participants may not be a very time-conserving or conflict-reducing option, for example, if policy positions are already well established. Case studies of siting debates have shown that injecting technical expertise into the process often is an incentive for greater citizen involvement; it may also serve to legitimize, on technical grounds, these entrenched positions. But these same studies have demonstrated that access to technical expertise provides a certain equity in the decisionmaking process. That is, "equal access to knowledge, and the resulting ability to question data used to legitimize decisions, may at least permit political forces to operate on a more equitable basis" (Nelkin, 1974:36; see also Gerlach, 1978).

The barriers to implementing such support strategies are also well known: the issue of direct financial aid to citizens particularly threatens established government rules and routines. Despite widespread theoretical acceptance of the principle of citizen involvement, recognition of the substantial costs of participation has resulted in Congress not outlining any federal position on the subject. Thus, the issue of financial aid has been left to the discretion of individual regulatory agencies. Caught in a web of legal and procedural conflicts regarding its authority and the desirability of using appropriations to fund public participation, as well as the problems of determining which interests require financial support, each agency responds in a different fashion. The most determined opposition to financial aid has come from agencies with highly complex and technological missions, such as the Nuclear Regulatory Commission (NRC) and the FPC (Paglin and Shor, 1977: 147). Both these bodies have refused to fund intervenors in adjudicatory or rule-making procedures. Thus, while congressional debates continue on proposals for a more comprehensive and liberal policy of reimbursement, there remains significant resistance to providing financial assistance to citizens' groups involved in the siting of energy facilities.

Fortunately, there are other strategies to support public participation that are more flexible than providing financial aid. Many agencies have acted to provide "facilitative assistance." These measures include: liberalizing the rules for public admission to proceedings, relaxing some financial and procedural standards, providing access to internal data, and supplying technical staff assistance or information services. Other agencies have implemented an "in-house" public counsel or office of citizen advocacy (Paglin and Shor, 1977:146–47). Taken together, these alternatives provide a good foundation for increasing public representation in siting decisions.

Summary and Comparison of Alternatives

Siting energy facilities presents several difficulties and contradictions. There is a tremendous need for better information about the trade-offs of technology-site combinations, yet such information can be used to delay development regardless of the relative advantages of a particular project. Delay and uncertainty have become increasingly unmanageable problems in siting, yet the pressures to increase public involvement in the process continue. And although many of the consequences of development can be mitigated, existing institutions often fail to provide assistance soon enough or to distribute it broadly enough.

The alternatives we have considered in this chapter will not, in themselves, be adequate to resolve these difficulties. However, they can help reduce the impact of siting decisions and improve the nature of the process. The conclusions for each of these alternatives are summarized in Table 11-8.

The most comprehensive choice is technology-site evaluation, which can be used to improve the information available to all participants. If this evaluation is successful in generating credible information—and if participants accept the fact that at least some negative consequences will emerge from *any* combination of the technologies and locations considered here—then some of the most serious consequences may be avoided.

The siting process can also be improved by providing impact assistance and by enhancing citizen involvement. Both of these choices can help reduce opposition to a particular development. There are two critical questions in this regard: whether the interested parties perceive the process as open and fair in allowing diverse opinions to be presented; and whether the process is successful in guaranteeing assistance to areas before large-scale effects are felt. If the process is not perceived as open and fair, citizen participation will almost certainly mean delays, and uncertainties about development will increase. This seems likely regardless of either the quality of information generated or the quantity of impact assistance provided.

Each of the alternatives that has been identified in this chapter is feasible—the knowledge exists to identify affected areas, and a variety of mechanisms exists to reduce the barriers to public participation. In addition, the alternatives are flexible in the sense that they can be applied to a variety of conditions found all across the eight-state study area. However, the general weakness of all of these choices is that each will require the successful implementation of either new or improved institutional mechanisms. This will present the biggest barrier to adopting technology-site evaluations because such a variety of infor-

Table 11-8: *Summary and Comparison of Energy-Facility Siting Alternatives*

Criteria	Technology-Site Evaluations	Impact-Assistance Methods	Citizen Involvement	Support for Public Participation
Effectiveness: Achievement of Policy Objective	Comprehensive evaluations can be used either to select the most appropriate sites or to eliminate sites, since negative consequences will be identified for most areas.	Impact assistance will help reduce opposition to some sites and expedite the siting process, but it must be provided with sufficient lead time to facilitate planning.	If the process of public participation is perceived as fair and equitable it can reduce delays in site selection.	Support for participation will increase interaction, but it may also make compromise more difficult by providing expertise to diverse and antagonistic groups.
Efficiency: Costs, Risks, and Benefits	If the least expensive locations are eliminated, overall development costs will increase. If the most controversial sites are eliminated, the risks of long-term public opposition can be reduced.	Administrative costs of impact aid could be high. However, the risks and uncertainties — and thus, the costs — associated with citizen resistance can be reduced.	Options for direct citizen involvement are relatively inexpensive, but data exchange and litigation can be costly.	Support options increase the efficiency of public participation in such areas as decision making.
Equity: Distribution of Costs, Risks, and Benefits	Comprehensive evaluations can help to determine accurately the winners and losers in siting decisions.	Impact assistance is an equitable choice if it is targeted to areas with major effects and if it is provided well in advance of the beginning of construction.	Public participation can help balance the risks of development, especially if "process support" (such as technical or financial assistance) is provided to interest groups.	Providing support to participants is equitable because it allows diverse groups to participate more equally in decision making.
Flexibility: Adaptability to Local Needs	Technology-site evaluations require frequent updating and modifications which are relatively expensive	Impact assistance can be applied flexibly, but it will require an on-going monitoring system at the state level.	Because participation can be achieved in a variety of ways (information exchange, interaction, voting, etc.), it is applicable to a wide range of situations.	Support options are applicable to many kinds of citizen involvement, but they can be very difficult to administer.
Implementability: Institutional Constraints and Acceptability	Acceptance of this mechanism by various participants will largely depend on perceptions of the credibility of information. Thus, evaluations will need to be done by independent, reputable groups.	Some groups will be opposed to development regardless of the degree either of assistance or mitigation.	Although citizen participation is generally an accepted part of the decision making process, it can be very difficult to administer.	Support options generally threaten those interests that are well established and have a stake in the status quo.

mation will be required. If technology-site evaluations are done at the state level, either new coordinating agencies or a reorganization of existing agencies will be required. Improving public participation is also likely to require new institutional processes. A critical question in this regard is representativeness—that is, establishing decisionmaking processes which reflect the diverse values of the broad range of interests typically affected by energy development.

Conclusion

Site selection is one of the most critical components of energy policymaking since it is at this point that most of the important questions and issues of resource development come together. However, there is little to suggest that the current siting system can convert these diverse values into reasonable policies. In many respects, the current system of site selection represents fragmented, incremental decisionmaking *ad absurdum.*

The value of technology-site evaluations is that they can provide something the present siting system largely lacks: a comprehensive consideration of the relative advantages and disadvantages of various siting strategies. While these evaluations can never remove all of the uncertainties associated with energy development, they can significantly improve the information upon which siting decisions are made. In particular, technology-site evaluations provide a means to reduce the risks of energy development by removing from consideration those areas where public opposition is most widespread.

While the effectiveness of technology-site evaluations partially depends on the quality and timeliness of the information produced, genuine success will also require more fundamental changes in the system. The most important of these appear to be providing equal access to decisionmaking for interested parties *in combination with* imposing some limits on how long and in what manner siting decisions can be contested after review processes are over. This does not imply that our knowledge of energy development technologies and locational factors is advanced enough to accurately predict the full consequences of development. It does suggest, however, that if the issues are given a fair hearing initially, then strict limits should be placed on how the decisions can be appealed. This approach to siting will probably lengthen, rather than shorten, the early stages of the process in order for various interests to be expressed. However, it could shorten the total time required to bring an energy facility into operation and could also reduce some of the uncertainty associated with its siting.

However, perhaps not even these revisions will improve the siting process. In the end, "reasonable" determinations of where to locate energy development facilities may depend on the willingness of interested parties to recognize the legitimacy of other values. If participants are not able to reach compromises, then we must either accept the uncertainties, inequities, and, at times, the paralysis of the existing system or adopt more centralized institutions.

REFERENCES

Arnstein, Sherry R. 1972. "Maximum Feasible Manipulation." *Public Administration Review* 32 (July/August):377–90.

Aron, Joan B. 1975. "Decisionmaking in Energy Supply at the Metropolitan Level: A Study of the New York Area." *Public Administration Review* 35 (July/August):340–45.

Baram, Michael S. 1976. *Environmental Law*

and the Siting of Facilities: Issues in Land Use and Coastal Zone Management. Cambridge, Mass.: Ballinger.

Beatty, Haradon. 1973. "Federal Water Pollution Control in Transition." In *Rocky Mountain Mineral Law Institute: Proceedings of the 18th Annual Institute,* pp. 493–530. New York: Matthew Bender.

Bishop, A. Bruce, Mac McKee, and Roger D. Hansen. 1978. *Public Consultation in Public Policy Information: A State-of-the-Art Report,* for Energy Research and Development Administration. N.p.: Intermountain Consultants and Planners, Inc.

Brown, David S. 1972. "The Management of Advisory Committees: An Assignment for the '70s." *Public Administration Review* 32 (July/August):334–42.

Business Week. 1977. "U.S. Indians Demand a Better Energy Deal." December 19, p. 53.

Calvert, J. R. 1978. "Licensing Coal-Fired Power Plants." *Power Engineering* 82 (January): 34–42.

Carter, Luther J. 1977. "Coal: Invoking 'the Rule of Reason' in an Energy-Environment Conflict." *Science* 198 (October 21):276–79.

Carter, Luther J. 1978a. "Virginia Refinery Battle: Another Dilemma in Energy Facility Siting." *Science* 199 (February 10):668–71.

Carter, Luther J. 1978b. "The Attorney General and the Snail Darter." *Science* 200 (May 12): 628.

Carver, John A. 1976. "Energy, Information, and Public Policy: Reflections of a Former Policy Maker." *American Behavioral Scientist* 19 (February):279–85.

Cavanaugh, H. A. 1977. "Utility Decisions: What Voice for the Public?" *Electrical World* 188 (December 15):89–92.

Cirillo, Richard R., et al. 1976. *An Evaluation of Regional Trends in Power Plant Siting and Energy Transportation.* Argonne, Ill.: Argonne National Laboratory.

Clean Air Act Amendments of 1970, Pub. L. 91-604, 84 Stat. 1676.

Clean Air Act Amendments of 1977, Pub. L. 95-95, 91 Stat. 685.

Coal Mining and Processing. 1978. "Navajos Reject $1-Billion Coal Gasification Plant." 15 (May):20–21.

Coastal Zone Management Act Amendments of 1976, Pub. L. 94-370, 90 Stat. 1013.

Council on Environmental Quality (CEQ). 1976. *Environmental Quality Seventh Annual Report.* Washington, D.C.: Government Printing Office.

Crittenden, Ann. 1978. "Coal: The Last Chance for the Crow." *New York Times,* January 8, sec. 3, p. 1.

Davis, Thomas P. 1976. "Citizen's Guide to Intervention in Nuclear Power Plant Siting: A Blueprint for Alice in Nuclear Wonderland." *Environmental Law* 6 (Spring):621–74.

Doerksen, Harvey R., and John C. Pierce. 1975. "Citizen Influence in Water Policy Decisions: Context, Constraints, and Alternatives." *Water Resources Bulletin* 11 (October):953–64.

Endangered Species Conservation Act of 1969, Pub. L. 91-135, 83 Stat. 275.

Endangered Species Preservation Act of 1973, Pub. L. 93-205, 87 Stat. 884.

Federal Power Act of 1935, Title II of Public Utility Act of 1935, Pub. L. 74-333, 49 Stat. 838.

Federal Land Policy and Management Act of 1976, Pub. L. 94-579, 90 Stat. 2743.

Federal Water Pollution Control Act Amendments of 1972, Pub. L. 92-500, 86 Stat. 816.

Fradkin, P. L. 1977. "Craig, Colorado: Population Unknown, Elevation 6,185 feet." *Audubon* 79 (July):118–27.

Frauenglass, Harvey. 1971. "Environmental Policy: Public Participation and the Open Information System." *Natural Resources Journal* 11 (July):478–96.

Garvey, Gerald. 1975. "Environmentalism Versus Energy Development: The Constitutional Background to Environmental Administration." *Public Administration Review* 35 (July/August):328–33.

Gerlach, Luther P. 1978. "The Great Energy Standoff." *Natural History* 87 (January):22–37.

Greider, William. 1977. "Indians Organize Own Energy Combine." *Washington Post,* July 17, p. A–3.

Hall, Timothy A., Irvin L. White, and Steven C. Ballard. 1978. "Western States and National Energy Policy: The New States' Rights." *American Behavioral Scientist* 22 (December):191–212.

Heberlein, Thomas A. 1976. "Some Observations on Alternative Mechanisms for Public Involvement: The Hearing, Public Opinion

Poll, the Workshop, and the Quasi-Experiment." *Natural Resources Journal* 16 (January):197–212.

Heintz, H. Theodore, Jr. 1977. "The Government Role in Mitigating Impacts of Energy Development." Paper presented at the Symposium on State-of-the-Art Survey of Socioeconomic Impacts Associated with Construction/Operation of Energy Facilities, St. Louis, Missouri, January.

Historic Preservation Act of 1966, Pub. L. 89-665, 80 Stat. 915.

Holden, Constance. 1977. "Contract Archeology: New Source of Support Brings New Problems." *Science* 196 (June 3):1070–72.

Horowitz, Donald L. 1977. "The Courts as Guardians of the Public Interest." *Public Administration Review* 37 (March/April): 148–54.

Ingram, Helen. 1973. "Information Channels and Environmental Decision Making." *Natural Resources Journal* 13 (January): 150–69.

Johnson, Haynes. 1975. "The Last Round-Up." *Washington Post*, August 3, p. C-5.

Joskow, Paul L. 1974. "Approving Nuclear Power Plants: Scientific Decisionmaking or Administrative Charade?" *Bell Journal of Economics and Management Science* 5 (Spring): 320–32.

Keeney, Ralph L., and Keshavan Nair. 1977. "Nuclear Siting Using Decision Analysis." *Energy Policy* 5 (September):223–31.

Kirschten, J. Dicken. 1977a. "The Clean Air Conference—Something for Everybody." *National Journal* 9 (August 13):1261–63.

Kirschten, J. Dicken. 1977b. "Watch Out! The Great Coal Rush Has Started." *National Journal* 9 (October 29):1683–85.

Lamm, Richard D. 1976. "States Rights vs. National Energy Needs." *Natural Resources Lawyer* 9 (No. 1):41–47.

McFarland, Andrew S. 1976. *Public Interest Lobbies: Decision Making on Energy.* Washington, D.C.: American Enterprise Institute for Public Policy Research.

Metz, William D. 1977. "Nuclear Licensing: Promised Reform Miffs All Sides of Nuclear Debate." *Science* 198 (November):590–93.

Mills, Jon L., et al. 1975. *Energy: The Power of the States.* Gainesville, Fla.: University of Florida, Holland Law Center, Center for Governmental Responsibility.

Monaghan, James E. 1977. "Managing the Impacts of Energy Development: A Policy Analysis from a State Government Perspective." In *Project Interdependence: U.S. and World Energy Outlook Through 1990.* U.S. Congress, House Committee on Interstate and Foreign Commerce, Subcommittee on Energy and Power; and Senate Committee on Energy and Natural Resources and the National Ocean Policy Study of the Committee on Commerce, pp. 608–29. Washington, D.C.: Government Printing Office.

Montana Major Facility Siting Act of 1975, Montana Revised Codes Annotated §§70–801 through 70–823 (Cumulative Supplement 1975).

Muntzing, L. Manning. 1976. "Siting and Environment: Towards an Effective Nuclear Siting Policy." *Energy Policy* 4 (March): 3–11.

Murray, Francis X., ed., 1978. *Where We Agree: Report of the National Coal Policy Project,* 2 vols. Boulder, Colo.: Westview Press.

Myhra, David. 1977. "Fossil Projects Need Siting Help Too." *Public Utilities Fortnightly* 99 (September 29):24–28.

National Environmental Policy Act of 1969, Pub L. 91-190, 83 Stat. 852.

Nelkin, Dorothy. 1974. "The Role of Experts in a Nuclear Siting Controversy." *Bulletin of the Atomic Scientists* 30 (November):29–36.

Nelkin, Dorothy. 1977. *Technological Decisions and Democracy: European Experiments in Public Participation.* Beverly Hills, Calif.: Sage Publications.

Nelkin, Dorothy, and Susan Fallows. 1976. *The Politics of Participation in Energy Policy.* Ithaca, N.Y.: Cornell University, Program on Science, Technology and Society and Department of City and Regional Planning, Appendix 2.

North Dakota Energy Conversion and Transmission Facility Siting Act, North Dakota Century Code, Chapter 49-22-10.

O'Hare, Michael. 1977. "'Not on My Block You Don't': Facility Siting and the Strategic Importance of Compensation." *Public Policy* 25 (Fall):407–58.

O'Riordan, Timothy. 1976. "Policy Making and Environmental Management: Some Thoughts on Processes and Research Issues." *Natural Resources Journal* 16 (January):55–72.

Paglin, Max D., and Edgar Shor. 1977. "Regulatory Agency Response to the Development of Public Participation." *Public Administration Review* 37 (March/April):140–48.

Peirce, Neal R. 1976. "Northeast Governors Map Battle Plan for Fight Over Federal Funds Flow." *National Journal* 8 (November 27): 1695–1703.

Rivers and Harbors Act of 1899, 30 Stat. 1121.

Rosenbaum, Walter A. 1977. *The Politics of Environmental Concern.* 2nd ed. New York: Praeger.

Rycroft, Robert W. 1979. "Energy Policy Feedback: Bureaucratic Responsiveness in the Federal Energy Administration." *Policy Analysis* 5 (Winter):1–19.

Sive, David. 1971. "The Role of Litigation in Environmental Policy: The Power Plant Siting Problem." *Natural Resources Journal* 11 (July):470–96.

Southern Interstate Nuclear Board. 1976. *Power Plant Siting in the United States.* Atlanta, Ga.: Southern Interstate Nuclear Board.

Spangler, Miller B. 1974. "Environmental and Social Issues of Site Choice for Nuclear Power Plants." *Energy Policy* 2 (March):18–32.

The Energy Center, Inc. 1977. "Issues in Power Plant Siting: The 94th Congress and the States." In *FEA Energy Facility Siting Workshops,* U.S. Federal Energy Administration, pp. 364–461. Springfield, Va.: National Technical Information Service.

U.S. Congress, Committee on Interior and Insular Affairs. 1977. *Congress and the Nation's Environment.* Washington, D.C.: Government Printing Office.

U.S. Department of the Interior (DOI), Bureau of Reclamation (BuRec), Upper Missouri Region. 1977. *ANG Coal Gasification Company, North Dakota Project: Draft Environmental Impact Statement.* Billings, Mont.: BuRec.

Valeu, Robert L. 1977. "Financial and Fiscal Aspects of Monitoring and Mitigation." Paper presented at the Symposium on State-of-the-Art Survey of Socioeconomic Impacts Associated with Construction/Operation of Energy Facilities, St. Louis, Missouri, January.

Verity, Victor, John Lacy, and Joseph Geraud. 1974. "Mineral Laws of State and Local Government Bodies." In *The American Law of Mining,* vol. 2, edited by Rocky Mountain Mineral Law Foundation, pp. 627–38. New York: Matthew Bender.

Wall Street Journal. 1976. "The Law's Delay: Huge Plant's Demise Signals Trouble Ahead for Energy Expansion." September 7, p. 1.

Ward, P. S. 1977. "Water Act Amendments: Deja Vu in the Halls of Congress." *Water Pollution Control Federation Journal* 49 (October):2055–61.

Western Interstate Nuclear Board. 1977. *Regional Factors in Planning and Siting Electrical Energy Facilities in the Western States.* Washington, D.C.: Nuclear Regulatory Commission.

White, Martin. 1977. "Colstrip Power Project—Its Monitoring and Mitigation Program." Paper presented at the Symposium on State-of-the-Art Survey of Socioeconomic Impacts Associated with Construction/Operation of Energy Facilities, St. Louis, Missouri, January.

Wolpert, Julian. 1976. "Regressive Siting of Public Facilities." *Natural Resources Journal* 16 (January):103–15.

CHAPTER 12

Conclusions and Recommendations

In the preceding chapters we have focused on the most significant problems and issues likely to result from western energy development. As we considered each of the major consequences of development, we have indicated how alternative policies will distribute the costs, risks, and benefits. Our purpose here is to present a series of conclusions and recommendations for policymakers associated with western energy. Following a brief description of key characteristics of the policymaking system, we summarize our findings for five sets of interests: local governments, state governments, the federal government, Indian tribes, and energy developers. For each of the five we have made policy recommendations that aim at achieving the benefits of developing western energy resources while minimizing the costs and risks.

Key Characteristics of the Policymaking System

Several key characteristics of this nation's political system must be recognized when addressing the issues that surround western energy development. First, policy is most often made in substantive categories (such as those used to organize our analyses in chapters 3 through 11). Thus, "western energy development" is not a focal point around which policies are established. For example, air- and water-quality policies are not for-

mulated specifically with regard to western energy development. As a consequence, the structure that affects the development of energy resources in the West is fragmented. Policies are made and implemented by various levels, branches, and agencies of government and by a range of participants in the private sector. A fundamental issue is whether certain problems fall in the public or private sector, and if public, whether they should be dealt with by federal or state agencies. Rapid development of western energy is now disrupting, and will continue to disrupt, previous relationships between the various sectors and levels of government. New rules will have to be worked out.

Although the policy system is continually changing, many of the decisions that will determine the outcome and consequences of western energy development (such as the choice of location and technology) will continue to be primarily private-sector decisions. Given present policies and responsibilities, the private sector will largely decide such things as whether to "strip and ship" coal or convert it near the mine-mouth; whether to convert coal to electricity or synfuels; whether to minimize water consumption; and whether to go to commercial-scale oil shale development. Private-sector decisions are, of course, affected by siting laws, leasing and land-use policies, air- and water-quality regulations, tax and pricing policies, certifications of public convenience and necessity, and other government policies. But, just as there is no unified western energy policy, there is no unified system for making and implementing the public policies that affect development. Not only are multiple levels of government involved, but at each level the executive, legislative, and judicial branches have a role. Furthermore, within the executive, more than one agency frequently has administrative, rule-making, and/or regulatory responsibilities; and responsibilities among agencies often overlap or conflict.

For these reasons, it is not easy for policymakers systematically to address energy resource development in the West. An important exception exists for the large land areas which are owned either publicly or by Indian tribes. There, comprehensive planning is a more feasible alternative. Our study indicates that the most direct way to control consequences is to control the allowable combinations of technologies and locations. This conclusion suggests that where public and/or Indian ownership is involved, comprehensive facility siting and land-use planning and management could directly control the consequences of development.

The fact that we address our recommendations to five sets of interests should not imply that policymakers do or should act independently of each other. As noted throughout this study, our system of government is characterized by shared, overlapping, and often ambiguous responsibilities. These interactions are noted in the following discussion; and we pay particular attention to those areas where more intergovernmental coordination and more cooperation between the public and private sectors is needed if the problems are to be dealt with adequately.

Local Governments

Local areas will benefit substantially from energy development, yet municipal governments will face immediate and potentially overwhelming growth-management prob-

lems. Most municipal governments have little independent capability to respond to the broad range of problems they will face, and institutional mechanisms for channeling assistance from other levels of government or from the private sector either do not exist or are inadequate. Although communities will not be able to solve their problems without outside assistance, they can improve their capacity to manage growth. In this regard, we recommend that:

- *County governments develop cooperative arrangements with municipal governments to enhance the equitable distribution of revenues;*
- *Officials of municipal governments, county governments, and multicounty agencies work to overcome their traditional antagonisms to planning and pool their resources to provide improved planning capacity in the abscence of state, federal, or developer assistance; and*
- *Municipal and county governments cooperate to improve the quality of mobile home parks through zoning and land-use controls.*

Virtually all the consequences of developing western energy resources create issues of concern to local governments. However, local officials have very little say in when, where, and how such development occurs. Moreover, they lack the resources and the authority to control most of these consequences and, therefore, are usually dependent on other levels of government and energy developers for assistance. For example, the federal and state governments generally regulate the impact on air, land, and water resources; and decisions by energy developers largely determine the population-related costs and benefits.

Local governments are creatures of the states: their taxing, zoning, public health and safety, and other powers are defined by the states. Although local governments may use their planning and zoning powers to affect the physical and environmental consequences of development, they can do very little to impose direct environmental controls on energy development. In response to such problems as the impact on air and water quality and the disturbance of surface land, local governments are largely limited to attempting to influence state and federal policies and programs. For most of the small, predominantly rural communities affected by energy development, this is not an effective means of control. The same point can be made for the socioeconomic consequences faced by municipal governments. Even though they lack adequate resources and the authority to control these problems, municipal governments are expected both to create an environment favorable to economic growth and to provide residents—old and new—with an opportunity to achieve the quality of life they desire.

The most serious dilemma facing local policymakers is how to balance the real economic benefits of energy-related growth against the limited capability to meet the demands of rapid and large increases in the population. The social and economic advantages for local areas include expanded employment opportunities, higher salaries, and the more indirect benefits of expanded retail trade and increased social and cultural opportunities. Throughout our eight-state area, both residents and local officials are generally among the most enthusiastic supporters of energy development. In large part this reflects the fact that most of these areas are economically underdeveloped in terms of job opportunities, median household incomes, and general economic diversity.

The potential economic benefits of energy development must be weighed against the serious growth-management problems that local officials will face. As elaborated in Chapter 8, municipal services and facilities quickly become inadequate for current residents as well as for those newcomers who move into the area because of energy-related jobs. To meet the needs of a larger population, new facilities will have to be constructed for such services as water supply, sewage treatment,

detention facilities, and medical care. The cost of public facilities is high—capital costs range from $750,000 to $1.15 million, with operating costs of $65,000 annually for every 500 new residents.[1]

The fundamental problem is that municipalities have few financial or planning resources. Therefore, they must depend heavily on other units of government and on energy developers. While county governments and school districts face similar problems, theirs are generally fewer and less severe; and they both will have larger tax revenues than municipalities with which to respond.

Energy-impacted communities will face serious imbalances between revenues and expenditures. Although the effects are felt as soon as development starts, energy facilities do not pay taxes until they start operation, some five to eight years later. In addition to this "front-end" financing problem, an energy facility is typically located outside the legal jurisdiction of nearby towns. However, many of the workers employed there will usually live in the towns, increasing the strain on public services and facilities. These communities typically receive very little revenue from county sources; thus, they must rely on local sales taxes, local property taxes, and other charges and fees which are nearly always inadequate to meet the needs of rapidly expanding populations.

In addition, in each of the eight western states we studied, local financial difficulties are intensified by the state ceilings on bonded debt. These ceilings limit the capacity of towns, counties, and school districts to finance and construct public facilities prior to the influx of large numbers of new residents.

---

[1] These estimates include fire protection, law enforcement, water, sewage treatment, garbage service, health care, jails, juvenile treatment and custody facilities, county and municipal courts, recreation facilities, and administrative space. They are based on capital costs of $1,500 to $2,300 per capita and on annual operating costs of $130 per capita (1975 dollars); see Chapter 8 for details.

The financial problems of municipalities are also magnified by inadequate planning. A professional planning capability is nonexistent in most of the western towns likely to be affected by energy resource development. This is because these towns are typically very small (under 5,000 people); in addition, traditional antagonism to planning and land-use control, and the lack of information concerning the schedule and the magnitude of energy development, contribute to the inability to anticipate and plan locally for rapid growth.

Another problem faced by communities in the vicinity of energy developments is the inadequate quantity and quality of housing (see Chapter 7). Housing is typically in short supply because of uncertain long-term housing needs, financing constraints in the private market, and the absence of governmental programs to address housing problems in rapidly growing areas. As a result, mobile homes have become the dominant type of housing for the majority of newcomers. However, mobile homes (which provide little property tax benefit to local governments because they are taxed as personal rather than real property) are usually restricted by local zoning to the less desirable areas. Hence, mobile home parks typically cluster in unincorporated areas on the edges of a town, typically receive poor public services (e.g., streets and parks), and are generally unattractive.

The picture that emerges for local governments is bleak, at least in the short term. Municipal governments face immediate and serious growth-management problems over which they have very little control. As discussed in the following section, the responsibility for dealing with these problems rest primarily with state governments. To the extent that local governments have any power, their options are essentially to try to make a bad situation a little better rather than to "solve" problems.

Among local governmental units—municipalities, counties, and school districts—*county governments should assume a large share of the responsibility for growth management.*

This conclusion reflects the nature of the problem of revenue distribution discussed above: county governments can anticipate substantially greater revenues and substantially smaller problems from energy development than can municipalities. Essentially two models exist for interaction between counties and municipalities. The more preferable is one of cooperation in which county governments offer assistance to municipalities by providing services and facilities. In addition to ensuring some degree of equity, this approach should make it easier to deal with the problems associated with mobile homes, problems which affect both counties and municipalities.

However, the more likely model, to judge from recent experience, is one of a lack of cooperation—or even conflict—between counties and municipalities. In the face of inadequate county assistance, the capacity of a municipality to deal with these problems would be improved by annexing county land so as to include the new energy facility in the town's tax base. Wyoming's Joint Powers Act (1974), for example, encourages cities and counties to cooperate in providing services and facilities, but like most state legislation, it requires county consent for any cooperative arrangements. Thus, in the absence of county willingness to assume more responsibility for growth management, the annexation option requires state government to assume responsibility.

Thus, traditional antagonisms to planning frequently leave local areas dependent on energy developers, state governments, and the federal government. To some extent, these antagonisms result from a historical relationship with the federal government in which "planning" has implied federal interference. However, throughout our study area resistance to any kind of planning frequently exists. This problem will not be easily resolved; it will require recognition on the part of states and the federal government of the need for local control of planning as well as acceptance on the part of local governments of the importance of planning to mitigate the impact of energy development.

Local governments have virtually no control over the availability of housing. However, they can attempt to improve the quality of housing by alleviating the mobile home situation. By encouraging better quality mobile home parks and county-wide land-use controls, the population density can be reduced and more amenities, such as paved streets and neighborhood parks, can be provided. Since these services will require additional public expenditures, classifying mobile homes as real, rather than personal, property could increase the needed tax revenues. Again, state governments would have to enact legislation authorizing this change.

State Governments

The western states are in a pivotal position to deal with the consequences of energy development. States will be affected by virtually every issue discussed in this book—from environmental protection to growth management and transportation. State governments obviously will not exert complete control in any single area; for example, their options for protecting the environment are greatly circumscribed by the federal government. Nevertheless, state governments have considerable control over the problems they face, either through application of their own laws and regulations or through cooperative arrangements with other governmental units and with the private sector.

Indeed, in general, the most important questions facing the region's state governments over the next few decades are, first, how well they will be able to adapt to new situations associated with energy development and, second, how well they will be able to develop working arrangements with the federal government, energy developers, Indian tribes, each other, and other interested parties. In particular, three areas stand out as being of primary concern to state governments:

water-resource management, growth management, and environmental protection.

Water-Resource Management

Managing increasingly scarce water resources will be the most serious, long-term problem associated with western energy development for which states will have primary responsibility. Although western states have successfully managed many water problems in the past, resolution in the future will be more difficult. In the absence of a concerted and timely management response by the individual states, pressures at the national level are likely to increase the demand for expanded federal control. In order to deal with water-resource problems, we recommend that state governments:

- *Develop better mechanisms for considering water availability and water quality jointly, including coordinated water planning and regulation within the state's administrative framework;*
- *Develop better management mechanisms, including improved information systems and basin-wide approaches; and*
- *Develop more comprehensive and flexible definitions of "beneficial use" to recognize a broad range of water uses in the public interest.*

The management of water resources is already a difficult issue for western states even without energy development, largely because of the scarcity of water in the region and the exceedingly complex system of water laws. Despite these factors, however, the policy system has been historically successful: there has been enough water to supply the basic activities of the region—agricultural, municipal, and industrial growth. In fact, a basic philosophy of water policy in the West has been that users have an inherent right to the resource. Hence, the policy system has traditionally responded to water-resource management in two ways: first, by damming, storing, and diverting water in order to make it available for users; and second, by developing a complex legal and administrative system primarily designed to identify and protect water rights.

However, it is increasingly likely that this system will be inadequate to deal with questions of water availability and water quality in the future, particularly in the Colorado River Basin (CRB). This does not mean that the current institutional system has failed. Rather, we must recognize that *fundamental* changes in water use are occurring that will require new approaches to water management, not merely incremental changes in past laws, regulations, and institutions.

The most important change is that increasingly water use will approach the limit of the physically available supply in river systems or in particular streams or stream segments. This situation will be worsened both by expanding demands from traditional users, primarily irrigated agriculture and municipal uses, and by several other new demands. The most significant of these are: (1) energy resource development, particularly coal conversion facilities; (2) reserved water rights of Indians; (3) salinity control; and (4) environmental uses, particularly instream values, which have increasingly been recognized as legitimate water uses in the West. The essential fact for future water-resource management in the West is that *it will not be possible to meet all of these legitimate demands.* This is particularly evident in the CRB, which has already, to a large extent, reached the limit to which physical controls can be used to augment and transfer water supplies in many areas. And it is also becoming a reality in energy development areas in the Yellowstone River Basin of the Upper Missouri system.

Three critical deficiencies exist in the states' present policies for managing future water problems. First, state appropriation systems *settle disputes over water use after damage (or impairment) of a water right has*

occurred. Although states record the holder, date, seniority, and use to which a water right is granted, conflicts among users are usually settled only after one user claims a right has been impaired by another user. Thus, the current system as practiced in most states is designed primarily to react, after the fact, to questions concerning the legality of rights and uses and not to manage the resource.

Second, *water policy is overly fragmented* in the sense that relationships among various components of water-resource management typically are treated as if they were unrelated. Although fragmentation characterizes American policymaking in general, it is overwhelmingly evident in questions of water use, making comprehensive water management difficult, if not impossible. The most obvious example of the lack of coordination is that the water-allocation system in most western states is not tied directly to issues of water quality. For many of the problems and issues considered in this report, both the quantity and quality of water are involved. Salinity, for example, is caused both by water withdrawals, which tend to concentrate salinity levels, and by runoff from irrigated agriculture, which picks up salt as the flows return to the stream. However, there are few regulatory or administrative links which allow policymakers to consider water use and water quality together. Thus, decisions about water allocation too often ignore the consequences for water quality.

Third, *the current system of establishing water policies largely discourages change.* This can be attributed primarily to its long and complex history and to the reliance on court rulings to determine water policy and to define and clarify how water is used. As a result, "water policy" is an extensive series of judicial decisions which almost never address issues beyond the very narrow questions of a given case. Therefore, this incremental legal and regulatory system is virtually devoid of any integrated or holistic framework, is very difficult to change, and often creates incentives for water use which add to, rather than subtract from, existing problems. This situation is exemplified in doctrines such as "use or lose" and "nonimpairment" which often create disincentives for individual users to minimize their use of water or to protect its quality.

These criticisms do not suggest that the federal government should have an expanded role in western water management; our knowledge of the current policy system and of the issues likely to be associated with the expanding demands for water (including energy development) leads us to conclude that states should continue to be the level of government primarily responsible for managing water resources. This position is based on two primary factors which are evident in the discussions throughout our report. First, a variety of values and problems exist across the eight states in our study area, and water resources are directly related to state priorities for growth and development. It is very difficult to imagine a centralized and/or uniform federal water policy that would effectively address these diverse situations. Second, federal interests in the region are also diverse and complex; they include encouraging energy development, preserving the region's environmental quality, subsidizing a large fraction of its agricultural development, and regulating its water quality. Thus, an expanded federal role in western water management would require, at a minimum, both a better consensus about priorities and a more coherent bureaucratic system. The recent conflicts and failures of federal intervention in the region, as evidenced by the Carter water policy reforms, suggest how difficult such a reorganization would be.

Despite this conclusion, it seems almost certain that pressure for an expanded federal role in managing the West's water resources will increase. This pressure can be traced to the high percentage of federally owned land in the West and to the variety of federal activities being promoted in the region, including increased energy production and environmental protection. To the extent that these priorities are affected by water-related conflicts in the future—for example, if energy facilities are unable to obtain water rights—the chances will increase for federal assumption of state prerogatives. In fact, there is already evidence of this trend in the form of recent claims to federal "non-reserved" rights made by the solicitor general of the Department of Interior (DOI). It is uncertain how far and how quickly this trend will develop. However, it is clear from both past environmental legislation and proposals for a national water policy that federal policymakers frequently use any perceived weakness in state policy to justify a stronger federal role.

Dealing with the problems of water resources will require a combination of technological, legal, and institutional changes. Among the most attractive technological choices is water conservation by energy facilities—for example, up to 77 percent of the water required for a 3,000-megawatt electric generating plant can be saved by using water-conserving cooling technologies (see Chapter 4). State governments should adopt policies to encourage the use of these technologies on a regional basis, for example, by states of the CRB. Rather than require the uniform application of a particular technology, however, ceilings on water use for power plants could be developed depending on the severity of local water problems. Thus, energy industries would be required to adopt a level of water conservation corresponding to the local situation, including state and local feelings about how water should be used. This policy would both save water and enhance water quality, since it would require reduced withdrawals from streams. The necessary legal and institutional changes in the present system would almost certainly generate substantial opposition, and many of the changes would not have a realistic chance of enactment. However, as water-resource problems become more serious, many alternatives are likely to become more feasible politically.

One of the biggest difficulties in managing the use of water is the general lack of adequate information on streamflows, on return flows from agriculture, and, in some cases, on actual withdrawals. Thus, states should develop more accurate and up-to-date ways to monitor water availability and use. Such programs would require close coordination with the Water and Power Resources Service (formerly the Bureau of Reclamation), the Department of Commerce, and the U.S. Geological Survey (USGS), since these agencies currently collect substantial information about streamflow, rainfall, etc. However, the point of this recommendation is for states to develop better capacities to manage the use of water on a day-to-day basis. Thus, while federal assistance will be needed, primarily to collect data, states would be responsible for applying this information to their own needs.

States should also initiate and enforce a criterion of "reasonableness" for allocating water so that various legitimate interests are reflected in the decisionmaking processes. While it is likely that no single standard of reasonableness could be developed, policies that encourage legal recognition of various water uses would be a marked improvement. The current system makes it very difficult for new uses to achieve legitimacy and thus encourages many participants to pursue their interests in the courts. At the present time, uncertainty over water rights is becoming increasingly problematical, and this uncertainty is worsened both by the reluctance of states to recognize environmental values as "beneficial" and by the inability of states and Indian tribes to resolve their conflicts. This statement is not meant to appreciate or depreciate any set of water interests; it is intended to suggest that waters of the Colorado and Upper Missouri River basins legitimately have multiple uses. The rapid increase in demands for multiple uses is the most fundamental change occurring today in western water-resource management.

Growth Management

Areas close to energy facilities and state governments generally can anticipate substantial economic benefits from energy development. However, local communities will also suffer serious growth-management problems such as inadequate housing, water supply, medical care, police and fire protection, and other social services. Although the federal government and private developers will have a role to play, the primary responsibility for dealing with these problems rests with state governments. Thus, we recommend that state governments:

- *Develop institutional mechanisms specifically designed to provide local impact assistance;*
- *Make financial and technical assistance available to local governments on a timely basis; and*
- *Require energy developers to play a larger role in providing impact assistance and planning information about development.*

Growth-management issues present two stark contrasts. The first is that local and state governments can anticipate substantial economic benefits from energy over the long-term, yet communities will face serious and potentially overwhelming short-term consequences. The most important economic benefit for state governments, of course, will come from increased tax revenues, increased personal incomes, and secondary industrial growth. To the extent that state governments impose severance taxes on coal or oil shale, a substantial percentage of a state's revenue can be supplied by energy development (see Chapter 7). On the other hand, many public-sector costs of energy development will be paid almost immediately by the affected communities. As discussed in the section on local governments, the underlying cause of adverse socioeconomic effects is the rapidly fluctuating population growth—the boom-and-bust cycle. For example, a standard-sized coal gas-ification plant (250 million cubic feet/day) requires approximately 4,700 workers during the construction phase but only about 590 workers after construction is completed. Few local areas will be able, by themselves, to deal effectively with the planning, front-end financing, revenue distribution, and housing problems caused by these population changes.

The second aspect of growth management is that although most growth problems can be managed, few institutional mechanisms for doing so exist. We are not suggesting that dealing with growth-management issues will be straightforward. However, compared to most of the issues associated with energy development, growth management is easier in several respects: (1) the causes of impacts are generally well known—the major uncertainties are the timing, location, and size of development; (2) the technical solutions to the problems generally exist—that is, it is generally known how to deal with most housing, public-service, and public-facility problems; and (3) although local communities will be unable to afford them, most solutions will not require large investments compared to the ultimate economic benefits generated.

The problem presented by growth management, then, is establishing the procedures and institutions to provide assistance to local areas. The federal government can and should play a role in this area, especially when the leasing of federally owned resources is involved. Also, energy industries can and should help improve the quality and availability of information about development and should help provide housing and general impact assistance. However, because state governments stand to gain many direct economic benefits from development, and because states directly control the type of responses a local government can make to development, *the primary responsibility for growth management should rest with state governments.*

To deal effectively with growth, many states will have to address three characteristics of the policy system that contribute to the se-verity of the problems faced by local governments:

- Information available to states and local governments concerning the size and scale of energy projects is usually too late or is insufficient to accurately estimate population changes;
- Sparsely populated rural communities affected by energy development have inadequate planning and zoning capabilities; to the extent that local areas want improved planning capabilities, they strongly oppose attempts by "outsiders" (including state agencies) to plan for them; and
- Financial resources are seldom available at the beginning of construction of an energy facility or within the proper governmental jurisdiction to provide the needed facilities and services.

The most immediate need for states is to establish institutions specifically mandated to provide financial and technical assistance to energy-impacted areas. While these institutions could take a variety of forms, the critical point is that they be specifically targeted in order to address the particular kinds of problems faced by energy-impacted areas.

Several alternatives have been identified in this report for generating the revenue needed for front-end financing. The most direct and effective options available are to change state and local tax structures. Such changes include requirements for energy industries to prepay property taxes to eliminate the "lead-time" problem; tax incentives to encourage, or laws to require, impact assistance by the private sector; and increased severance taxes, to be distributed to the affected communities.

As discussed in chapter 7 and 8, several states have instituted programs to mitigate the socioeconomic impacts of energy development. For example, Colorado, Montana, and North Dakota earmark a percentage of their severance-tax receipts for local governments.

The key variable is how these taxes are distributed. If a significant portion of severance tax revenues is allocated to the local government, the jurisdictional imbalances between energy-related revenues and expenditures could be overcome. However, the trend in the West is for states to retain most of the severance-tax revenues for state purposes. Furthermore, the institutions for distributing tax revenues to local governments will have to include methods for identifying the needy communities without relying on outdated information, such as census data. This should be a state rather than a local responsibility.

States also need to develop institutional mechanisms for improving the quality of impact assistance that is provided by the energy companies. State options in this area include regulations to make siting permits conditional on the developer's taking certain actions: for example, developers should be required to provide timely information on their development plans. At a minimum, information on the construction and operating work forces, on land use, and on the likely environmental consequences should be provided directly to local governments. State governments should also require developers to help provide adequate housing for the expected work force. Options here include construction of housing units, guarantee of loan funds, and support for private housing projects. Requiring developers to provide impact assistance is especially attractive because it is a flexible solution: that is, specific measures can be tailored to meet the conditions that result from a particular technology at a particular site.

Although such siting requirements are not yet common in the West, the Wyoming Industrial Development Information and Siting Act of 1975 is an example of this type of policy alternative. This act makes the approval of siting permits for major energy and industrial facilities contingent upon the mitigation of socioeconomic consequences. As a result of this act, Basin Electric Power Cooperative agreed to provide 1,900 homes and other impact assistance in return for permission to build a plant near Wheatland, Wyoming (see Chapter 7).

Environmental Protection

States will face numerous environmental threats from energy development, yet their authority over environmental issues is greatly circumscribed by the federal government. Although federal control will make it more difficult to respond to concerns about environmental quality than to problems of managing water resources and growth, state governments do have several options for balancing environmental protection and energy development. We recommend that state governments:

- *Develop policies to control environmental problems not now covered by federal regulations—in particular, waste disposal, reclamation of non-coal surface-mined lands; and lessening the local impact of coal unit-train traffic;*
- *Increase the flexibility of implementation procedures in current regulations to protect environmental quality; and*
- *Improve the siting process to reduce barriers to participation and to plan more comprehensively for energy development.*

States will face numerous environmental issues as a result of the development of energy resources. The most serious of these issues identified in our study are:

- Reduced air quality, including reduced visibility;
- Ecological effects of reduced streamflows;
- Potential pollution of both surface water and groundwater;
- Land impacts from surface mining and waste disposal;
- Local disruption of groundwater flows;

- Aesthetic impact of surface mining and conversion facilities; and
- Noise and hazards from unit trains.

Some western states have been frustrated in their attempt to deal with environmental quality over the past two decades. Frustration has generally taken two forms. First, the federal government has preempted or assumed primary responsibility for environmental protection—even in areas, such as surface-mine reclamation, where several western states had developed strict regulatory policies in advance of federal legislation. Second, federal policies have typically established uniform standards and regulations which lessen a state's flexibility in balancing the competing goals of economic growth, energy development, and environmental quality. It is easy to understand why state governments in the West frequently rebel against what they view as unnecessary federal intervention.

However, it is clear that states still have considerable authority in protecting the environment. Their authority exists in three general areas. First, several potential environmental problems are not covered by current federal regulations. Second, in areas that are covered by federal regulations, states typically have the authority to impose stricter standards and are responsible for implementing federal regulations. Third, states have primary responsibility for siting energy facilities on non-federal lands; this authority can be used to encourage energy development and/or to improve environmental controls.

The environmental problems not adequately covered by existing federal and state regulations include land rehabilitation for surface mining, waste disposal,[2] and mitigating the impact of increased unit-train traffic. State action on land rehabilitation is needed for two reasons. First, the recent federal law on strip-mine reclamation covers only coal min-

ing; and land reclamation laws in several western states either do not apply to resources other than coal (e.g., oil shale, uranium, and geothermal energy) or are inadequate. Second, the existing federal and state regulations governing waste disposal may be inadequate for the West, especially in the cases of spent shale and uranium mill tailings. Strong state action now in these areas could preclude the need for federal action.

States can also exercise control through the implementation of environmental regulations. A consistent finding of our study is that local differences are often critical in determining the magnitude of an environmental impact. In fact, the severity will vary dramatically from state to state and from site to site within a state. Thus, *state policies and implementation procedures should be designed to maximize the flexibility of environmental regulations.* For example, although the federal government has set national ambient air-quality standards, the states are responsible for developing State Implementation Plans to achieve or maintain compliance with the standards. Since states can set standards that are stricter than the federal ones, they can use this flexibility to restrict energy development in some areas in order to protect the environment. On the other hand, states can redesignate some areas as Class III under Prevention of Significant Deterioration (PSD) regulations (CAA, 1977:§108). Because this classification allows the largest increase in pollution, it can be used to promote energy development.

Water-quality control offers another illustration of how states can improve the regulatory system within the federally established boundaries. For example, states must meet federal salinity standards for the Colorado River. Until recently the states depended almost totally on federal funds to meet these standards. Now, however, the states in the CRB have established the Colorado River Basin Salinity Control Forum to recommend standards and to develop mechanisms for meeting these goals. The forum is the first

---

[2]Comprehensive regulations for waste disposal are being developed under the 1976 Resource Conservation and Recovery Act.

attempt by western states to address the water-quality problems of an entire river basin through a coordinated decisionmaking process rather than by relying on traditional physical controls. While the success of this body is still undetermined, it could be a useful model for interstate cooperation in other areas.

Perhaps the most comprehensive, yet largely unused, area of state control is the siting process—a policy area in which states have primary responsibility. *We recommend that state governments develop agencies and policies to coordinate the siting process, plan more comprehensively for energy development, and reduce barriers to participation in the siting process by interested citizens.* The objectives of this recommendation are to select areas that will help avoid or reduce the impact of energy development and to minimize costly delays once major investments are made.

Information provided in this study begins to provide the basis for more comprehensive technology-site evaluations in the siting process.[3] This includes a range of specific strategies, such as elimination of some sites from consideration, designation of appropriate sites, elimination of certain technologies at given sites, and determination of the size, environmental controls, and/or socioeconomic safeguards that would be required for a particular technology at a given site. These strategies represent an extension of present zoning practices on a larger scale. North Dakota's Energy Conversion and Transmission Facility Siting Act (1975) represents an example of designating certain locations as unacceptable. Under the provisions of this law, energy facilities are excluded from state parks, from areas where animal or plant species unique or rare to the state would be irreversibly damaged, from prime farm land, and from irrigated land. More flexible methods include site screening to evaluate each technology-location configuration individually: for example, a

power plant utilizing wet cooling may be prohibited from a particular area although less water-consumptive technologies might be acceptable.

The determination of acceptable combinations of technology and site could be implemented by formal governmental regulatory and economic policies, or it could be carried out in a more informal manner, such as creating an ad hoc siting task force made up of representatives of government agencies, industry, and other interest groups. Such an informal approach has been successfully used in Utah to select acceptable areas for coal-fired power plants in the central part of the state, thus protecting the more scenic, southern area which contains many national parks (see Chapter 5).

Siting procedures frequently fail to include adequate participation by interested parties, which often leads interest groups to rely on delaying tactics to slow or block development. Thus, states should reduce barriers to citizen involvement by setting up programs to improve notification procedures, establishing broadly based advisory committees, creating review boards for siting agencies, and providing financial support for intervenors. The objectives of these programs would be both to improve the substance of siting decisions by incorporating more pluralistic values and interests and also to enhance the participatory process itself by allowing interested parties to be decisionmakers rather than outsiders. None of these options, however, is likely to avoid the conflict inherent in siting an energy facility. Nevertheless, they can help facilitate a compromise at an early decisionmaking stage and avoid some of the long-term delays in siting.

The success of these changes in the siting process depends in part on the quality and timeliness of the information produced. However, success may ultimately require more fundamental changes in the system, including some aspects of the newly proposed Energy Mobilization Board. The two most important changes that are needed are equitable access

[3] See Chapter 11 for a general discussion of this approach.

to decisionmaking by interested parties and limits on how long and in what manner siting decisions can be contested after the review processes are over. This does not imply that our knowledge of energy development technologies and locations is advanced enough to predict, once and for all, the consequences of development. It does suggest that if issues are given a fair hearing, stricter limits should be placed on how they can be appealed. This approach to siting will probably lengthen, rather than shorten, the initial hearing processes in order for various interests to be expressed. However, the value of this approach is in its potential to shorten the total time required to get a facility into operation and to reduce some of the uncertainty associated with siting.

However, perhaps not even these revisions will improve the siting process. In the end, "reasonable" determinations of where to locate energy development facilities may depend on the willingness of interested parties to recognize the legitimacy of diverse values. If participants are not able to reach compromises, the two major alternative approaches appear to be either to accept the uncertainties, inequities, and, at times, paralysis of the existing system or to adopt more centralized and controllable policymaking institutions.

Federal Responses

A variety of policymakers at the federal level will have an important influence on western energy development. In addition to the president, the most important of these are (1) the Congress, particularly those sub committees dealing with environmental and energy policy; (2) the Environmental Protection Agency (EPA); (3) the Department of Energy (DOE); and (4) the DOI, particularly the Bureau of Land Management, the Water and Power Resources Service (formerly Bureau of Reclamation), the Bureau of Indian Affairs (BIA), and the USGS.

Federal policymakers will be increasingly

responsible for dealing with the consequences of the development of western energy resources. Pressures for federal responsibility originate from many sources, but include primarily the following:

- The federal government owns about 35 percent of the total land area in our eight-state study area and substantial percentages of the energy resources, including about 50 percent of the coal, uranium, and geothermal energy and 80 percent of the oil shale resources.

- Environmental problems have steadily come under federal jurisdiction during the past two decades. Federal agencies now have primary authority in formulating air-quality, water-quality, waste-disposal, and reclamation policies.

- National energy policies, including the National Energy Plan[4] and recent administration proposals for a major program to develop synthetic fuels, emphasize development of domestic resources including western energy. Since energy produced in the West will be used primarily to meet energy demands outside the region, solutions for many of the problems caused by development will require national policies. These problems include the equitable distribution of costs and benefits of development across the nation.

- Some problems appear to be beyond the capacity of other levels of government, either because their financial resources are inadequate (e.g., supporting demonstration plants) or because federal laws

[4]The National Energy Plan was set forth in an April 20, 1977, speech by President Carter (see U.S. President, 1977; U.S. President, Office of the White House Press Secretary, 1977). After more than a year of debate, controversy, and revision, Congress passed a five-part energy legislation package in October 1978, consisting of: the Public Utility Regulatory Policies Act, the Energy Tax Act, the National Energy Conservation Policy Act, the Powerplant and Industrial Fuel Use Act, and the Natural Gas Policy Act.

and regulations discourage or prevent them from assuming control (e.g., in financing sewage-treatment plants).

On the other hand, although the federal government can intervene in many policy areas, comprehensive national control of the problems associated with western energy development is limited by several factors—especially the fragmented nature of energy policymaking at the federal level. Also, a limited role is consistent with the explicit desire of Congress to restrict federal involvement in some areas of energy policymaking. For example, although the federal government has assumed primary responsibility for environmental protection, major air- and water-quality legislation expressly reserves to the states primary planning, implementation, and enforcement responsibilities.[5] Another constraint on the federal role is increased demands by western states for control over the public lands within their boundaries, a conflict which has been popularized in the media as the "sagebrush rebellion." As discussed in previous chapters, these demands have been particularly strong regarding water-resource management and land use and reclamation. Hence, it may be politically difficult for Congress and the various regulatory agencies to expand federal control over issues that are perceived to be primarily regional or local. Furthermore, federal assistance and expertise are limited. In an era of increasing concern about tax levels, inflation, balanced budgets, and the appropriate role of the federal government, comprehensive federal management of energy development in the West will be politically and economically controversial.

Despite these constraints, as the previous eleven chapters have amply demonstrated, the federal government is already involved and will likely continue to be involved in most, if not all, of the problems associated with

[5] For an elaboration of the federal-state relationship in air and water quality, see Jones, 1974; and Lieber, 1975.

energy development. These chapters have discussed a variety of policy options available to federal lawmakers and regulators. Three broad areas of policy action can be identified: environmental regulations, financial and technical assistance to impacted communities, and research and development for new technologies. In the following pages we summarize our major conclusions for each of these areas.

Environmental Regulation

To a large extent, the basic policies exist at the federal level to safeguard air quality, water quality, and the land. Nevertheless, significant environmental risks exist over the long term both because of the continuing uncertainty over how well the standards will be enforced and because of gaps in some regulations. Furthermore, the complex and poorly coordinated regulatory system often leads to unnecessary delays in siting energy facilities. Therefore, we recommend that the federal government:

- *Expand its environmental monitoring and research program to reduce uncertainties about impacts and to monitor the success of regulations;*
- *Expand its regulatory programs to fill existing gaps particularly regarding, first, water-quality protection and land reclamation for both oil shale and uranium production and, second, the long-range transport of air pollution;*
- *Improve the procedural requirements of regulations to reduce the costs and risks of siting energy facilities—for example, by coordinating the many permits required.*

For air quality, water quality, and land use the basic policy-management system exists in most cases to limit adverse effects. For air quality the policy system consists of a set of emission limits, ambient air-quality standards, and regulations to prevent significant deterioration of the existing air quality, especially in scenic or highly valued recreation areas. Water-quality regulations establish permit procedures which limit the direct discharge of pollutants into surface waters. The risk of water pollution to surface runoff and ground-water from solid waste disposal has been inadequately regulated in the past, but this appears to be changing with the passage of several state and federal laws, most notably the Resource Conservation and Recovery Act (RCRA) of 1976. Also, surface water and groundwater pollution from coal mining is now covered by the 1977 Surface Mining Control and Reclamation Act. This legislation is also intended to ensure that mined lands are returned to a productive state and that mining is prohibited on lands that cannot be adequately reclaimed.

However, a significant limitation to informed environmental policy and regulation is uncertainty about how given actions will affect the environment over the short and long terms. For example, even if perfect knowledge existed about the emission rates of various pollutants from a proposed coal-fired electric power plant, there can be errors on the order of 100 percent in our scientific ability to predict the peak or average annual concentration of any pollutant downwind of the plant. There is an even greater uncertainty about the ultimate effects of air pollutants on human health, ecology, etc. Both experience and research will help reduce these kinds of uncertainties and identify appropriate ways to control the adverse consequences (including choice of technology, scale of development, or siting).

Similar conclusions apply to water and land resources. In the case of water quality, the long-term effects of solid waste disposal are still uncertain. Research in this area is only beginning, and the RCRA regulations are still being formulated. In land reclamation, there is still considerable uncertainty about the long-term potential for revegetation of lands in arid and highly grazed areas. The regulatory machinery exists to minimize or mitigate the impact of energy development once we learn what effects need to be controlled. As our knowledge expands, or as society's goals change concerning the balance between environmental and economic factors, the standards and regulations can be modified.

To reduce uncertainty, we recommend an increase in selected environmental monitoring and research programs. For example, one important unknown is groundwater quality prior to energy resource development. Therefore, predevelopment measurements of groundwater quality can provide the baseline data essential to determining appropriate controls, standards for permits and monitoring, and liability for subsequent changes. A variety of federal research programs already exist within the DOE, EPA, and the USGS. However, these programs do not provide sufficient and timely data. A wider range of studies is needed to cover several energy resources in a variety of environments and include potential risks and controls for mining, *in situ* recovery, and waste disposal. Although federal funds will be needed, such studies should be carried out cooperatively with state agencies and energy companies in order to enhance their effectiveness.

Two policy deficiencies stand out regarding environmental issues. The first concerns gaps in present regulations. For example, water-quality protection and land reclamation requirements for oil shale and uranium mining are not adequately covered under existing federal law. In air quality, the Clean Air Act (1977) is inadequate to deal with long-range transport of pollutants and pollution across state lines. These gaps should be a top priority for attention in Congress and in the EPA's regulatory programs.

The second deficiency is more subtle, more complex, and probably more difficult to correct. It is the fragmented and poorly coordinated nature of the regulatory and administrative systems which set and enforce environmental standards. For example, our analysis shows that the choice of emission controls, plant size, or plant location is usually sufficient to meet any one of the individual air-quality standards (such as the Class I PSD standards). However, when all of the various environmental standards are considered together, with their continual changes and with the patchwork of government regulatory agencies involved, it is easy to understand the difficulties typically encountered in constructing a new energy facility.

Siting new facilities becomes such a lengthy and uncertain process that accurate planning is esentially impossible. Thus, costs are increased, capital investments in new domestic energy facilities are eventually restricted, and energy imports are increased. *Lawmakers in Congress and regulators, principally in EPA and the DOI, seem to account properly for the substantive requirements of their individual decisions, such as the direct economic costs associated with certain levels of sulfur dioxide (SO₂) control. However, they seldom account for the procedural requirements and consequences of their collective actions.* In essence, this is because no formal mechanism currently exists for weighing this kind of consideration.

It is difficult to advance specific proposals to correct this inadequacy. Overcentralization of regulatory authority poses its own set of threats. However, it seems clear that the regulatory system must be improved. As discussed in the previous section, some changes are being initiated at the state level. There also have been attempts to make improvements at the federal level, but little real progress has yet to occur. Concern with the government's regulatory apparatus resulted in President Carter's creating two new executive level institutions in 1978: the Regulatory Analysis Review Group and the U.S. Regulatory Council (currently headed by EPA Administrator Costle). As an example of attempted actions in the environmental arena, EPA has begun to improve coordination and to simplify various regulatory activities concerning air and water pollution and waste disposal. EPA is also working with other fed-

eral agencies to develop a single-permit application package for new industrial facilities. This effort is not the same as the so-called one-stop siting process, since, due to specific requirements in some of the environmental legislation, it is not possible to combine all federal permit review and approval procedures for siting without legislative changes. To be truly effective, Congress would probably have to rewrite the procedural requirements for siting in much of the environmental legislation. However, the single-permit application package will eliminate much of the time and expense that now goes into providing the same information in many different forms for different permit applications.

Assistance to Impacted Communities

Although states should take the primary responsibility for aiding local communities in dealing with the boom-and-bust cycles of energy development, the federal government must also be involved, especially when energy projects are the direct result of federal initiatives or when impacts overlap state boundaries. The federal government should:

- *Provide direct and timely financial aid—primarily to impacted towns rather than counties or school districts—using existing programs where possible; and*
- *Provide technical and planning assistance to local communities in a manner that allows local communities to develop their own management capabilities.*

As was discussed in the previous section, the primary responsibility should lie with the states for mitigating the adverse social and economic impacts of energy development on surrounding communities. Nevertheless, there are several justifications for federal assistance in some areas: (1) the federal government already provides financial resources to states and communities through such programs as revenue sharing; (2) much of the rapid growth in the development of western energy resources is the result of federal policies; (3) many energy developments will occur on federal lands; and (4) federal policies contribute to the needs of states and localities for additional revenue (for example, requirements for wastewater treatment). We recommend two basic approaches for the federal government — to provide financial aid and to provide technical assistance to local officials.

A number of federal impact-assistance programs are already operating or have been proposed, although not all are applicable to our study area. These include programs to share royalties and other revenues from the energy development that takes place on federal lands with states and localities,[6] a variety of loan programs,[7] and assistance programs targeted for specific impacts.[8] Even when the existing laws are applicable, however, they are often inadequate for western communities affected by energy development because the laws are not directed towards small communities or because they don't provide sufficient and timely assistance.

When circumstances justify a direct federal role (for example, when the developments are the result of accelerated federal coal-leasing programs), financial assistance should take several forms:

- *Funds targeted to specific problem areas,* such as street construction and housing, which could go directly to impacted communities, rather than indirectly through the states;
- *General impact assistance* channeled directly to impacted communities, which could be used at the discretion of town governments to provide planning, services, and facilities;
- *Assistance to private-sector lenders and builders* in order to increase the supply and mix of housing (apartments, single-family homes, and townhouses).

Such approaches could reduce many of the problems experienced by most energy development areas. Direct federal assistance to localities will probably be resisted by most state governments, however, since these problems are traditionally a state responsibility. Nevertheless, as suggested above, many circumstances of western energy development suggest that a federal role is appropriate; the extent of that role should largely be determined by the success of state programs in dealing with the problems.

Successful use of impact assistance requires securing "front-end" money before large increases in the population occur, targeting to impacted communities rather than counties or school districts, and local administering of grants so that each community can respond to its particular problems and needs. The process of developing federal programs to meet these requirements will be difficult for three reasons. First, it is necessary to have accurate identification of the communities and the anticipated effects; thus, the success of impact assistance will depend, in part, on technical assistance, particularly improved information systems (discussed below). Second, coordination among local, state, and federal governments will be required to avoid delays that could make assistance ineffective. Third, and probably most importantly, assistance will be impeded by budgetary constraints, especially in these times of high inflation. However, at least in the case where federal resources are involved, portions of lease bonuses and royalties can be used, as is already being done in several cases (identified above).

As discussed earlier in this chapter, many local governments in the West have little or no planning capacities for dealing with energy

[6] These include the Public Lands–Local Government Funds Act (1976), which include the In Lieu of Tax Payment Act, and the Federal Land Policy and Management Act (1976).

[7] One of the more significant of these was the proposed Energy Impact Assistance Act of 1978, which was not passed.

[8] Such as the Coastal Zone Management Act of 1972.

development and typically have inadequate information about development. Thus, policies are needed which create better, up-to-date information on the potential impact and provide assistance in using the information. These two considerations were included in the unsuccessful Energy Impact Assistance Act of 1978 (the so-called Hart-Randolph bill). This bill would have provided impact-assistance teams composed of federal, state, and local officials who would assess the likely effects of energy facilities and identify sources of assistance. Teams would also prepare a mitigation plan for dealing with the immediate consequences and procedures for establishing a continuing planning capability.

As stated above, we do not believe that technical assistance is solely a federal responsibility—indeed states and energy developers can do much to alleviate weaknesses in information and planning. However, federal programs do appear necessary to improve the quality of information about development and the capacity of local policymakers to use that information and obtain financial assistance. On the other hand, it is also clear that local areas want local people for these kinds of jobs, largely because planning for energy-related impacts is inherently a political process that helps determine how the costs and benefits of development will be distributed. Thus, policies that provide "outside" experts—whether they be federal, state, or from other areas within the same state—are not likely to be successful. And even alternatives that try to mix federal, state, and local teams, such as the proposed Hart-Randolph bill, are less likely to succeed than options that allow local governments to improve, but ultimately control, their own ability to plan and respond to their energy-related problems.

Research, Development, and Demonstration Programs

Improvements in existing processes to produce energy and commercialization of new *technologies can reduce the adverse consequences of energy development. Although the energy industry should be expected to take the lead in such technological innovations, the federal government should play an active role in encouraging the development of new technologies and providing incentives for their commercialization.*

Although energy developers should be expected to play the major role in research, development, and demonstration (RD&D),[9] this is also one of the most important long-term strategies that can be pursued by the federal government. Joint federal, industry, and state programs for RD&D are necessary to improve energy production and conversion systems as well as specific environmental control.

Several new energy production and conversion technologies considered in this study are not yet commercially viable but have features which may make them more attractive than existing processes. For example, on the questions of air quality and water availability, two of the biggest issues associated with western energy development, coal synfuels appear to be less problematic than electric power plants. Commercial-scale demonstration programs should be established for a wide range of new technologies (such as coal synfuels, oil shale, and geothermal power) in order to better understand the costs and benefits of these options. Since such demonstration projects can be quite costly (e.g., $1 billion for a coal synfuel plant), federal participation is probably required. Regarding environmental controls, important options for minimizing the impact of western energy development are technologies for reducing air pollution (e.g., improved flue-gas desulfurization systems, fluidized beds, and coal cleaning), minimizing land impacts (e.g., improved reclamation methods in arid climates), and

---

[9]For a further discussion, see the section on responses by energy developers.

minimizing water-quality threats (e.g., improved waste-disposal methods).

While such options are long-term in nature and provide no certain solutions, they would reduce uncertainties about energy hardware and its impact on the social, economic, physical, and ecological environment. They would also provide more choices about how to increase our energy supply. Of course, a major constraint is the large financial outlays required of the federal government and other participants, including private industry and, possibly, the states. However, we believe that the costs are worth paying.[10]

Indians

Indians will be increasingly important in energy policymaking both as mineral owners and as major influences on water availability and air quality. Indians will face the decision of whether to encourage of discourage development on their own lands. Although considerable diversity among Indian tribes exists over this question, it is apparent that they will seek more control over development decisions on and near their lands as well as increased independence from the BIA. Perhaps the most significant effect of energy development for Indians is that traditional federal-Indian relationships are becoming obsolete. In addition, Indians will need to improve their management capabilities in order to deal with the impacts of energy development on the reservation and must forge new relationships with various levels of government and energy industries.

Indian lands in our eight-state study area contain abundant and accessible energy resources, particularly coal and uranium. As Indians face continuing demands on their resources, they also will have expanding opportunities to make energy policy. Furthermore, Indians can significantly affect the development of non-Indian resources, primarily because of the water rights they own and because of their ability to redesignate Indian land to either a Class I or Class III status under the PSD regulations.

Energy development will create many of the same problems for Indians as it does for state and local governments. Growth management is likely to be the most serious of these, particularly the need to provide adequate housing, water and sewer facilities, and public services. In addition, Indians will face problems of air- and water-quality degradation and of land rehabilitation. Most of these will create pressures for better relationships among Indian tribes, various levels of government, and energy developers. The most visible and contentious issue in this regard is water rights, particularly the question of whether Indian water rights will be resolved in state or federal courts. Questions of state enforcement of air-quality standards on Indian lands and of the responsibility for land reclamation raise similar issues of control and responsibility.

In many respects, Indian tribes are similar to state governments in terms of their responsibilities and the solutions they may pursue. However, the trust relationship with the federal government also places Indian tribes in a unique position within the political system; their responses to energy development will depend both on how relationships with interests in the public and private sectors evolve and on their own values regarding development and self-determination.

Issues of culture and lifestyle are among the most critical for Indians in regard to energy development. In fact, some tribes, such as the Northern Cheyenne, have decided not to develop their resources rather than to accept what they consider to be the likely negative consequences for their culture and lifestyle. In other tribes—for example, the Navajos—these potential changes have increased tensions among different segments

[10] For a comprehensive discussion of the policy importance of energy research, development, and demonstration programs, see Kash et al., 1976.

of the tribe. In this regard, Indians can influence mineral-leasing arrangements since both tribal approval and concurrence by the secretary of the interior are required before Indian lands can be leased. Although the extent of Indian control in this process varies, the trend is to give tribal leaders more control, including allowing them to consider joint ventures, production-sharing agreements, and other legal arrangements with developers.

In fact, Indians have already been exercising this kind of control, in some cases without approval of the DOI. For example, the Jicarilla Apaches in Arizona have successfully developed joint ventures in oil and gas development; and other tribes have proposed agreements for variable—rather than fixed—royalty rates, production-sharing agreements, and service contracts. Several tribes have either petitioned or filed suit against the DOI to negate previously negotiated leases, alleging procedural irregularities or failure of the secretary of the interior to uphold his trust responsibilities.

Furthermore, the number and strength of Indian interest groups has been increasing. Consequently, in 1975 some 25 tribes banded together to form an organization known as the Council of Energy Resource Tribes (CERT). CERT has since effetively mobilized and organized Indian tribes toward the goal of becoming full participants in the process of energy development on Indian lands. Its early efforts included securing premium prices for Indian energy products, intensifying mineral inventory work, and restructuring minerals taxation to benefit Indians. More recently, CERT has fought implementation of the "windfall profits" tax on Indian oil and lobbied for more federal aid to Indian areas impacted by large-scale energy development (*New York Times,* 1979).

These cases are important indicators that Indians have been gaining power to control their resources; the cases also demonstrate that mineral-leasing arrangements provide Indians with an opportunity to require de-

velopers to help mitigate the impact of development. For example, tribes can stipulate environmental-protection measures, growth-management assistance, and local Indian employment as part of the lease arrangement. By imposing such requirements, Indians take the same risk states do when they institute such measures as high severance taxes: that is, developers may decide to develop in less costly areas.

Indians can also control development through environmental regulation. Both tribes and energy developers on Indian lands must comply with federal environmental statutes, including those regulating air quality, noise, water pollution, solid wastes, reclamation, and safe drinking water. In addition, other legislation specifically addresses the development of minerals on Indian lands, including preserving cultural, recreational, scenic, and ecological values. Implementation of these regulations is the responsibility of the BIA superintendent. However, the role of Indians in establishing and implementing environmental standards and regulations is still ambiguous. This is largely the case because Indian tribes are often not explicitly identified in federal statutes,[11] including those that call for implementation by state governments and those that provide funding assistance to state and local governments. For example, one difficulty arises because states are responsible for implementing federal regulations (e.g., under the Clean Air Act), and yet generally, the states do not have jurisdiction on Indian lands.

Thus, the current approach to environmental protection—particularly EPA's regulatory approach, which has largely excluded Indians from the federal-state-local system of establishing and implementing environmental protection regulations—increases the diffi-

[11]Indians are mentioned more frequently in recent environmental legislation: for example, the 1977 CAA specifically establishes Indian authority in redesignation procedures under Prevention of Significant Deterioration.

culty for Indian tribes in planning for, attracting, and/or controlling energy development options. Because Indian authority is often unclear in federal law and largely unspecified in state law, this adds to conflicts among Indians and with state and federal agencies, since each may suspect the other's environmental and energy-planning objectives.[12] Indian responses are also affected by the limited authoirty they have over non-Indian activities on Indian lands. For example, Indian tribes have no authority to enforce tribal laws on non-Indians on the reservation without specific congressional authority (*Oliphant* v. *Suguamish Indian Tribe,* 1978), and such authority does not exist in present environmental laws.

In spire of these general limitations, Indians have been able to influence policies governing water availability and air quality. Although the full extent of Indian influence on water availability in the West is still not clear, their potential influence is considerable because of the reserved rights doctrine (see Chapter 3). Indians have already claimed the right to substantial quantities of water in the CRB. If these claims are granted, or if Indian water rights in general are quantified, energy development on non-Indian lands in the CRB could be restricted, or Indians at least would gain more control over development by controlling water supplies. Indians can also use air-quality regulations to restrict development, as we have seen in the Northern Cheyenne's redesignating their reservation to a Class I status. Indians could also use these regulations to encourage development by reclassifying their lands as Class III.

Although Indians' external powers were terminated when the trust relationship was established with the federal government, tribes still possess powers of self-government, except as modified by treaty or express provisions of federal legislation. These powers include in-dependence from state governments unless Congress has expressly granted state authority.

Ironically, although this situation grants some degree of authority to Indian tribes, it also serves to increase their dependence on other governmental units or on the private sector. This is because, like local governments, Indian tribes generally do not have the taxing powers necessary to respond adequately to growth-management problems. Thus, although tribes may receive substantial economic benefits from development, they usually require support from the BIA, state governments, or developers unless impact assistance has been made part of the lease arrangement. As with local governments, assistance will be needed primarily in the form of "front-end" money to provide health care, water, sewers, and other public services.

Each of these potential responses by Indians to energy development raises fundamental questions regarding Indian values and the role of the federal government in Indian affairs. There seems to be considerable diversity among Indian tribes regarding the attractiveness of development and the relative advantages of self-determination (compared to the protection of the trust relationship). For most tribes, complete isolation is not an option: when faced with the question of whether to develop their energy resources, the choice is not whether tribal members will be exposed to different cultures and lifestyles, but whether to increase the existing exposure. If the tribes choose to intensify their exposure, current trends suggest that they can realize many economic benefits from development, yet they will also increase their dependence on other policymakers for growth-management assistance and for environmental protection. Furthermore, increased Indian participation in energy policymaking will probably make traditional federal-Indian relationships obsolete in the sense that the BIA will be incapable of protecting Indians and deciding what is best for them.

[12] This thesis has been developed in part by Schaller, 1978.

Responses by Energy Developers

Energy developers—and this includes corporations such as coal and oil companies and the regulated gas and electrical utilities—have a major role in shaping energy development plans. They will decide, for example, whether to "strip and ship" coal or to convert it at the mine-mouth to electricity or synthetic fuels; whether to use conventional or solution mining of uranium; and whether to build commercial-scale oil shale production facilities. The principal criterion for making such choices will be economics; energy companies alone have both the expertise and the capital to undertake most of the developments considered in this study.

In estimating economic benefits, developers must be concerned with the variety of problems and issues which energy development creates. For example, requirements for siting and for mitigating the environmental impact, as well as the possibility of adversary actions, are some of the consequences that will directly affect the decisions of developers—either by increasing the cost of producing the energy or by increasing uncertainty about where and how to proceed.

Since energy developers are prime actors in western energy development, they have a great opportunity to avoid or mitigate the potential problems. Of course, the developers' willingness and ability to act is often limited by practical economic considerations and by regulatory uncertainties. Thus, financial incentives (e.g., investment tax credits for pollution-control equipment) and/or government regulations (e.g., emission standards) will often be required if energy developers are to respond adequately. Nevertheless, such policies aimed at avoiding conflicts will be in the long-term self-interest of the energy industry if they help minimize costly delays in the start of construction. While policies such as impact assistance will increase the cost of a project, the shorter time between planning and start-up may more than compensate. In the following pages, we offer recommenda-

tions for energy developers in three areas: social and economic impact mitigation, site and technology selection, and research and development.

Mitigation of Social and Economic Consequences

Energy companies should assume a greater responsibility for reducing the adverse effects of "boomtown" growth by announcing project plans at an early stage, providing direct impact assistance, and reducing the peak influx of population.

The "boomtown" conditions that result from rapid population growth in small western communities can mean a substantially lower quality-of-life for some residents: inadequate medical care, housing, and recreational facilities are probable results. All of these factors can result in worker dissatisfaction, leading to lowered productivity and increases in the rate of worker turnover. For these reasons, it may be in the long-term interest of developers to take actions which minimize these problems, even though the immediate costs may be considerable.

We recommend that energy developers play a larger role in mitigating these social and economic consequences. While the appropriate response will depend on many factors specific to the particular project, alternatives include early announcement of project plans, direct impact assistance, and programs to reduce the peak influx of population during construction.

Boomtown problems are often magnified because local government officials are simply not given adequate advanced warning about new projects and therefore have little time to plan for rapid growth. In many cases, energy developers could alleviate some of the resulting problems, with very little increase in cost, simply by providing local officials with estimates of personnel requirements and other important information such as the likely tax benefits. Because industry's plans often change in response to government regulations and bureaucratic red tape, as well as because of its own choices, both developers and local governments will want to weigh these uncertainties to avoid making large investments until the project plans are made final. However, these uncertainties do not preclude cooperative arrangements between industry and local government to plan for housing, retail services, sewage-treatment facilities, and other service needs. The responsibility for developing these cooperative arrangements rests largely with the energy companies.

In addition to supplying planning information as early as possible, industry should take more direct action by providing housing for its workers. There are several examples of industries that have responded this way, including ARCO's mobile home park near its Black Thunder Mine in Wyoming and the "new town" built by Montana Power at Colstrip, Montana. In several instances, companies have been willing to provide guarantees to home builders for a certain number of new homes. Industry could also provide for some of the community's expanded infrastructure by paying property taxes in advance, but this approach is presently discouraged by Internal Revenue Service rules which do not allow for deduction of the full amount in the year in which the prepayment is made.

Energy companies could reduce the peak influx of population for small communities in several ways, including encouraging construction workers to commute long-distance from outlying population centers. Energy companies can facilitate this option by providing fleets of mass-transit vehicles or by subsidizing travel costs for workers using their own vehicles. Of course, the success of this approach depends on the location of the construction site relative to larger towns. Where feasible, it may offer the added advantage of allowing some workers' families to maintain their current residence.

Selections of Site and Technology

Selecting the combination of site and technology for a new energy facility is the most critical decision affecting the impact of energy development. Companies have often experienced costly and debilitating delays, and even prohibitions, due to opposition to their siting decisions. To help lessen the impact and to reduce the risk of opposition, energy developers should:

- *Select combinations of technology and location that are likely to minimize the undesirable consequences of development;*
- *Work actively with government entities and interested parties to establish procedures that aim at building a consensus on future energy development sites.*

Siting is one of the most critical components of energy decisionmaking, since it is an occasion at which most of the important questions and issues about resource development come together. Although site and technology selections are influenced by a great many government policies and regulations, energy developers have the primary initiative. *We recommend that companies make better use of information on the role of technological and locational factors in order to help avoid or minimize the adverse consequences of development.* That is, energy developers should adopt the "best practicable location-technology combinations," even if they are not required by current regulations to do so. While this may increase the short-term costs, it can be a long-term economic benefit by helping to avoid conflicts and delays. One example would be the use of water-saving cooling technologies for coal conversion facilities in some areas of the West. Water availability is one of the key issues associated with western energy development, especially in the CRB. If water-saving cooling technologies are used, water requirements can be reduced by about 65 to 75 percent for power plants and by about 20 to 40 percent for synthetic coal fuel plants. It is true that in many parts of our eight-state study area, the price of water is low enough so that it is not economically attractive to use these water-conserving technologies. (In many situations, however, the economic penalty for using them would be small [see Chapter 3].) If they were employed, opposition to energy development from agricultural, environmental, and other interests might decline, and a higher overall level of energy development might be possible.

Of course, defining the "best practicable location-technology combination" is a political as well as a technical decision. For this reason, the second part of our recommendation is that *energy developers should take more initiative in cooperating and coordinating with governmental agencies and interested parties in order to reach a consensus on sites and technologies acceptable for energy facilities.* The goal could be to avoid the most problematic sites, thus avoiding court suits and construction delays. An example of this approach is the ad hoc siting task force established in Utah, which is discussed above in the section on state responses.

Research, Development, and Demonstration Programs

The energy industry should take the leading role—in cooperation with states and the federal government—in research, development, and demonstration programs to commercialize new technologies that can reduce the adverse consequences of western energy development.

As discussed earlier in the section on federal responses, a potentially effective policy option is research, development, and demonstration programs aimed at developing or improving energy technologies that would avoid many of the problems created by present-day processes. Examples include commercially viable processes for coal liquefaction, im-

proved air pollution controls on coal-fired power plants (e.g., regenerable SO_2 scrubbing), and more efficient dry or wet/dry cooling systems. With only a few notable exceptions,[13] private industry is now, and will continue to be, the major source of energy supply. For that reason, *we believe that industry should play a major role in planning and conducting energy research to help establish the necessary bridge between theory and actual use.*

While the industry might undertake these programs without government participation, either within individual companies or within broader research organizations such as the Electric Power Research Institute or the Institute of Gas Technology, there are three important factors that suggest government involvement is required. First, the high cost and risks of some projects (especially commercial-scale demonstrations) are beyond the capacity of all but the largest firms; thus, joint efforts by government and industry will probably be required for financial reasons. Second, much of the required work will be basic research, which has traditionally been supported largely by the federal government and carried out in universities and government laboratories. Finally, much of the required research is in the area of environmental effects; if such studies are funded and carried out solely by the industry, they would have little credibility.

The interaction between the public and private sectors is curcial to the development of a successful energy RD&D program. The key questions concerning the private versus the public role are: (1) what are the proper levels of risk that should be assumed by each; and (1) how can effective management procedures be structured? It is clear that for this recommendation to be implemented, *the federal government must develop stable and consistent policies for supporting technological innovations in the private sector, policies*

---

[13] For example, the Tennessee Valley Authority and the Bonneville Power Administration.

that will protect the public interest and yet ensure that the results will be used (see Kash et al., 1976).

Summary

The United States is under considerable pressure both at home and abroad to reduce imports of foreign oil. To do so will require success both in reducing energy consumption and in increasing the domestic production of energy from a variety of sources. If domestic energy supplies are to be expanded, it is clear that a significant portion of the increase must come from the West, which is rich in coal, oil shale, oil, natural gas, uranium, and geothermal energy resources. The purpose of this study has not been to predict or recommend levels of energy production. Rather, we have tried, first, to anticipate and assess a broad range of environmental, social, and economic consequences if large-scale energy development occurs and, second, to analyze policy options for addressing the problems and issues that are likely to arise.

The picture which emerges from this study is of a region full of conflicts, many of which are unavoidable. For example, the West is rich in natural resources, but it is poor in one of the most essential—water. It is rich in energy resources, but development of these will come directly into conflict with the desire to protect the relatively pristine air and natural beauty of much of the area. And there is a great diversity of opinion among residents concerning the most desirable course to be followed for future energy development.

Because of these and other conflicts, the process of developing appropriate policy responses will be difficult, and this difficulty is worsened by many uncertainties. Among the most significant are: the feasibility of several technologies for energy conversion and environmental control; the inherent errors in current methods for identifying the

costs, risks, and benefits of energy development; and the complexity of policies affecting energy development.

Despite the uncertainties, much is already known about the problems that will be created and the trade-offs that can be expected if alternative policies are chosen. The challenge for policymakers will be to develop and implement improved mechanisms for enhancing the economic benefits of increased energy development while minimizing its adverse environmental and socioeconomic impacts.

In this chapter we have identified only a few of the options available to policymakers at various levels of government and in the private sector. Several technological options appear promising, particularly in the area of protecting the air and water. For example, water-saving cooling techniques can reduce conflicts over the appropriate use of scarce water resources; and stringent sulfur controls can reduce air-quality problems. However, in addition to technological fixes, the resolution of conflicts associated with large-scale energy development in the West will also require a variety of legal, institutional, and informational solutions. This is particularly true for growth-management problems, which will necessitate changes in state tax structures, federal impact assistance, and improved relationships between the public and private sectors.

Resolving conflicts over air and water will be particularly difficult because of the complexity of the regulatory system and the uncertainty associated with predicting the impact of development. In the case of water, this is further complicated by the problem of accurately describing the quantity and quality of the resource. Thus, policymakers will need to expand research programs and develop new institutional mechanisms to accommodate the diverse interests and values at stake *before* large-scale consequences are felt. While this conclusion applies to literally all of the problems studied, it is particularly true for water-resource management. While many institutions presently exist for administrating water resources, these institutions date from and largely reflect an era in which widespread agreement existed about how such a scarce resource should be allocated. This study shows that such agreement no longer exists, and questions about the appropriate use of western water have created serious intergovernmental and public- private-sector conflicts. Thus, new institutional mechanisms will be needed to better integrate water-quality and water-availability concerns, to increase the day-to-day management capacity of states, and to provide incentives for efficient use of the resource.

Recommendations for new and more comprehensive institutional mechanisms do not suggest that uniform approaches to these various conflicts are appropriate. In fact, differences among the western states, variations in impact from different energy development technologies, and the uncertainty associated with these impacts suggest that flexible policy responses will be needed. This is perhaps best exemplified by land-use problems, which are likely to require a diversified approach rather than a strict requirement for reclamation to predevelopment use.

Western energy development will provide substantial benefits to local areas, to the western states, and to the nation. However, development will also create many new problems as well as worsen existing ones in the region. While many of these issues are manageable, they will require a mix of technical, legal, and institutional responses.

REFERENCES

Clean Air Act Amendments of 1977, Pub. L. 95-95, 91 Stat. 685.

Coastal Zone Management Act of 1972, Pub. L. 92-583, 86 Stat. 1280.

Energy Conversion and Transmission Facility Siting Act (1975), North Dakota Century Code, Chapter 49-22-10.

Energy Impact Assistance Act of 1978, S. 1493 (Hart-Randolph Bill).

Energy Tax Act (1978), Pub. L. 95-618, 92 Stat. 3174.

Federal Land Policy and Management Act (1976), Pub. L. 94-579, 90 Stat. 2743.

Industrial Development Information and Siting Act of 1975, Wyoming Statutes §§35-502.75–35-502.94 (Cumulative Supplement 1975).

Joint Powers Act (1974), Wyoming Statutes §§9-18.3–9-18.20 (Cumulative Supplement 1975).

Jones, Charles O. 1974. "Local-State-Federal Sharing in Air Quality Control." *Publius* 4 (Winter):69–85.

Kash, Don E., et al. 1976. *Our Energy Future: The Role of Research, Development, and Demonstration in Reaching a National Consensus on Energy Supply.* Norman: University of Oklahoma Press.

Lieber, Harvey. 1975. *Federalism and Clean Water.* Lexington, Mass.: Lexington Books.

National Energy Conservation Policy Act (1978), Pub. L. 95-619, 92 Stat. 3206.

Natural Gas Policy Act (1978), Pub. L. 95-621, 92 Stat. 3350.

New York Times. 1979. "Western Governors and Indians in Pact on Energy." December 9, p. 72.

Oliphant v. *Suguamish* Indian Tribe, 435 U.S. 191 (1978), 98 S. Ct. 1011.

Powerplant and Industrial Fuel Use Act (1978), Pub. L. 95-620, 92 Stat. 3289.

Public Lands-Local Government Funds Act (1976), Pub. L. 94-565, 90 Stat. 2662.

Public Utility Regulatory Policies Act (1978), Pub. L. 95-617, 92 Stat. 3117.

Resource Conservation and Recovery Act of 1976, Pub. L. 94-580, 90 Stat. 2795.

Schaller, David A. 1978. "An Energy Policy for Indian Lands: Problems of Issues and Perception." *Policy Studies Journal* 7 (Autumn): 40–49.

Surface Mining Control and Reclamation Act of
U.S. President, Office of the White House Press
   Secretary. 1977. "President Carter's Energy
   Message: White House Fact Sheet, April 20,
   1977." *Energy Users Report,* Current Report
   No. 193 (April 21):31–40.

1977, Pub. L. 95-87, 91 Stat. 445.
U.S. President. 1977. "Energy Message to Joint
   Session of Congress, April 20, 1977." *Energy
   Users Report,* Current Report No. 193 (April
   21):28–31.

INDEX

Heritage Conservation and Recreation
Bureau of Outdoor Recreation Act of 1963: 172
Bureau of Reclamation (BuRec): 40, 45-46, 55,
91, 95, 104. *See also* Water and Power
Resources Service
Burlington Northern Railroad: 254, 274(box),
276(box), 277, 283-84

Campbell County, Wyo.: 84, 197(box), 218, 220
Canyonlands National Park: 173(box)
Capital investment: 142, 247-50, 271, 276,
283-84, 340
Capital Reef National Park: 129, 148
Carbon County, Wyo.: 161
Carbondale, Colo.: 49
Carbondale, Utah: 225
Carter, James E.: 9, 11(box), 12, 52-53,
53(box), 146n., 172, 238(box), 331, 337n.,
340
Carter Administration: 52-53, 64, 166, 275,
286, 299, 318
Casper, Wyo.: 197
Certificate of public convenience and necessity:
276, 294, 301
Chaco Canyon National Monument: 159n.
Chicago and North Western Railroad: 277, 283
Citizen participation: *see* Public participation
Clean Air Act (CAA) of 1970: 9, 127
Clean Air Act (CAA) of 1977: 9, 12, 124,
130-31, 133-37, 141-42, 151-52, 251,
296, 300, 309-10, 339, 344, 344n.
Clean Water Act (CWA) of 1977: 9, 77, 85-87,
90, 96n., 102. *See also* Federal Water
Pollution Control Act Amendments of 1972
Coal: 18-23, 37, 86, 113-14, 165, 168, 223,
247, 251, 254-56, 261-63, 266, 299,
299(box); demand for, 137-38; reserves, and
ownership of, 3, 4; transportation, 269-70,
273-77, 279-81. *See also* Slurry pipelines;
Unit trains
Coal-fired power plants: *see* Electric power plants
Colorado: Air Pollution Commission, 130(box);
Department of Health, Water Quality Control
Division, 89; Division of Wildlife, 49; General
Assembly, 287; Mined Land Reclamation
Board, 180n.; Public Utilities Commission,
Railroad-Highway Grade Crossing Protection
Fund, 285; Senate Committee on Health,
Environment, Welfare and Institutions,
130(box); Water Conservation Board, 49
Colorado Highway User Trust Fund: 285
Colorado River flow estimates: 40-42

Colorado River Basin (CRB) Project Act of
1968: 55
Colorado River Basin (CRB) Salinity Control
Act of 1974: 95, 103-104, 106-107, 110
Colorado River Basin (CRB) Salinity Control
Forum: 95-96, 110, 335-36
Colorado River Compact of 1922: 40, 43
Colorado River Conservation District: 49
Colorado River Water Quality Improvement
Program: 95, 105-106
Colstrip, Mont.: 4(box), 13, 43, 119, 130, 198,
294, 298(box), 347
Common Cause: 297
Community Development Block Grant Program:
199
Comprehensive Planning Grants ("701")
Program: 228
Consolidated Coal Co.: 181
Construction work force: 145, 194, 197, 201,
205-206, 210-11, 220, 239, 254, 333-34.
See also Operation work force
Contaminants: *see* Pollutants
Continental Oil Co.: 256
Cooling systems: 20, 22, 28, 39, 60-64, 67, 72,
142, 142n., 144, 336, 348
Cooling towers: *see* Cooling systems
Corps of Engineers: 181, 296
Costle, Douglas M.: 299(box), 340
Council of Energy Resource Tribes (CERT):
13, 253, 298, 344
Council on Environmental Quality: 172
Councils of governments: 228
Criteria pollutants: 118-20, 122, 124, 128;
definition of, 118n.
Crow Reservation: 129
Crow Tribe: 253
Curry, Robert R.: 161

Davis, Grant: 161
Dean, Ernest H.: 129-30
Decker, Mont.: 161, 219
Demonstration plants: 146-48, 151, 250, 337,
342, 348
Dempsey, William: 274
Denver, Colo.: 53, 67, 200, 207, 228
Department of Agriculture (USDA): 238-39
Department of Commerce: 229, 239, 270, 313,
332
Department of Energy (DOE): 101, 147, 161,
250, 263, 274(box), 286, 294, 313, 337, 339
Department of Housing and Urban Development
(HUD): 199, 210, 228-29, 313; Section 235